Lean Combustion

Lean Combustion
Technology and Control

Edited by

Derek Dunn-Rankin
Department of Mechanical and Aerospace Engineering
University of California
Irvine, CA, USA

ELSEVIER

AMSTERDAM • BOSTON • HEIDELBERG • LONDON
NEW YORK • OXFORD • PARIS • SAN DIEGO
SAN FRANCISCO • SINGAPORE • SYDNEY • TOKYO

Academic Press is an imprint of Elsevier

Front cover photograph: Lean premixed low swirl burner
operating on propane (courtesy of S. Abbilian, University of California, Irvine).

Academic Press is an imprint of Elsevier
30 Corporate Drive, Suite 400, Burlington, MA 01803, USA
525 B Street, Suite 1900, San Diego, CA 92101-4495, USA
84 Theobald's Road, London WC1X 8RR, UK

This book is printed on acid-free paper. ☺

Library of Congress Cataloging-in-Publication Data
Lean combustion : technology and control / edited by Derek Dunn-Rankin.
    p. cm.
  ISBN-13: 978-0-12-370619-5 (acid-free paper)
  ISBN-10: 0-12-370619-X (acid-free paper)
  1. Combustion. 2. Combustion engineering. I. Dunn-Rankin, Derek.
  QD516.L353 2008
  621.402'3—dc22

                                                    2007025533

British Library Cataloguing-in-Publication Data
A catalogue record for this book is available from the British Library.

ISBN: 978-0-12-370619-5

For information on all Academic Press publications
visit our web site at www.books.elsevier.com

Printed and bound by CPI Group (UK) Ltd, Croydon, CR0 4YY

Transferred to Digital Print 2011

Dedication

To the memory of Professor James H. Whitelaw whose energy and enthusiasm initiated the workshops on which this volume is based and who was the founding editor of the Combustion Treatise series in which it is published

Contents

Chapter 4

Lean-Burn Spark-Ignited Internal Combustion Engines   95

Robert L. Evans

Chapter 5

Lean Combustion in Gas Turbines   121

Vince McDonell

Chapter 6

Lean Premixed Burners   161

Robert K. Cheng and Howard Levinsky

Chapter 7

Stability and Control   179

Sivanandam Sivasegaram

Chapter 8
Lean Hydrogen Combustion   213
Robert W. Schefer, Christopher White, and Jay Keller

Contributors

Derek Bradley, University of Leeds, England, UK

Antonio Cavaliere, Università Federico II, Naples, Italy

Robert K. Cheng, Lawrence Berkeley National Laboratory, Berkeley, CA, USA

Mariarosaria de Joannon, Instituto di Ricerche sulla Combustione, Consiglio Nazionale delle Ricerche Naples, Italy

Derek Dunn-Rankin, University of California, Irvine, CA, USA

Robert L. Evans, University of British Columbia, Vancouver, Canada

Jay Keller, Sandia National Laboratories, Livermore, CA, USA

Howard Levinsky, University of Groningen and Gasunie Engineering and Technology, Groningen, The Netherlands

Vince McDonell, UCI Combustion Laboratory, University of California, Irvine, CA, USA

Matt M. Miyasato, South Coast Air Quality Management District, Diamond Bar, CA, USA

Trinh K. Pham, California State University, Los Angeles, USA

Raffaele Ragucci, Instituto di Ricerche sulla Combustione, Consiglio Nazionale delle Ricerche Naples, Italy

Robert W. Schefer, Sandia National Laboratories, Livermore, CA, USA

Sivanandam Sivasegaram, University of Peradeniya, Sri Lanka

Christopher White, Sandia National Laboratories, Livermore, CA, USA

Chapter 1

Introduction and Perspectives

Derek Dunn-Rankin, Matt M. Miyasato, and Trinh K. Pham

Nomenclature

BACT	Best available control technology
EGR	Exhaust gas recirculation
FGR	Flue gas recirculation
HCCI	Homogeneous charge compression ignition
IC	Internal combustion
LNB	Low NO_x burner
PCV	Positive crankcase ventilation
RQL	Rich-quench-lean
SCR	Selective catalytic reduction
λ	Relative air–fuel ratio
ϕ	Equivalence ratio

1.1. INTRODUCTION

Lean combustion is employed in nearly all combustion technology sectors, including gas turbines, boilers, furnaces, and internal combustion (IC) engines. This wide range of applications attempts to take advantage of the fact that combustion processes operating under fuel lean conditions can have very low emissions and very high efficiency. Pollutant emissions are reduced because flame temperatures are typically low, reducing thermal nitric oxide formation. In addition, for hydrocarbon combustion, when leaning is accomplished with excess air, complete burnout of fuel generally results, reducing hydrocarbon and carbon monoxide (CO) emissions. Unfortunately, achieving these improvements and meeting the demands of practical combustion systems is complicated by low reaction rates, extinction, instabilities, mild heat release, and sensitivity to mixing. Details regarding both of these advantages and challenges appear in the various chapters in this book. As a whole, therefore, the volume explores broadly the state-of-the-art and technology in lean combustion and its role in meeting current and future demands on combustion systems. Topics to be examined include lean combustion with high levels of preheat (mild combustion) and heat recirculating burners, novel IC engine lean operating modes, sources of acoustic instability in lean gas turbine engines and the potential for active control of these instabilities, lean premixed combustion using both gaseous and prevaporized liquid fuels, and the potential role of lean hydrogen combustion (particularly in comparison to fuel cells) in meeting future power generation needs.

This book is an outgrowth of two international workshops on the topic of lean combustion. The first workshop, held in Sante Fe, New Mexico in November 2000, identified the range of applications in which lean burn could be and is used, along with the tools available to help design the combustion processes for these systems. The second conference, held in Tomar, Portugal in April 2004, focused on the role of lean combustion technology in an energy future that is increasingly constrained by concerns over emissions (both criteria pollutants and greenhouse gases) and excessive fossil fuel use. This second conference centered discussions on which applications of lean combustion are likely to be the most effective contributors to rational energy utilization and power generation. Both of these conferences included participants with backgrounds in a wide range of industries as well as combustion researchers focused on different aspects of lean combustion fundamentals. One goal of the workshops was to try and identify fundamentals, processes, design tools, and technologies associated with lean combustion that could be used across application boundaries. This book is created in that same spirit.

1.2.   BRIEF HISTORICAL PERSPECTIVE

Studies of lean combustion are among the oldest in the combustion literature because its extreme represents the lean limit of inflammability, which was a well-recognized hazard marker from the inception of combustion science. In fact, Parker (1914) argues that the first useful estimates for the lean limit of methane/air mixtures were reported by Davy (1816) in his efforts to prevent explosions of methane gas (called "fire-damp") in coal mines (incidentally, this is the same paper where Davy published his famous explosion-safe lantern that used wire gauze walls to allow air and light to pass but prevent flame propagation). Davy reported limits of inflammability between 6.2% and 6.7%. In modern terminology, this represents an equivalence ratio range for methane between 0.68 and 0.74. Parker also reports a three-fold variation in the limits reported by the early literature (with Davy's near the upper end), which he attributes to the fact that the limit of inflammability depends on the vessel used for the test, among other experimental variations. This recognition led eventually to standard inflammability measurements based on the upward propagation of a flame through a mixture indefinitely. However, Parker further complicated the concept of the lower limit of inflammability by examining mixtures of oxygen and nitrogen rather than using the standard ratio of these molecules in air. His findings are shown in Figure 1.1, with a minimum at 5.77% methane using a 25% oxygen mixture. The exact values are not that important but these results show clearly that the limits of lean combustion depend not simply on an equivalence ratio, but on the oxidizer and the diluent composition. For example, if the 5.77% methane is considered relative to a stoichiometric mixture of methane and normal air, the equivalence ratio would be $\phi = 0.61$, but if the stoichiometry is taken relative to the slightly oxygen-enriched mixture reported, the equivalence ratio at the

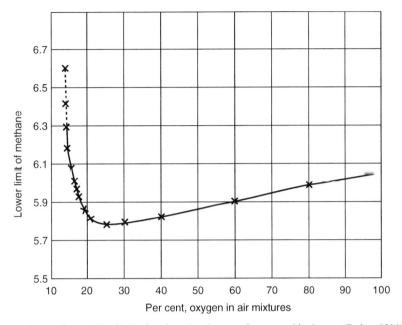

Figure 1.1 Inflammability limit of methane in mixtures of oxygen with nitrogen (Parker, 1914).

lean limit is 0.47. In 1918, Mason and Wheeler also complicated the picture of lean limits by demonstrating conclusively that the temperature of the mixture affected dramatically the limits of inflammability, as is illustrated in Figure 1.2. Their finding was not surprising, even to them, but it showed that combustion limits depended on the chemical and physical properties of the reactant mixture, and on the details of the combustion vessel and the ignition method. Although these dependencies can make global predictions of lean combustion behavior (and accidental explosion) difficult, they can also be quite advantageous because they allow technologies to manipulate the process over a wide range of the parameter space in order to achieve a desired performance outcome. Essentially, all of the technologies reported in this book take advantage of temperature and dilution control to manipulate the power and emission output from the lean combustion process.

Lean combustion was considered only with regard to explosion hazards until the late 1950s, when lean flames were introduced as useful diagnostic tools for identifying detailed reaction behavior (Kaskan, 1959; Levy and Weinberg, 1959). However, it was not until the late 1960s that lean combustion began to be discussed as a practical technology, particularly for trying to improve fuel economy (Warren, 1966) and reduce emissions from spark-ignited reciprocating IC engines (Lee and Wimmer, 1968). The latter was in response to the 1965 amendment of the revised 1963 US Clean Air Act that set, for the first time, federal emission standards beginning with the 1968 model year. These standards called for 72% reduction in hydrocarbons, 58% reduction in CO, and complete capture of crankcase hydrocarbons over the 1963 model year vehicles. Note that based on this regulation, the lean combustion approach was being used initially to reduce HC and CO only, not NO_x. The 1970 amendments to the Act controlled NO_x for the first time, requiring a 90% reduction from the 1971 levels beginning with the 1978 model year. These emission requirements kept lean combustion a viable and important

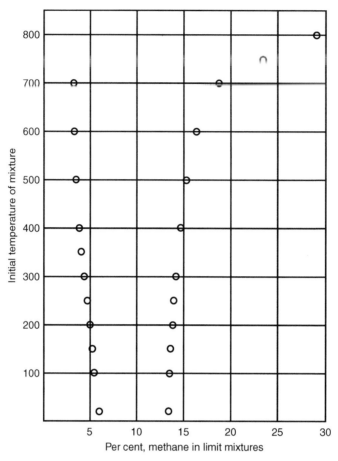

Figure 1.2 Limits of inflammability for methane/air mixtures with various initial temperatures in °C (Mason and Wheeler, 1918).

technology for IC engines for almost two decades (Anonymous, 1975, 1977, 1978, 1984a, b). Eventually, however, more stringent emission standards, coupled with the advent of a three-way catalyst technology that required near-stoichiometric operation, moderated interest in lean combustion for spark-ignited IC engines. A dilute (lean) combustion retrofit for pipeline compressor engines was proposed more recently (Balles and Peoples, 1995) in response to the NO_x reduction requirements promulgated in the 1990 Clean Air Act amendments, but catalyst technology essentially obviated most further developments in IC engines until just before the new millennium. A recent paper (Döbbeling _et al._, 2005) shows a very similar 25-year technology trend for stationary gas turbines, again in response to the Clean Air Act emission regulations, except that there was no hiatus. Lean technology has contributed continuously in this arena since the early 1970s.

As indicated by this brief history, developing an understanding of lean combustion that crosses traditional boundaries requires first that the definition of lean burn be examined in the context of the various applications in which it is used and the primary driver (i.e., emission regulation) for them.

1.3. DEFINING LEAN COMBUSTION

Because lean combustion got its start assuming atmospheric air as the oxidizer in premixed combustion systems, the concept of fuel lean referred originally to a condition related simply to the level of excess air provided to the reaction. Since different fuels require different amounts of air for stoichiometric conditions, and because the stoichiometry can be defined in terms of mass or number of moles, a normalized value is more appropriate for comparison across fuel types. The equivalence ratio, often denoted ϕ (fuel/air by mass or mole divided by the same quantity at stoichiometric conditions), and its inverse, the relative air–fuel ratio, often denoted λ, are two common examples of such normalized quantities. In this circumstance, lean combustion occurs when $\phi < 1$ or $\lambda > 1$. The concept is, essentially, that lean combustion refers to conditions when the deficient reactant is fuel, and the more deficient, the more fuel lean.

In many ways, however, the excess air during lean combustion, including the oxygen contained therein, acts primarily as a diluent. There is little difference, therefore, between providing excess air and providing a stoichiometric level of oxygen with the additional excess made up with an inert (such as nitrogen). This approximate equivalence can be seen in Figure 1.1 from Parker, where there is little difference in flammability limit when increasing the oxygen fraction beyond its stoichiometric proportion in methane/nitrogen/oxygen combustion. Hence, while the reaction might be described as lean because the deficient reactant is still the fuel (though perhaps only slightly), the level of leanness is governed by the level of diluent (of any kind). Under these conditions, it can be reasonable to describe dilute mixtures as lean even if they nominally contain enough oxygen for reaction that is very close to stoichiometric conditions. Exhaust gas and flue gas are used commonly to dilute mixtures without quenching the reaction, and it is this dilution-based form of lean combustion that characterizes the highly diluted combustion discussed in Chapter 3.

Under standard premixed flame propagation conditions, that is, when heat evolved from the reaction front is responsible for heating the incoming reactants to combustion temperatures, the equivalence ratio is usually a very good parameter for classifying flame behavior (such as laminar burning velocity). Generally, as the equivalence ratio falls and the mixture becomes dilute, its capability for sustaining a reaction front decreases. By preheating the reactants, it is possible to extend the region of mixture inflammability (where there is a definable flame front), as seen in Figure 1.1. In addition, it is possible to react the mixture even without a recognizable flame front, if the preheat is sufficiently high. In this case, the mixture is clearly lean, but the concept of equivalence ratio as characterizing the strength of inflammability loses some of its correlating capability since a majority of the enthalpy is not coming from the reaction directly. Characterizing such situations with highly recirculated exhaust heat, homogeneous compression ignition, and the like must use new relevant parameters to describe combustion behavior.

When the reactant mixture is not uniform, further complications in defining lean combustion domains can arise. In a diesel engine, for example, where the primary reaction might occur in the locally fuel rich zone surrounding the fuel spray, the overall fuel/air ratio is lean. It is this non-uniform combustion, in fact, that produces the features characteristic of typical diesel engines, namely soot and NO_x from the rich combustion, and high efficiency from the overall lean reaction that follows (though some of the efficiency gain is from the reduction in throttling losses as well). A similar phenomenon occurs for stratified charge gasoline engines, where the mixture near the spark is enriched intentionally to enhance stability, but the flame kernel is then able to consume the remaining lean reactants. How then is lean combustion to be defined for partially premixed or non-premixed systems? In this book, we include systems that are overall lean, even if the primary reaction zone is not, because their goal is to achieve the characteristics of lean burn behavior (i.e., low NO_x and high efficiency). In gas turbine engines, for example, the rich-quench-lean (RQL) approach to lowering NO_x emissions relies on creating a mixing environment that is rapid relative to reaction times. In this case, the staged reaction sequence is part of a lean burn system even if one stage is fuel rich. When defining the critical processes operating in each stage, however, it is important to assess the level of leanness in terms of the local temperature, equivalence ratio, and other properties.

When reading the remaining chapters of this book, it is useful to keep in mind the above range of meaning in the umbrella term "lean combustion."

1.4.   REGULATORY DRIVERS FOR LEAN COMBUSTION TECHNOLOGY DEVELOPMENT

At the second lean combustion workshop, participants from across the technology spectrum were asked to assess the principal drivers for lean combustion in their respective fields, as well as the research and technology needs for achieving the performance goals. The conclusion of this discussion, essentially uniform for all technologies and all industrial sectors, was the recognition that the primary driver for lean combustion was to reduce emissions (primarily NO_x) to meet externally imposed regulations. That is, national and international regulations for stationary and mobile sources, as well as more aggressive regulations, such as those established by the state of California, have largely been "technology forcing," especially with respect to mobile sources. According to the National Academy of Sciences (2006),

> *"'Technology forcing' refers to the establishment by a regulatory agency of a requirement to achieve an emissions limit, within a specified time frame, that can be reached through use of unspecified technology or technologies that have not yet been developed for widespread commercial applications and have been shown to be feasible on an experimental or pilot-demonstration basis"*

As an illustration of technology forcing in the United States, the number of emission control related patents concerning power plants, motor vehicles, as well as lean combustion technologies are plotted with respect to time in Figure 1.3. Critical air pollution regulations have been superimposed to show the connection between regulation and technology. Specifically, the release of the Clean Air Act amendments of 1977 and 1990 and the implementation of Tier I and Tier II air emission standards of 1991 and 1999 are shown. Although there are many other factors that likely drive the number of patents, including political events and world oil prices, there is a recognizable increase in the technology activity following regulatory events.

The prolonged surge in patents being issued after the 1990 amendment resulted because, unlike its counterpart in 1977, the 1990 amendment not only increased the regulatory limits on emissions but also made provisions to foster the growth of renewable power generation technologies (Wooley and Morss, 2001).

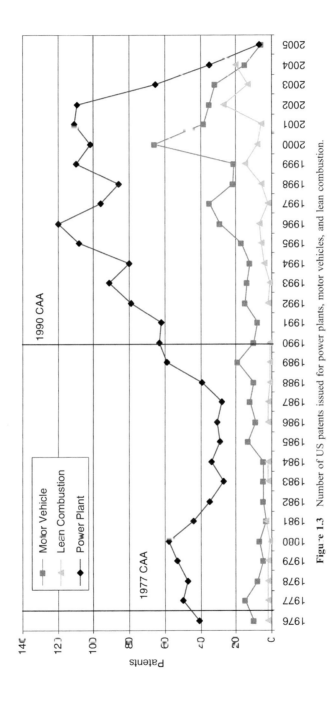

Figure 1.3 Number of US patents issued for power plants, motor vehicles, and lean combustion.

Motor vehicles were not specifically targeted in the 1977 Clean Air Act amendment, but Title Two of the 1990 Clean Air Act amendment addresses the issue of vehicle pollutants. Hence, the issuing of the 1990 Clean Air Act amendment was followed by an increase in motor vehicle pollution control patents, and motor vehicle patents rose again following the implementation of the 1991 Tier I air emission standards and again, even more steeply, after the Tier II standards of 1999.

As described in the previous section, interest waned in lean combustion for motor vehicles as catalytic converters provided sufficient exhaust clean up performance to meet regulations. Therefore, the number of patents for lean combustion did not increase until after the 1990 Clean Air Act amendment. As the Tier I and Tier II air emission standards came into play, automakers turned again to lean combustion as a way to meet the new regulations. The greatest number of patents for lean combustion were issued after 1999, reflecting an enthusiasm for strategies like homogeneous charge compression ignition (HCCI) (e.g., Aceves *et al.*, 1999).

Regulation not only increased the number of patents related to air pollution technology, but also the number of academic papers published. A plot of papers pertaining to "lean burn" or "lean combustion" per year indicates an increasing frequency following the Clean Air Act amendment of 1990. Although not normalized to account for what might be an increase in total papers published, Figure 1.4 shows that the number of both academic automotive journal articles from The Society of Automotive Engineers and academic combustion journal articles in Combustion and Flame increased following the 1990 amendment.

Section 1.4.2 describes a survey of responses from technology providers about how they address these regulations. The survey demonstrates that lean combustion is often

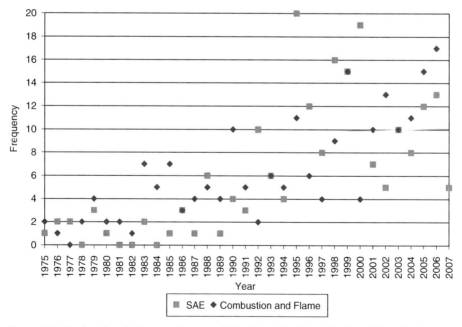

Figure 1.4 Number of academic journal papers published involving "Lean Burn" and "Lean Combustion" for *The Society of Automotive Engineers* and *Combustion and Flame*.

Table 1.1

Technology strategies considered to represent "Lean Combustion"

Strategy	Driver	Methods
Oxidation	Reduced hydrocarbon and CO emissions	Better mixing complete combustion $\phi \leq 1$ (lean)
Dilution (Lower Temperatures)	Reduced NO_x emissions	Flue gas recirculation staging premixed radiant

the first or most favored control strategy attempted. The success of lean combustion, however, appears dependent on the pollutant to be controlled (e.g., hydrocarbons, CO, or oxides of nitrogen) and on the application (e.g., light-duty vehicle, boiler, or heater). Furthermore, as discussed earlier (and throughout this book), lean combustion can represent two strategies: oxidation and dilution, as shown in Table 1.1. Providing excess air to complete combustion through the oxidation of hydrocarbon and CO was not only a means of improving efficiency but also of reducing harmful emissions. As focus shifted to reducing oxides of nitrogen (NO_x) emissions, lower combustion temperatures were achieved through dilution using leaner mixtures, exhaust gas recirculation (EGR), and staged combustion. The relative importance of these pollutants associated with particular combustion applications produced slightly different technological responses to the regulations from the mobile and stationary source communities.

1.4.1. MOBILE SOURCES

Southern California, due to its unique atmospheric conditions, large vehicle population, and dense urban centers, experiences the worst air quality in the United States. As a result, California has led the way in developing and implementing strict air quality regulations. As shown in Figure 1.5, during the late 1950s and through the 1960s, peak ozone (smog) levels in Los Angeles reached 0.6 ppm (South Coast Air Quality Management District, 2006). This is a level greater than six times the state 1-hour standard (0.09 ppm) established to protect public health.

In 1959, the State Legislature created the California Motor Vehicle Pollution Control Board to test and certify emission control devices. Regulations initially focused on controlling unburned hydrocarbons and eventually the efforts of the California Motor Vehicle Pollution Control Board led to the passage of legislation in 1961 requiring positive crankcase ventilation (PCV) valves on all passenger vehicles starting with the 1963 model year. The PCV valve was designed to capture any unburned fuel–air mixture that escaped past the piston ring, commonly termed "blow-by gas," and force it back into the combustion chamber through the intake manifold. This technology reduced hydrocarbon emissions by diluting the combustion charge with the blow-by gas, which was previously vented to the atmosphere. Stricter hydrocarbon and CO tailpipe regulations were enacted in California in 1966, which again compelled the manufacturers to produce more complete combustion; the result was leaner carburetors to more fully oxidize the fuel/air mixtures and reduce hydrocarbon and CO concentrations in the exhaust.

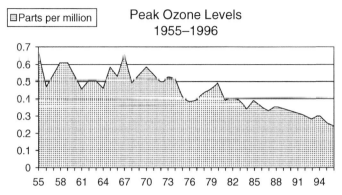

Figure 1.5   Ozone trend in South Coast air basin (SCAQMD, 2006).

Table 1.2

Lean combustion technology responses to mobile source regulations

Regulated Emissions	Technology Response	California Model Year Requirement	Federal Model Year Requirement
Crankcase emissions	Postitive crankcase ventilation	1963	1968
First CO and VOC standards	Leaner carburetors, air pumps, and retarded timing controls	1966	1968
NO$_x$ at 2.0 g/mile	Increased EGR	1974	1977
VOC and CO	Lean mixtures and oxidation catalysts	1975	1975

Peak ozone levels, however, were not reduced significantly with just hydrocarbon controls. It became clear by the late 1960s that emission of both smog precursors, hydrocarbons and NO$_x$, needed to be mitigated. In order to address NO$_x$ emissions, flue gas recirculation (FGR), a well-known stationary source technique, was employed for vehicles. For the automotive application, the strategy relied on diluting the intake charge with exhaust gas to reduce peak combustion temperatures and thermal NO$_x$. EGR is still used today in conjunction with three-way catalysts to control automotive emissions. Table 1.2 summarizes the targeted emissions to be reduced, the technologies used to address the emissions, and the implementation, by the California and Federal governments.

As described earlier, although modern gasoline, light-duty vehicles rely almost exclusively on near-stoichiometric fuel–air ratios and three-way catalysts, early strategies to control NO$_x$ favored lean combustion. Honda Motor Corporation, for example, used lean combustion technology to certify the first ultra low emission vehicle in 1974. The lean combustion Civic was commercialized the following year. Honda has continued to rely on lean combustion for their next generation diesel light-duty engine, which uses lean combustion as a part of the complex emissions controls system that includes selective catalytic reduction (SCR) (Honda Motor Corporation, 2006).

1.4.2.   STATIONARY SOURCES

For stationary sources, lean combustion technologies were identified through a review of the 2006 best available control technologies (BACT) database for the South

Coast Air Quality Management District (SCAQMD). The database includes 21 combustion categories, shown in Table 1.3, with 92 different BACT.

The results of the BACT review, shown in Figure 1.6, illustrate that almost half of the technologies used for stationary applications utilize low NO$_x$ burners (LNBs), which is the generic term for some type of lean or dilution strategy in the combustion zone. Further analysis showed a mixture of external FGR, SCR, a combination of both SCR and FGR, or lean premixed radiant burners. The categories, or applications, which employ these LNB technologies are represented in Figure 1.7, which illustrates that boilers and dryers or ovens are the most prevalent application for LNBs.

In order to identify the perspective of combustion practitioners associated with these technologies, an anonymous survey, shown in Table 1.4, was sent to the members of the American Flame Research Council (AFRC). The AFRC is the US affiliate of the International Flame Research Foundation, which is a cooperative research and technical organization serving industry through collaborative research and networking among industry, academia, and government members. The 28 responses (25% response rate) provide an insight into the industrial community's view of the potential for lean combustion technologies.

The responses are represented graphically in Figures 1.8 and 1.9. Most of the respondents were burner manufacturers or consultants. Consistent with the survey of

Table 1.3

Combustion BACT categories from the South Coast Air Quality Management District

Absorption chiller	Flow Coater, dip tank and roller coate	ICE – stationary, non-emergency
Aluminum melting furance	Gas turbine	ICE – digester
Asphalt batch plant	Gas turbine – landfill	ICE – emergency; diesel
Boilers	Heater – refinery	ICE – emergency; spark ignited
Refinery heaters	ICE – fire pump, spark ignition	ICE – fire pump; compression ignition
Dryer or oven	ICE – portable, compression ignition	Metal heating furnace
Catalyst regeneration – fluidized catalytic cracking unit	ICE – landfill gas-fired	Process heater

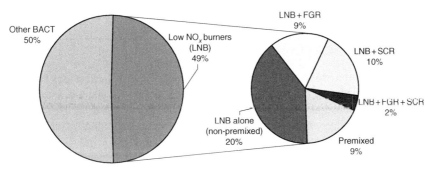

Figure 1.6 Listing of the SCAQMD's BACT categories for combustion sources utilizing low NO$_x$ burners and other strategies.

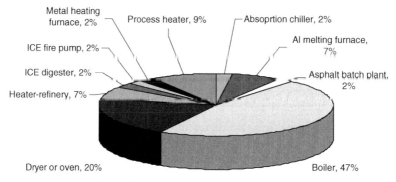

Figure 1.7 Low NO$_x$ burner BACT by category.

Table 1.4

Anonymous survey submitted to the American Flame Research Council members

Question	Responses
1. Please select your organization's main function.	Burner manufacturer
	Government
	Research
	Production
	Academia
	End-user
	Other (consultant)
2. Please select the main driver for your organization in terms of combustion processes (high, medium, low priority).	Emissions regulations
	Energy efficiency
	Greenhouse gas emissions
	Market penetration
	Intellectual property
	Research
3. Please rank the following technologies in the order of highest potential to address your organization's needs (please select only three).	Lean combustion (premixed or rapid mixing)
	Catalytic aftertreatment (e.g., SCR)
	Particulate aftertreatment (e.g., baghouse)
	Flue gas recirculation
	Staged combustion
	Pulsed combustion
	Hydrogen injection
4. Please add any other thoughts on drivers or technologies you believe were not addressed in the previous two questions.	Open-ended response

the lean combustion workshop participants, the respondents list emissions regulations and energy efficiency as the primary drivers for their organizations, with rapid mixing or premixed combustion, staged combustion, and FGR identified as the most likely technologies to address these drivers. All three of these technologies represent lean combustion as identified in Table 1.1.

Due to the theoretical simplicity of achieving complete combustion and low emissions simultaneously, lean combustion is prevalent in both mobile and stationary applications. However, in practice, the goals of complete fuel oxidation and reducing

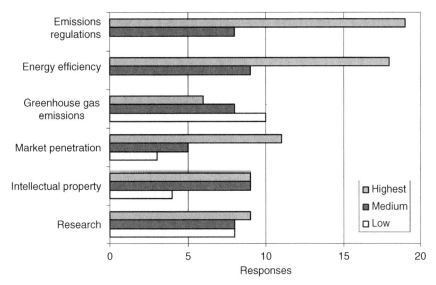

Figure 1.8 Responses to survey Question 2: "Please select the main driver for your organization in terms of combustion processes."

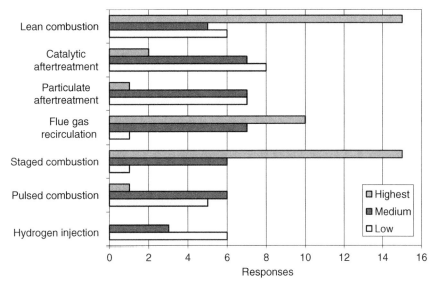

Figure 1.9 Responses to survey Question 3: "Please rank the following technologies in the order of highest potential to address your organization's needs (please select only three)."

combustion temperatures to reduce NO_x are often in conflict. Generally, therefore, there are few applications that use lean combustion alone; most applications rely on a combination of lean combustion and additional techniques, including FGR and catalytic after treatment. Based on previous experience, however, and depending on the application, lean combustion will continue to be one important factor in a portfolio of strategies utilized to address inevitable regulatory requirements.

1.5. LEAN COMBUSTION APPLICATIONS AND TECHNOLOGIES

In part, because fuel lean combustion can be defined in the diverse ways described above, this concept appears across a wide range of heat and power generating systems. In order to examine scale dependence, operating condition variation (e.g., pressure and temperature in the combustion zone), and to separate transient from continuous processes, we have divided these systems, somewhat arbitrarily, into (a) reciprocating IC engines, (b) moderate-scale burners for process heat, (c) large-scale burners for power station boilers, (d) stationary gas turbine engines, and (e) aircraft gas turbine engines. We have also identified a wide range of lean combustion approaches that include: (a) traditional lean premix, (b) high preheat or dilute combustion, (c) partial premixing or stratified charge, followed by dilution, (d) catalytically assisted combustion, and (e) homogeneous charge compression combustion. Naturally, there are several overlaps and gray boundaries between these approaches, particularly when they refer to combustion systems with very different character, and one goal of this book is to help identify these overlaps by laying the various systems and approaches side-by-side for examination. A third dimension of importance in defining lean combustion behavior is the fuel properties. Not only are there large differences across applications (e.g., Jet-A liquid fuel to natural gas) and lean burn approaches, but there are also fuel sensitivities within these categories since real fuels are not pure hydrocarbons but some mixture of components that exhibit crudely the same heat release behavior. In lean combustion, these subtle fuel composition fluctuations can lead to large changes in combustion performance. The potential role of hydrogen combustion in the energy future is another example of the fuel dimension of lean burn.

With the complexity, breadth, and dynamic nature of the lean combustion field, it is impossible for any book to claim a complete state-of-the-art representation since the information base grows daily. HCCI is one example where recent advances have spurred enthusiasm in research and development. Although the concept is not entirely new, the term HCCI entered the combustion vernacular in around 1999 (e.g., Aceves *et al.*, 1999), and this topic now would require its own volume alone (as do many of the other subjects discussed). Therefore, there are several lean burn approaches and technologies that this volume unfortunately addresses only cursorily. HCCI and catalytic combustion are just two examples. There is no implication that their limited role in this volume represents in any way their significance or potential. Rather, we have focused somewhat on the slightly more static elements of lean combustion, hoping by so doing to provide a foundation for discussion and understanding.

1.6. BRIEF HIGHLIGHTS OF THE CHAPTERS

The remaining chapters of the book examine the following:

Chapter 2: Fundamentals of Lean Combustion – This chapter presents the foundations of lean combustion including the roles of turbulence, flame front instabilities, and flame speed in controlling the robustness of the reaction. An important aspect of this chapter is the recognition that real fuels (i.e., practical blends like gasoline) do not always behave as their idealized counterparts (e.g., iso-octane). This chapter shows that when turbulent stretch is very high and the mixture is very lean, flameless, autoignitive combustion can be achieved but may require both heat and hot gas recirculation. Although recirculation both dilutes and increases the temperature of the mixture, the temperature effects tend to dominate in autoignitive burning as compared to flame propagation, resulting in the general need for autoignitive burning for very lean mixtures. For non-premixed fuel jets burning in air, surrounding the jet with hot recirculated gases is shown to allow very lean mixtures to burn when the flame would otherwise blow off. In this situation, combustion is governed by low temperature chemical kinetics rather than high temperature kinetics, which is associated with propagating flames.

Chapter 3: Highly Preheated Lean Combustion – As shown in Chapters 1 and 2, with sufficiently high preheat temperatures it is possible to burn arbitrarily lean pre-mixtures even when the enthalpy released in the reaction is a small fraction of the total. This chapter discusses the concepts and application of combustion in these highly preheated, highly diluted, and excess enthalpy systems, including their taxonomy based on whether the dilution and/or preheating occurs in the fuel and/or oxidizer stream. The authors also describe the simple processes that define and characterize MILD combustion systems, and compare them to the properties of more traditional premixed and diffusion flames. This chapter also includes a conceptual application of MILD combustion to a gas turbine engine. Specifically, a power plant analysis is presented whereby combinations of Rankine, Brayton, Ericsson, and Hirn cycles are employed in an attempt to increase plant efficiency and reduce pollutant emissions. Final thoughts are given regarding the possible categorization of MILD combustion as a prototypical clean combustion process since pollutant species containing carbon, hydrogen, and nitrogen are suppressed, destroyed, or reduced to less harmful by-products.

Chapter 4: Lean Burn Spark-Ignited Reciprocating Engines – As discussed above, lean burn spark-ignited IC engines were explored (and mostly rejected) decades ago. This chapter describes and demonstrates some relatively minor modifications to standard engines that would allow robust lean burning using stratified charge concepts. In particular, special piston designs that induce turbulent mixing and direct injection fueling near the spark plug are featured. These engine concepts use conventional spark-igniters to control ignition timing. This chapter leaves the discussion of HCCI engines for another volume, concentrating instead on strategies that can be implemented or retrofitted into current engines without difficulty. The HCCI autoignitive approach appears, however, in Chapters 2, 3, and 8, and for further details regarding this concept the reader is referred to Aceves *et al.* (1999), Flowers *et al.* (2001), and Yeom *et al.* (2007).

Chapter 5: Lean Burn Gas Turbine Engines – This chapter discusses gas turbines for both stationary power and aircraft propulsion applications, as well as the aircraft auxiliary power unit (APU), which has features of each. Aircraft propulsion engines focus first on safety and reliability, while ground-based systems can relax the demands in these categories slightly and push toward higher efficiency and lower emissions by operating close to static stability limits. Hence, as mentioned in Chapters 2 and 6, sensitivity to fuel variability is an important element of lean turbine design. Since many gas turbine engines operate in a non-premixed mode, the concept of lean combustion here refers primarily to a locally lean condition since overall dilution is always required to meet the thermal constraints imposed by material properties. As discussed in Chapter 7, the recirculation zones typical of high-swirl non-premixed injectors can be susceptible to combustion oscillation. It may be possible, therefore, to implement a low swirl approach similar to that described for burners in Chapter 6. This chapter does not cover catalytic combustion concepts for gas turbines in much detail, and so for further information, the reader is referred to Mantzaras (2006) and Reinke *et al.* (2007) as a starting point. Overall, this chapter shows that progress in lean gas turbine combustion has been substantial, but that the major driver remains NO$_x$ reduction, and avoiding material limits remains a major challenge.

Chapter 6: Lean Premixed Burners – The challenges associated with lean combustion in burners are often underappreciated because these ubiquitous combustion devices have been operating lean and with low emission reliably, efficiently, and economically for decades. However, as emission requirements tighten, the demands on lean burners increase. Because these systems are usually fan or blower driven, they are susceptible to acoustic feedback; and as the mixtures approach their lean limits, the heat release becomes sensitive to fluctuations. In addition, relatively small changes in fuel composition can produce substantial changes in local stoichiometric conditions because the airflow is rarely controlled relative to the instantaneous fuel content. This combination of challenges means that turndown windows narrow if traditional high swirl injection is used to provide flame stabilization. This chapter also discusses a reconsideration of the need for recirculation-stabilized flames by presenting a relatively new concept for lean premixed combustion, the low swirl burner. Many of the factors discussed in this chapter, including sensitivity to fuel composition, the role of high swirl in flame stabilization, and the potential for acoustic instability growth, cut across several lean combustion technologies and appear in other chapters as well.

Chapter 7: Stability and Control – Combustion oscillations have been studied in great detail in a wide range of circumstances owing to their importance and potential detrimental effects. Much of this prior work, however, has concentrated on oscillation in near stoichiometric combustion (either premixed or non-premixed). This chapter focuses on the issues of acoustic oscillations that dominate near the lean limit. These include the fact that under lean conditions the reaction zone can be spatially distributed rather than compact so that the effective pressure forcing is also broadly distributed. This distributed heat release can lead to lower intensity oscillations than occurs for stoichiometric systems, but it can also excite alternate combustor modes. In addition, lean combustion systems can exhibit cycles of extinction and relight near the lean limit that can span 5–100 Hz, which can couple into longitudinal modes, but rarely stimulates transverse or circumferential ones. This chapter offers detailed and comprehensive research results related to lean combustion oscillations as related to the gas turbine environment, and many of the issues identified in this chapter are mentioned as challenges in Chapters 5 and 6 as well. This chapter also discusses control strategies and the particular challenges to their implementation and practicality for suppressing lean combustion oscillations. For example, while it may be relatively straightforward for a control system to respond to a tightly defined dominant source, it can be more difficult when the source character is distributed broadly and somewhat unpredictably across the frequency spectrum. For additional perspective, the reader may be interested in the recent work on these topics by Cho and Lieuwen (2005), Coker *et al.* (2006), Li *et al.* (2007), Muruganandam *et al.* (2005), Taupin *et al.* (2007), and Yi and Gutmark (2007).

Chapter 8: Lean Hydrogen Combustion – It is difficult to imagine any modern exposition on combustion technology without considering the implications of including hydrogen in the fuel stream. This chapter discusses lean burn hydrogen combustion in gas turbines and reciprocating IC engines. It explains that while utilizing hydrogen fuel poses some challenges, operating hydrogen combustion engines have already been demonstrated. The special features of hydrogen for lean burn include its high flame speed, low flammability limit, high mass-specific energy content, lack of carbon, and high molecular diffusivity. These properties, when properly exploited, can enhance flame stability at the low temperatures capable of limiting NO_x formation and, depending on the source of the hydrogen, reduce or eliminate the emission of the greenhouse gas carbon dioxide. With these benefits, come some significant challenges including compact fuel storage, potential explosion hazards, pre-ignition limited peak power in IC engines, and autoignition and flashback in gas turbines. This chapter discusses some advanced engine concepts for overcoming these challenges and ends with the optimistic view that lean hydrogen combustion could, with relatively little effort, provide an effective bridge to future carbon neutral fuel scenarios.

Based on the above highlights, it is clear that lean combustion has several cross-cutting features. Lean burn can provide low emissions, particularly those of NO_x, in a wide range of combustion configurations. The challenging behaviors of lean flames within these configurations include sensitivity to fuel composition and relatively weak reaction fronts in highly dynamic fluid flows. In some situations, high levels of preheat can take lean combustion beneficially away from a flame mode and toward a distributed reaction. In any case, it seems likely that as emission regulations continue to tighten, clever implementations of lean combustion technology will continue to be needed.

ACKNOWLEDGMENTS

The authors appreciate very much the assistance of Jesse Pompa and Peter Therkelsen for identifying emission regulations and their results on the technology. Professor Frank Schmidt of the Pennsylvania State University and the Engineering Foundation was instrumental in assembling the lean combustion workshops mentioned in this chapter.

REFERENCES

Aceves, S.M., Smith, J.R., Westbrook, C.K., and Pitz, W.J. (1999). Compression ratio effect on methane HCCI combustion. *J. Eng. Gas Turb. Power – Trans. ASME* **121**(3), July, 569–574.

Anonymous (1975). Chryslers electronic lean – burn engine. *Mach. Des.* **47**(17), 24–26.

Anonymous (1977). What limits lean combustion – spark or propagation. *Auto. Eng.* **85**(1), 48–51.

Anonymous (1978). Homogeneous mixtures may not be optimal for lean combustion. *Auto. Eng.* **86**(6), 66–70.

Anonymous (1984a). Lean combustion – a review, 1. *Auto. Eng.* **92**(2), 49–54.

Anonymous (1984b). Lean combustion – a review. *Auto. Eng.* **93**(3), 53–58.

Balles, E.N. and Peoples, R.C. (1995). Low-cost NO_x reduction retrofit for pump scavenged compressor engines. *J. Eng. Gas Turb. Power – Trans. ASME* **117**(4), October, 804–809.

Cho, J.H. and Lieuwen, T. (2005). Laminar flame response to equivalence ratio oscillations. *Combust. Flame* **140**(1–2), 116–129.

Coker, A., Neumeier, Y., Zinn, B.T., Menon, S., and Lieuwen, T. (2006). Active instability control effectiveness in a liquid fueled combustor. *Combust. Sci. Tech.* **178**(7), 1251–1261.

Davy, H. (1816). On the fire-damp of coal mines, and on methods of lighting the mines so as to prevent its explosion. *Phil. Trans. R. Soc. London* **106**, 1–22.

Döbbeling, K., Hellat, J., and Koch, H. (2005). Twenty-five years of BBC/ABB/Alstom lean premix combustion technologies. *J. Prop. Power* (ASME IGTI GT2005–68269?).

Flowers, D., Aceves, S., Westbrook, C.K., Smith, J.R., and Dibble, R. (2001). Detailed chemical kinetic simulation of natural gas HCCI combustion: Gas composition effects and investigation of control strategies. *J. Eng. Gas Turb. Power – Trans. ASME* **123**(2), 433–439.

Honda Motor Corporation (2006). Civic/CVV Fuel Economy Draws Praise in the US, http://world.honda.com/history/challenge/1972introducingthecvcc/text07/index.html.

Kaskan, W.E. (1959). Excess radical concentrations and the disappearance of carbon monoxide in flame gases from some lean flames. *Combust. Flame* **3**(1), 49–60.

Lee, R.C. and Wimmer, D.B. (1968). Exhaust emission abatement by fuel variations to produce lean combustion. *SAE Trans.* **77**, 175.

Levy, A. and Weinberg, F.J. (1959). Optical flame structure studies – examination of reaction rate laws in lean ethylene – air flames. *Combust. Flame* **3**(2), 229–253.

Li, H., Zhou, X., Jeffries, J.B., and Hanson, R.K. (2007). Sensing and control of combustion instabilities in swirl-stabilized combustors using diode-laser absorption. *AIAA J.* **45**(2), 390–398.

Mantzaras, J. (2006). Understanding and modeling of thermofluidic processes in catalytic combustion. *Catal. Today* **117**, 394–406.

Mason, W. and Wheeler, R.V. (1918). The effect of temperature and of pressure on the limits of inflammability of mixtures of methane and air. *J. Chem. Soc.* **113**, 45–57.

Muruganandam, T.M., Nair, S., Scarborough, D., Neumeier, Y., Jagoda, J., Lieuwen, T., Seitzman, J., and Zinn, B. (2005). Active control of lean blowout for turbine engine combustors. *J. Prop. Power.* **21**(5), 807–814.

National Academy of Sciences (2006). *State and Federal Standards for Mobile-Source Emissions*. The National Academy Press, Washington, DC.

Parker, A. (1914). The lower limits of inflammation of methane with mixtures of oxygen and nitrogen. *Trans. J. Chem. Soc.* **105**, 1002.

Reinke, M., Mantzaras, J., Bombach, R., Schenker, S., Tylli, N., and Boulouchos, K. (2007). Effects of H_2O and CO_2 dilution on the catalytic and gas-phase combustion of methane over platinum at elevated pressures. *Combust. Sci. Tech.* **179**(3), 553–600.

South Coast Air Quality Management District (SCAQMD) (2006). The southland's war on smog: Fifty years of progress toward clean air. http:///www.aqmd.gov/news1/Archives/History/marchcov.html with an access date of December.

Taupin, B., Cabot, G., Martins, G., Vauchelles, D., and Boukhalfa, A. (2007). Experimental study of stability, structure and CH^* chemiluminescence in a pressurized lean premixed methane turbulent flame. *Combust. Sci. Tech.* **179**(1–2), 117–136.

Warren, G.B. (1966). Fuel-economy gains from heated lean air–fuel mixtures in motorcar operation. 65-WA/APC-2. *Mech. Eng.* **88**(4), 84.

Wooley, D.R. and Morss, E.M. (2001). The Clean Air Act Amendments of 1990: Opportunities for promoting renewable energy. National Renewable Energy Laboratory Paper NREL/SR-620–29448.

Yeom, K., Jang, J., and Bae, C. (2007). Homogeneous charge compression ignition of LPG and gasoline using variable valve tifigloadming in an engine. *Fuel* **86**(4), 494–503.

Yi, T.X. and Gutmark, E.J. (2007). Combustion instabilities and control of a multiswirl atmospheric combustor. *J. Eng. Gas Turb. Power – Trans. ASME* **129**(1), 31–37.

Chapter 2

Fundamentals of Lean Combustion

Derek Bradley

Nomenclature

a	Acoustic velocity
A	Arrhenius constant, and area of elemental surface
A'	Empirical constant = 16, in relationship between λ and l
C_p	Specific heat at constant pressure
D	Diffusion coefficient of deficient reactant
E	Global activation energy
fn_s	Largest unstable wave number
F	Ratio of flame speeds
k	Thermal conductivity
K	Weighting factor
K	Turbulent Karlovitz stretch factor, $(\delta_\ell/u_\ell)(u'/\lambda)$
$K_{\ell c}$	Laminar Karlovitz stretch factor, $\alpha_c(\delta_\ell/u_\ell)$, for flame curvature
$K_{\ell s}$	Laminar Karlovitz stretch factor, $\alpha_s(\delta_\ell/u_\ell)$, for aerodynamic strain
l	Integral length scale of turbulence
Le	Lewis number, $k/\rho CpD$
Ma_c	Markstein number for flame curvature, based on reaction zone surface
Ma_s	Markstein number for aerodynamic strain rate, based on reaction zone surface
MON	Motor octane number
n	Inverse exponent for pressure dependency of τ_i, and wave number
n_l	Smallest unstable wave number
OI	Octane index
ON	Octane number
p	Pressure
Pe	Peclet number
Pe_{cl}	Critical Peclet number
r	Engine compression ratio, and radial distance in hot spot
r_o	Hot spot radius

R	Universal gas constant
R_l	Turbulent Reynolds number based on integral length scale of turbulence, $u'l/v$
RON	Research octane number
s	Dimensionless stretch rate, $\alpha\tau_\eta$
s_{q+}	Dimensionless positive stretch rate for laminar flame extinction, $\alpha_{q+}(\lambda/u')15^{-1/2}$
s_{q-}	Dimensionless negative stretch rate for laminar flame extinction, $\alpha_{q-}(\lambda/u')15^{-1/2}$
t	Time
T	Temperature (K)
u'	Rms turbulent velocity
u_a	Autoignition front propagation velocity
u_ℓ	Unstretched laminar burning velocity
u_n	Stretched and unstable laminar burning velocity
u_t	Turbulent burning velocity
α	Total flame stretch rate, $\alpha_s + \alpha_c$
α_c	Contribution of flame curvature to flame stretch rate
α_{q+}	Positive stretch rate for laminar flame extinction
α_{q-}	Negative stretch rate for laminar flame extinction
α_s	Contribution of aerodynamic strain to flame stretch rate
δ_ℓ	Laminar flame thickness, v/u_ℓ
ε	Aerodynamic/excitation time
λ	Taylor length scale of turbulence
Λ	Normalized wavelength
Λ_s	Smallest normalized unstable wavelength
v	Kinematic viscosity
ξ	Critical temperature gradient ratio
ξ_l	Lower detonation limit of ξ
ξ_u	Upper detonation limit of ξ
ρ	Density
τ_e	Excitation time
τ_i	Ignition delay time
τ_η	Kolmogorov time
ϕ	Equivalence ratio

2.1. COMBUSTION AND ENGINE PERFORMANCE

Historically, stoichiometric combustion has been regarded as the norm. One reason for this was that in many combustors the fuel and oxidant were introduced separately, in order to prevent any flash-back of the flame along the duct conveying the flammable premixture. With separate fuel and air streams, it was supposed that subsequent reaction would occur at the fuel/oxidant interface in a diffusion flame. In the case of laminar flow, the seminal paper of Burke and Schumann (1928) defined the flame as "the locus of those points where the rate of diffusion of combustible gas outward, and the rate of diffusion of oxygen inward, have the ratio required by the stoichiometric equation for complete combustion of the combustible gas." This concept of a stoichiometric flame front was widely adopted in descriptions of non-premixed fuel jet flames, even when they were turbulent.

However, with fuel jet flames discharging into air, premixing of fuel and air occurs at the base of the jet, in a region where the aerodynamic straining rates are usually sufficiently high to inhibit combustion. As a result, a high degree of mixing is possible before the strain rate relaxes sufficiently for combustion to occur. Consequently, there can be a wide range of local equivalence ratios, ϕ, in the mixture, some of which are very small (Bradley *et al.*, 1998a). A high jet velocity and small jet diameter skew the distribution of equivalence ratios towards lean mixtures. In the case of diesel engine fuel injectors with high injection pressures, this effect can eliminate sooting, if fuel droplet diameters are reduced to between 50 and 180 µm (Pickett and Siebers, 2002).

Another reason for near-stoichiometric combustion becoming the norm was that a slightly richer than stoichiometric mixture gave maximum power in gasoline engines. Later, the associated excessive emissions of carbon monoxide (CO), hydrocarbons, and oxides of nitrogen (NO_x), led to the introduction of three-way catalytic converters to reduce these emissions. For optimal operation, a sufficiently high temperature was necessary and, as a result, mixtures remained close to stoichiometric. However, increasing concerns about NO_x emissions led to the design of burners and engines for lean combustion at lower temperature. The advantages of lean combustion are not confined to possible reductions in emissions. In engines, the thermodynamic equilibrium properties of the lean reactants and products are such as to give increased thermal efficiencies.

Other factors also influence thermal efficiency and some general principles are illustrated by reference to the computed efficiencies of Otto (constant volume combustion) cycles at different work outputs, shown in Figure 2.1. These ideal performance characteristics are based on isentropic compressions and expansions, with constant volume combustion. Real thermodynamic equilibrium properties of iso-octane–air and its products of combustion are obtained from the Gaseq equilibrium code (Morley, 2007).

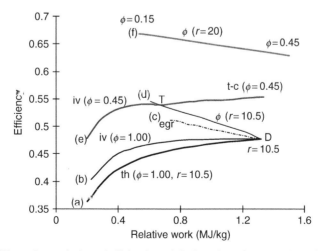

Figure 2.1 Different Otto cycle thermal efficiencies and single cycle work outputs. Control by (a) throttling, (b) inlet valve closure, (c) exhaust gas recirculation, (d) equivalence ratio, (e) inlet valve closure, iv, and turbo-charging, t-c, and (f) equivalence ratio at $r = 20$.

No account is taken of either the reciprocating engine speed or the burning rate of the mixture. The thermal efficiency is plotted against work done per kilogram per cycle, rather than engine power. Six different operational modes were computed. These show the effects, not only of decreasing ϕ, but also of a variety of different engine design details.

In ascending order of thermal efficiencies, for curve (a), labeled th, with $\phi = 1$ and an engine compression ratio, r, of 10.5, the amount of inlet charge is controlled by throttling the flow, so that the inlet pressure is less than atmospheric. "D" represents a datum point for these conditions and is for 1 kg of stoichiometric mixture initially at 0.1 MPa and 298 K. Increased throttling reduces both the work output and the efficiency. For curve (b), the output is controlled, not by throttling, but by varying the point in the cycle at which the inlet valve, iv, closes. This mode of control gives a superior efficiency to throttle control. The curves (c) and (d) show the effect of controlling the mixture. The highest output is at the datum condition, D. In curve (c), the output is reduced by an increasing amount of recirculated exhaust gas (egr) at 298 K, up to the point where its mole concentration is equal to that of air. In curve (d), the work is reduced by a decrease in ϕ from 1.0 to 0.45. In both cases, the efficiency increases as the work output decreases.

Still with $r = 10.5$, curve (e) is for lean burn with $\phi = 0.45$. The output is controlled by variable inlet valve closure at the lowest work output. This gives superior efficiency to curve (b) with $\phi = 1.0$. However, the maximum work with $\phi = 0.45$ is reduced and, to counter this, from the point T, higher outputs are achieved by turbo-charging. This increases up to a maximum charge pressure of 0.28 MPa. Isentropic efficiencies of 70% were assumed for compression and expansion. Finally, curve (f) exhibits the highest efficiencies and the advantage of both increasing r to 20, with an unchanged clearance volume, and decreasing ϕ from 0.45 to 0.15. Formally, it is the increase in *expansion* ratio, rather than the increase in *compression* ratio that increases the efficiency.

Fuel cells tend to have efficiencies that are higher than those on Figure 2.1 at low load, but which are lower at high load. The figure demonstrates not only the improved efficiency with leaner burning, but also the need to compensate for the decline in work output by pressurizing the charge. In general, increasing the pressure by turbo-charging followed by cooling improves both the cycle efficiency and the octane index (*OI*) of a gasoline fuel (see Section 2.3.3; Bradley *et al.*, 2004; Kalghatgi, 2005). Overall, the improvements in efficiency that arise from leaner burning are due to the increasing ratio of specific heats, as is apparent from the simple expression for the efficiency of the ideal air cycle in Chapter 4. In addition, for given octane numbers (*ON*s), a higher compression ratio is attainable with leaner mixtures before autoignition occurs.

With lean burn, CO_2 emissions are reduced by the improved efficiencies, and NO_x emissions are reduced by the lower temperatures of combustion. However, temperatures may become so low that the rates of oxidation of CO and unburned hydrocarbons, often formed in cylinder crevices, are seriously reduced. Aspects of practical engine design that lead to improved performance include high pressure fuel injection (aids evaporation, cools the charge), several fuel injections per cycle, variable valve timings, nitrogen oxide storage catalytic converters, and particulate filters. An increasingly large range of fuels is becoming available. This includes biofuels, reformed hydrocarbons, natural gas and hydrogen (H_2), liquid fuels synthesized from natural gas, H_2 and coal gas. This variable fuel situation necessitates an integrated design approach for matching the fuel blend and engine. Indeed, this applies to all combustion systems. The complex interplay of many controlling factors must be optimized to give high efficiency, sufficient power, and low, unwanted, emissions.

Lean combustion, of the type discussed above, has been variously termed dilute combustion, moderate and intense low oxygen dilution (MILD), flameless oxidation, homogeneous combustion, and low NO$_x$ injection. These are described in detail in Chapter 3. To help put the various elements of lean combustion into a fundamental framework, the present chapter covers aspects of flame propagation, flame quenching, flameless reaction, and the different modes of autoignition. Flame and flameless instabilities, and the different ways of achieving heat recirculation and flame stabilization also are discussed.

2.2. BURNING IN FLAMES

2.2.1. LAMINAR FLAMES AND FLAME STRETCH RATE

Within the limits of flammability, the reaction front of a premixed laminar flame involves many chain reactions that are sustained by thermal conduction and species diffusion at the leading edge, in the preheat zone. This is followed by a fuel consumption zone, or inner layer, in which the temperature rises rapidly and radical concentrations begin to fall. Finally, in hydrocarbon flames, H$_2$ and CO are oxidized and radical concentrations decay in an oxidation layer (Peters, 2000). The molecular transport processes combined with the chemical kinetics determine the unstretched laminar burning velocity, u_ℓ. This is the velocity at which the flame front propagates normal to itself, relative to any flow, into the unburned mixture. In principle, it can be computed from the detailed chemical kinetics and the relevant molecular transport coefficients (Warnatz *et al.*, 1996). Particular difficulties arise from uncertainties in the reaction rates for three body collisions, particularly at the higher pressures, and in the detailed chemical kinetics of rich hydrocarbon flames.

Two- or three-dimensional flames are stretched by the flow and this changes their burning velocity. The flame stretch rate, α, of a small surface element of area A is $(1/A)(dA/dt)$. Formal expressions for flame stretch rates, in terms of the strain rate and the stretch rate due to flame curvature have been presented by a number of workers (Bradley *et al.*, 1996). Both these terms, when multiplied by a chemical time given by the laminar flame thickness divided by the laminar burning velocity, δ_ℓ/u_ℓ, give the laminar Karlovitz stretch factors, $K_{\ell s}$ and $K_{\ell c}$, respectively.

The two contributions to the flame stretch rate (strain and curvature) change the value of the burning velocity. This change is conveniently expressed in dimensionless terms by the sums of the products of the appropriate K_ℓ and the associated Markstein number, Ma. The latter is a function of the density ratio and the product of the Zeldovich number (dimensionless activation energy for the burning velocity), and $(Le - 1)$, where Le is the Lewis number based on the diffusion coefficient for the deficient reactant (Clavin, 1985). The difference between the burning velocity of the unstretched flame, u_ℓ, and that of the stretched flame, u_n, when normalized by u_ℓ, is equal to the sum of the two products:

$$\frac{u_\ell - u_n}{u_\ell} = K_{\ell s} Ma_s + K_{\ell c} Ma_c. \qquad (2.1)$$

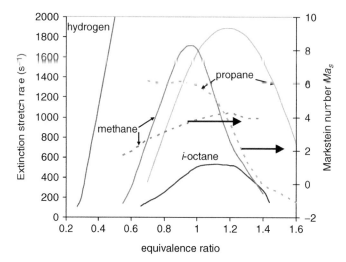

Figure 2.2 Positive stretch rates for extinction of CH$_4$-air and C$_3$H$_8$-air (Law *et al.*, 1986), hydrogen-air (Dong *et al.*, 2005), and *i*-octane-air (Holley *et al.*, 2006), under atmospheric conditions. Broken curves give Ma_s for C$_3$H$_8$-air (Bradley *et al.*, 1998a) and CH$_4$-air (Bradley *et al.*, 1996). (See color insert.)

The first term (strain rate) on the right of this linearized equation usually dominates over the second (curvature) term. At sufficiently high rates of stretch, the flame is extinguished. Five important physico-chemical parameters are necessary to characterize fully laminar burning: the unstretched laminar burning velocity, u_ℓ, two Markstein numbers and the extinction stretch rates for positive and negative stretch, α_{q+} and α_{q-}. Shown by the full line curves in Figure 2.2 shows some experimental values of α_{q+} for four different fuels at different equivalence ratios under atmospheric conditions. These were obtained from symmetrical counterflow flames of CH$_4$-air and C$_3$H$_8$-air by Law *et al.* (1986) and from flames of H$_2$-air and *i*-octane-air, counterflowing against an opposing jet of air or nitrogen, by Dong *et al.* (2005) and Holley *et al.* (2006), respectively. The broken curves give computed values of Ma_s for CH$_4$-air (Bradley *et al.*, 1996) and C$_3$H$_8$-air (Bradley *et al.*, 1998a).

Attempts have been made to generalize laminar burning velocity data in terms of the molecular structure of fuels. Bradley *et al.* (1991) found that u_ℓ for lean mixtures varies approximately linearly with the heat of reaction of a mole of mixture, but in different ways for different types of fuel. Farrell *et al.* (2004) studied burning velocities of a wide range of hydrocarbons and gave relative values for alkanes, alkenes, alkynes, aromatics, and oxygenates. There is a dearth of burning velocity data at high temperature and pressure with even bigger gaps in the data for α_{q+}, α_{q-}, and Markstein numbers. Such information is particularly critical for lean combustion because Figure 2.2 shows that these flames are susceptible to stretch extinction.

2.2.2. FLAME INSTABILITIES

Although ultimately capable of causing extinction, flame stretch also helps suppress instabilities. As the flame stretch rate decreases, so does its stabilizing effect, until eventually the flame becomes unstable and cellular structures develop through a

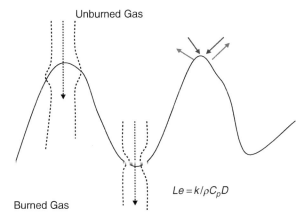

Figure 2.3   Darrieus–Landau and thermo-diffusive instabilities.

combination of Darrieus–Landau (D–L) instabilities with thermo-diffusive (T–D) effects. These wrinkle the flame, increasing the burning rate so that Equation (2.1) no longer holds. The mechanisms of these instabilities are illustrated in Figure 2.3, which shows a wrinkled flame and the streamlines through it. At the crest of the wave, the streamlines in the unburned gas diverge ahead of the flame surface. The velocity into the flame, therefore, decreases and, as a result, the burning velocity propagates the flame further into the unburned gas. There is an opposite effect at the valley bottom and this increases the flame wrinkling. This phenomenon is the D–L instability. With $Le < 1$, at the crest of the wave, the focused diffusion of enthalpy into the flames, shown by the arrows, exceeds the conductive energy loss and the local burning velocity increases. The opposite occurs in the valley and, as a consequence, the flame wrinkling increases. This is the T–D instability. Similar reasoning also shows the inverse and greater flame stability with $Le > 1$.

Spherical explosion flames demonstrate important aspects of flame instability. The flames are unstable between inner and outer wavelength limits, which can be derived from the theory of Bechtold and Matalon (1987). The ranges of the limiting unstable wave numbers that define a peninsula of instability are shown diagrammatically in Figure 2.4. The Peclet number, Pe, plotted on the x-axis, is the flame radius normalized by δ_ℓ. The instability begins to develop at a critical Peclet number, Pe_{cl}, from which a corresponding critical Karlovitz number can be derived. Plotted against Pe are the limiting wave numbers. The theory has been modified by Bradley (1999) to give values of Pe_{cl} corresponding to those found experimentally. The upper limit is for the largest unstable wave number, fn_s (smallest wavelength), and the lower line is for the smallest unstable wave number, n_l (largest wavelength). All intermediate wave numbers within the peninsula are unstable. The numerical constant, f, is selected to modify the theoretical value of the largest unstable wave number, n_s, to give the modified wave number fn_s that makes Pe_{cl} equal to the experimental value.

The wave numbers, n, are related to dimensionless wavelengths by

$$n = 2\pi Pe/\Lambda, \tag{2.2}$$

where Λ is the wavelength normalized by δ_ℓ. The gradient,

$$(dfn_s/dPe) = 2\pi/\Lambda_s, \tag{2.3}$$

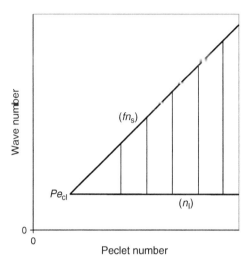

Figure 2.4   Instability peninsula with limiting wave numbers. Pe_{cl} is the critical Peclet number.

increases as Ma_s decreases, as must the flame instability. Experiments suggest that the smallest unstable wavelength, Λ_s, has a lower limit equal to about 50, with negative values of Ma_s (Bradley *et al.*, 2000).

Fractal considerations (Bradley, 1999) give the ratio of the fractal surface area with a resolution of the inner cut-off to that with a resolution of the outer cut-off as $(fn_s/n_l)^{1/3}$. The ratio, F, of flame speeds with instabilities to that without them is equal to the ratio of the surface areas:

$$F = (fn_s/n_l)^{1/3}. \tag{2.4}$$

Explosions at higher pressures are particularly prone to such instabilities, for two reasons: the smaller values of δ_ℓ imply Pe_{cl} is attained at smaller flame radii, and values of Ma_s tend to decrease with pressure. Unstable burning velocities can become several times greater than u_ℓ and this poses a problem for the measurement of u_ℓ at high pressure. Shown in Figure 2.5 are values of u_ℓ, obtained from explosions of *i*-octane-air at different values of ϕ, for five different p and T combinations with isentropic compression from 1 MPa and 358 K (Al-Shahrany *et al.*, 2005). D–L, T-D instabilities became quite pronounced as the flame radius and pressure increased. This was particularly so with the negative Markstein numbers of the rich mixtures, in which the burning velocities were increased almost seven-fold. Lean mixtures showed less variation with initial temperature and final compression. Values of F were derived, as indicated above, and this enabled the values of u_ℓ to be obtained from the measured unstable burning velocities, u_n. Similarly, derived values of u_ℓ at high pressure for lean mixtures of H_2-air have been presented by Bradley *et al.* (2007b).

The heat release rate of a propagating flame is the product of the heat of reaction, flame area, unburned gas density, and the burning velocity. A rapid change in this occurs with unstable explosion flames due to flame propagation and wrinkling can trigger pressure oscillations (Singh *et al.*, 2005). These can further wrinkle the flame surface due to Taylor instabilities. These arise when pressure fluctuations align non-orthogonally to the high density gradients of parts of the flame surface and generate vorticity through the baroclinic term, $\nabla(1/\rho) \times \nabla p$ (Batley *et al.*, 1996). As a consequence, the rate of change

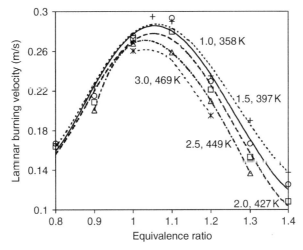

Figure 2.5 Values of u_ℓ for *i*-octane-air, at five different p and T in isentropic compression from 1 MPa and 358 K. Pressures in MPa followed by temperature.

of the heat release rate is increased still further by wrinkling and the pressure pulses are further strengthened. The underlying phenomenon was demonstrated by Markstein (1964), who showed how a smooth spherical laminar flame became severely wrinkled when it was struck by a shock wave. Shown in Figure 2.6 is the influence of a pressure pulse, originating from the opening of a vent during an explosion, upon the propagating flame front. The pulse, propagating to the left, hits the leading edge of the flame, propagating to the right, and accelerates the less dense burned gas backwards into that gas.

Acoustic waves can be further strengthened by a sufficient energy release at the flame, provided it feeds into the positive phase of the pressure oscillations. This Rayleigh (1878) instability is usually coupled with the Taylor instability. Not infrequently, particularly in explosions at high pressures, D–L, T-D instabilities act as a trigger for Rayleigh–Taylor instability. The consequent Rayleigh–Taylor instability,

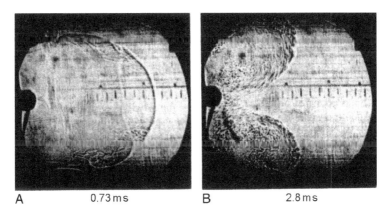

Figure 2.6 Schlieren ciné frames of flame propagating in closed vessel (A) before vent opening and (B) after vent opening, showing flame front deformation by resulting pressure pulse (Bradley and Harper, 1994).

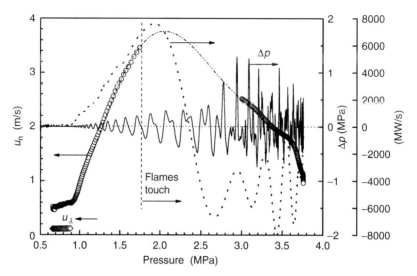

Figure 2.7 Changes in u_n and pressure oscillations during inward propagation of two near-simultaneous flames for iso-octane-air, $\phi = 1.6$, and initial conditions 0.5 MPa and 358 K (Al-Shahrany *et al.*, 2006). Dotted curve shows rate of change of heat release rate.

along with the associated strong feedback mechanism, can generate even higher burning velocities and large pressure amplitudes.

This sequence of events is illustrated in Figure 2.7, taken from Al-Shahrany *et al.* (2006). It shows the consequences of the changing burning velocity of two similar unstable imploding flames, within a spherical explosion bomb, as shown for another mixture in Figure 2.8 (Al-Shahrany *et al.*, 2005). For the conditions of Figure 2.7, a quiescent mixture of *i*-octane-air, $\phi = 1.6$, at initially 0.5 MPa and 358 K, which produced very unstable flames, was spark-ignited simultaneously at two diametrical opposite points, at the wall of the bomb. The flame was ciné-photographed and the pressure measured.

Figure 2.7 shows the measured values of the unstable burning velocity, u_n, plotted against the increasing pressure, p, by the circle symbols. Also shown, by the diamond symbols, are values of u_ℓ obtained by allowing for D–L, T–D instabilities, as described above. These latter values are close to those estimated from the direct measurements of Lawes *et al.* (2005). It, therefore, would seem that the elevation of burning velocities in this regime is entirely due to D–L, T–D instabilities. The rapid increase in u_n thereafter, from a value of about 0.6 m/s, could be a consequence of the rapid increase in the rate of change of heat release rate, shown by the dotted curve, to the high value of 7615 MW/s and the development of secondary acoustic (SA) oscillations of high amplitude, Δp, shown by the continuous curve. These further wrinkled the flame through Rayleigh–Taylor instabilities and it was no longer possible to derive u_ℓ from u_n. The values of u_n and Δp continued to increase by this positive feedback mechanism.

The figure also shows the two flames to have coalesced at the leading points before the maximum rate of change of heat release rate and the maximum value of u_n were attained. Because the method of deriving u_n after the two flames had made contact is less reliable than that employed before contact, the maximum value of u_n is uncertain, but it is estimated to be about 3.75 m/s. The reduction in the maximum wavelength of

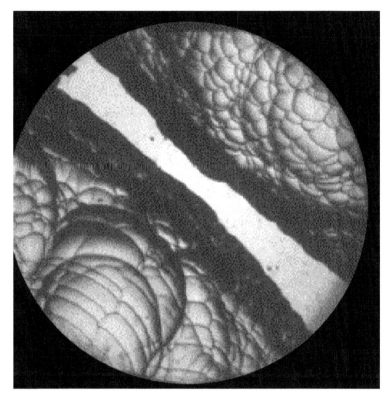

Figure 2.8 Schlieren image of closing twin flame kernels for CH_4-air at $\phi = 1$, initially at 0.1 MPa and 298 K.

the wrinkling caused u_n to decline sharply, although the now negative rate of change of the heat release rate maintained the strong pressure oscillations, with an amplitude of about 0.5 MPa. For the temperature and pressure at the estimated peak value of u_n, the estimated values of u_ℓ and Markstein number were 0.1 m/s and –34.

Shown in Figure 2.9 are similar records, derived for a comparable negative Ma_s, with two imploding lean H_2-air flames, $\phi = 0.4$, also initially at 0.5 MPa and 358 K. Again, D–L, T–D instabilities elevated u_n above u_ℓ, but an important difference with the rich *i*-octane-air flame was that, with this mixture, the rate of change of heat release rate was less than a twentieth of that in Figure 2.7. As a result, less energy was fed into the pressure pulses and there was no growth in the amplitude of the primary acoustic (PA) oscillations, Δp, comparable to that occurring in Figure 2.7. Consequently, there was no apparent further augmentation of u_n due to Rayleigh–Taylor instabilities.

On the basis of this assumption, values of u_ℓ could be derived allowing solely for D–L, T–D instabilities, and these are shown by the diamond symbols. These values are somewhat higher than those predicted by the chemical kinetic scheme of Ó Conaire *et al.* (2004), also shown in the figure, and somewhat lower than those measured by Verhelst *et al.* (2005). There is much uncertainty in both modeled and experimental values of u_ℓ for lean H_2 mixtures at high pressure and temperature.

Al-Shahrany *et al.* (2006) also studied the more stable flames of a stoichiometric iso-octane-air implosion. Results from all three mixtures are summarized in Table 2.1. It is difficult to establish values of Markstein number with any certainty, particularly for the

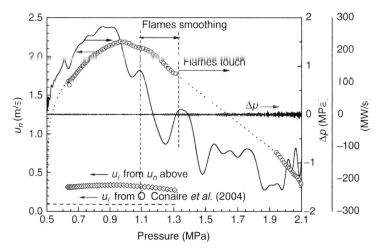

Figure 2.9   Changes in u_n and pressure oscillations during inward propagation of two near-simultaneous flames for H_2-air, $\phi = 0.4$, and initial conditions 0.5 MPa and 358 K (Al-Shahrany *et al.*, 2006). Dotted curve shows rate of change of heat release rate.

Table 2.1

Summary of experimental findings, particularly for maximum value of u_n. MRCHRR is maximum rate of change of heat release rate. PA is primary acoustic and SA secondary acoustic.

Mixture	u_ℓ at $(u_n)_{max}$(m/s)	Ma_s	Δp at $(u_n)_{max}$(kPa)	MRCHRR (MW/s)	u_n/u_ℓ at $(u_n)_{max}$	Type of instability
i-C_8H_{18}-air $\phi = 1.0$	0.295	2	0	190	2.24	D–L, T-D
H_2-air $\phi = 0.4$	0.34	−22	0.8	270	6.44	D–L, T-D, PA
i-C_8H_{18}-air $\phi = 1.6$	0.10	−34	300	7615	37.5	D–L, T-D, PA, SA

unstable flames. The ratio u_n/u_ℓ, at the maximum value of u_n, is indicative of the enhancement of the burning rate due to instabilities. The ratio increases with the decrease in Ma_s. Both the lean H_2 mixture and the rich *i*-octane mixture had very negative Markstein numbers, but perhaps the crucial difference between them, leading to the highest value of u_n/u_ℓ of 37.5, was the much higher rate of change of heat release rate for the rich *i*-octane mixture. This led to strong SA oscillations, with an amplitude, Δp, of 0.3 MPa at $(u_n)_{max}$ and increasing flame wrinkling due to Rayleigh–Taylor instabilities. The much lower value of Δp for the H_2 mixture is indicative of the absence of strong SA instabilities. There was even evidence of a degree of stabilization of D–L, T-D instabilities by the PA oscillations with this mixture.

These implosion phenomena parallel those observed during flame propagation along tubes. In the experiments with stoichiometric C_3H_8-air of Searby (1992), the secondary thermo-acoustic instability ultimately destroyed the cellular structure and u_n/u_ℓ reached a maximum value of about 18, in what appeared to be turbulent flow. Negative Markstein numbers were by no means essential for secondary instabilities, but an important difference with bomb explosions is that the tubes exhibit more coherent excitable resonant acoustic frequencies. The data from the tube explosions with CH_4-air of Aldredge and Killingsworth (2004) suggested a ratio of u_n/u_ℓ of about 6.4 at $\phi = 0.8$ and of about 2.9 at the larger Markstein number at $\phi = 1.2$.

Knocking combustion in engines usually is attributed to end gas autoignition, ahead of the flame front, and there is much photographic evidence to support this. However, the type of instabilities described here might also contribute to "objectionable noise," the definition of knock given by Miller (1947) (see Section 2.3.2). The instabilities of flames in continuous flow combustors and how they might be stabilized are discussed in Section 2.4.

2.2.3. TURBULENT FLAMES AND FLAME QUENCHING

Rather than relying on instability development, a more common way of increasing the overall burn rate in flames is to introduce turbulence into the flow field. The turbulent burning velocity increases with the root mean square (rms) turbulent velocity, u'. When u' is divided by the Taylor scale of turbulence λ for isotropic turbulence, it defines the rms strain rate on a randomly orientated surface (Bradley *et al.*, 1992). If this becomes sufficiently high it can extinguish the flame. Normalized by the chemical time, δ_ℓ/u_ℓ, it gives the turbulent Karlovitz stretch factor, K:

$$K = (u'/\lambda)(\delta_\ell/u_\ell). \tag{2.5}$$

The value of δ_ℓ is approximated by v/u_ℓ. For isotropic turbulence, λ is related empirically to the integral length scale, l, by $\lambda^2/l = A'v/u'$, where A' is an empirical constant, here assigned a value of 16. These relationships, when substituted into Equation (2.5), yield

$$K = 0.25(u'/u_\ell)^2 R_l^{-1/2}, \tag{2.6}$$

in which R_l is the turbulent Reynolds number based on l.

Flames in turbulent flow are exposed to a range of stretch rates, and probability density functions (pdfs) of these, $p(s)$, are shown in Figure 2.10. Here s is a dimensionless stretch rate, the product of α and the Kolmogorov time, $\tau_\eta, = (\lambda/u')15^{-1/2}$; s has

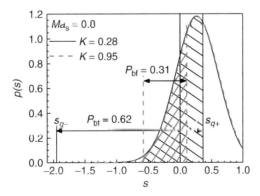

Figure 2.10   Ranges of stretch rates on $p(s)$ curve for $K = 0.28$ and 0.95. Positive and negative extinction stretch rates are s_{q+} and s_{q-}, and dotted line gives rms strain rate. (See color insert.)

been synthesized from the strain and the curvature contributions to the flame stretch rate, and $p(s)$ is a function of K, Ma_s and R_l (Bradley *et al.*, 2003). Two values of K are considered here, 0.28 and 0.95, with $Ma_s = 0.8$, applicable to *i*-octane–air, $\phi = 0.8$, at 1.0 MPa and 365 K. In this instance, with the high value of R_l, the two pdfs of s are almost identical. For $\alpha = u'/\lambda$, $s = 15^{-1/2} = 0.258$.

Dimensionless positive and negative extinction stretch rates, s_{q+} and s_{q-}, given by $\alpha_{q+}(\lambda/u')15^{-1/2}$ and $\alpha_{q-}(\lambda/u')15^{-1/2}$, are shown for the mixture at both values of K. Equation (2.5) gives values of λ/u'. The values of extinction stretch rate were obtained under steady state conditions, with values such as those depicted in Figure 2.2.

The horizontal, arrowed, lines indicate the corresponding range of stretch rates, which decreases as K increases, within which flamelets can propagate for the two values of K. The probability of a flame being able to propagate, unquenched, at a given position might be expressed by

$$P_{bf} = \int_{s_{q-}}^{s_{q+}} p(s)\ ds. \tag{2.7}$$

For $K = 0.28$ and 0.95, the respective shaded areas in Figure 2.10 give probabilities, P_{bf}, of 0.62 and 0.31, respectively. In fact, measured values of probabilities of complete flame propagation, p_f, in a fan-stirred bomb for these conditions were 1.0 and 0.6, higher than these values of P_{bf} (Bradley *et al.*, 2007a). One explanation of this is that under turbulent, non-steady conditions, there is insufficient time for extinction by positive stretch rates to occur (Donbar *et al.*, 2001). Consequently, ranges of stretch rates over which flames can be sustained are somewhat larger than those suggested by the arrowed spans in Figure 2.10. This time lag is particularly marked for positive stretch rates. Much less is known about the magnitude of negative stretch rates and their behavior, but Mueller *et al.* (1996) found that the time lag in the response to changing negative stretch rate was much less.

The correlation of a range of experimental observations on turbulent flame quenching for a variety of mixtures at atmospheric pressure, in different apparatus, suggested a dependence of quenching upon K and the Lewis number, Le. It was found that a 20% probability of total flame quenching ($p_f = 80\%$) occurred for $K \times Le = 2.4$ (Abdel-Gayed and Bradley, 1985). Increases of $K \times Le$ above this numerical value increased the probability of quenching. However, Markstein numbers are a more fundamental parameter in this context than Le. Later work in a fan-stirred bomb, in which u' was varied by changing the speed of rotation of four symmetrically disposed fans with additional mixtures at higher pressures and over a wide range of Ma_s, suggests that significantly higher values of K are sometimes necessary to induce quenching of an established flame. This is particularly so for negative values of Ma_s. The studies have led to the proposed expression for 20% probability of total quenching (Bradley *et al.*, 2007a):

$$K(Ma_s + 4)^{1.4} = 37.1, \quad \text{provided } -3.0 \le Ma_{sr} \le 11.0. \tag{2.8}$$

Turbulent flame quenching in engines usually becomes more probable as ϕ is reduced. It eventually sets a limit to lean burn and also to the associated increases in thermal efficiency that are suggested in Figure 2.1. Most hydrocarbon fuels burn erratically, if at all, when ϕ is reduced below about 0.7. With lean H_2–air mixtures, this limit is about $\phi = 0.2$. With $u' = 2$ m/s, the lowest values of ϕ measured in the fan-stirred bomb for 20% probability of total flame quenching at 1.0 MPa and 365 K are

Table 2.2

Comparisons of fan-stirred bomb limit ϕ for 20% probability of total flame quenching at 1.0 MPa and 365 K, $u' = 2$ m/s, with observed limits in engines. LNT indicates lean NO_x trap necessary.

Mixture	ϕ bomb	ϕ engine	Ma_s	u_ℓ (m/s)	NO_x (ppm)	Ind. thermal eff. (%)
i-octane-air	0.78	0.71	0.8	0.150	LNT	unknown
CH_4-air	0.57	0.62	−2.2	0.017	20	41.1
H_2-air	0.17	0.20	−2.5	0.030	0	33.2

given in Table 2.2, for three lean mixtures. The two lighter-than-oxygen fuel molecules have negative values of Ma_s and are less easily quenched than the i-octane–air mixture, for which the bomb lean limit is $\phi = 0.78$. This compares with a measured lean limit of $\phi = 0.71$ for stable running of a direct injection, spark ignition, and gasoline engine with a pressurized air supply (Stokes *et al.*, 2000). For isentropic compression to 1.0 MPa, this created a slightly higher temperature than that in the bomb. At this value of ϕ, the NO_x emissions were such as to necessitate a lean NO_x trap.

Similarly, the 20% probability of total flame quenching ($p_f = 80\%$) in the bomb occurred for CH_4-air at $\phi = 0.57$ compared with an engine limit of $\phi = 0.62$ for natural gas (listed as CH_4 in Table 2.2), which was observed by Van Blarigan (1996). He used a single cylinder Onan engine, with $r = 14$, running at 1800 rpm, to find the minimum value of ϕ at which the engine would operate with different fuels. At this minimum value of ϕ, the exhaust NO_x concentration was 20 ppm and the indicated thermal efficiency was 41.1%. For the very lean H_2 mixture, the limiting engine value of $\phi = 0.20$ compares with a bomb value of 0.17 for 20% quenching probability. The engine NO_x emission was effectively zero and the indicated thermal efficiency was 33.2% (Aceves and Smith, 1997). A higher thermal efficiency of 43.2% was achievable when ϕ was increased to 0.35, at which the NO_x emission was 2 ppm. The addition of H_2 to natural gas further lowered the observed lean engine limit from $\phi = 0.62$. With a composition of 67.5%, CH_4 and 2.5% C_2H_6 with 30.0% H_2, Van Blarigan measured a limit value of $\phi = 0.48$.

The addition of H_2 to hydrocarbon fuels generally increases laminar and turbulent burning velocities and reduces Ma_s (Mandilas *et al.*, 2007). On-board, catalytic partial oxidation of hydrocarbon fuels, such as gasoline or methanol, can generate mixtures of H_2 and CO. The presence of H_2 yields a value of u_ℓ for the partial oxidation products that is higher than that of the parent fuel. There is a consequent extension of the lean turbulent burning limit (Jamal and Wyszyński, 1994).

2.2.4. TURBULENT COMBUSTION REGIMES AND TURBULENT BURNING VELOCITY

The value of K influences not only flame quenching, but also the turbulent burning velocity and the nature of turbulent burning. The different regimes of premixed combustion can be defined diagrammatically, in terms of velocity and length scale ratios and this is done in Figure 2.11. Several diagrams of this type have been proposed (Peters, 2000). Figure 2.11 is of the type proposed by Borghi (1985), with the experimental regimes observed by Abdel-Gayed *et al.* (1989). Values of u'/u_ℓ are plotted logarithmically

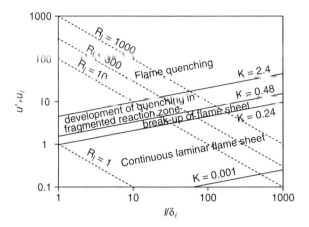

Figure 2.11   Regimes of premixed turbulent combustion.

against those of l/δ_ℓ, with lines of constant K and R_l. The different flame regimes, for Le close to unity, are described between limiting values of these lines. It is to be anticipated that appropriately normalized values of the turbulent burning velocity, u_t, would depend upon the other dimensionless parameters in the figure, as well as upon Le or Ma_s.

Expressions for u_t/u' and u_t/u_ℓ have been developed by several experimentalists and summarized by Peters (2000) and Bradley (2002), although not all of the expressions involved length scales. Allowance also must be made for Le, or Ma_s, as a decrease in these parameters increases u_t/u_ℓ and u_t/u' (Abdel-Gayed *et al.*, 1985). An expression that involves a length scale (contained within K, see Equation (2.6)) and Le is (Bradley *et al.*, 1992)

$$u_t/u' = 1.01(K \times Le)^{-0.3} \quad \text{for } 0.05 \leq K \times Le \leq 1.0, \text{ with } A' = 16. \qquad (2.9)$$

It shows u_t/u' to decrease progressively from a maximum value of 2.48 at the lower limit of $K \times Le$, to almost unity at the upper limit. An increase in u' increases u_t, but gives diminishing returns as flamelet extinctions develop. Fragmentation of the reaction zone becomes severe and the burning becomes erratic, so the limit for reasonably well controlled burning is close to $K \times Le = 1.0$.

It is rather more logical to correlate with $K \times Ma_s$ than with $K \times Le$ and this is done on Figure 2.12, taken from Bradley *et al.* (2005). The transformation of Equation (2.9) to Equation (2.10) is based on the data for Le and Ma_s for CH_4-air and C_3H_8–air mixtures, at atmospheric conditions and different ϕ. Equation (2.9) is shown, along with the correlation:

$$u_t/u' = U = 1.41(KMa_s)^{-0.43}. \qquad (2.10)$$

A limitation of many of the earlier expressions for u_t, including Equations (2.9) and (2.10), is that they were based on measurements at atmospheric pressure and with Le close to unity. More recently, Kobayashi *et al.* (1996, 1998, 2005), with pressures up to 3 MPa, and Lawes *et al.* (2005) at 0.5 MPa, have obtained high pressure measurements of u_t using a burner and fan-stirred explosion bomb, respectively. Kobayashi *et al.* correlated their results by expressing u_t/u_ℓ as a function of the pressure

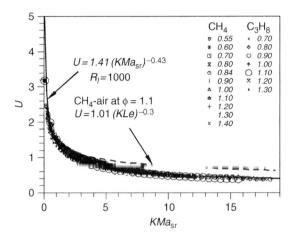

Figure 2.12 Variations of u_t/u' ($= U$) with KMa_s (Bradley *et al.*, 2005).

in atmospheres and u'/u_ℓ for different fuels. Higher pressures give higher values of u_t than do Equations (2.9) and (2.10). In a computational study, Bradley *et al.* (2005) attributed this to the smaller (and often negative) values of Ma_s at high pressure. These enhance the burning velocity in laminar flamelets (see Equation (2.1)). In addition, it was suggested that negative values of Ma_s tend to be associated with higher values of α_{q+}, the extinction stretch rate.

The transitional regime, $0 \leq K \leq 0.05$, behaves rather differently. As K is reduced to very low values, D–L instabilities become important and give rise to increasing values of u_t/u' (Cambray and Joulin, 1992, 1994; Kobayashi *et al.*, 1996, 1998; Bradley *et al.*, 2005), until at $K = 0$ the burning velocity is that of an unstable laminar flame. Fan-stirred bomb studies show that as K is increased in this transitional regime, the amplitudes of the pressure oscillations associated with any D–L, T-D, and Rayleigh–Taylor instabilities are progressively reduced. This has been attributed to the domination of flame wrinkling due to the imposed turbulence over that due to the original instabilities (Al-Shahrany *et al.*, 2006).

2.3. AUTOIGNITIVE BURNING

2.3.1. IGNITION DELAY TIME

When the value of K becomes so high that flames cannot be maintained, an alternative strategy is to employ flameless, autoignitive burning. This strategy has been adopted in "homogeneous charge" compression ignition (HCCI) engines and also for mild combustion in gas turbines (see Chapter 3). Unlike flame propagation, this type of burning is not driven by the molecular transport processes of diffusion and thermal conduction. It may occur under conditions where flame propagation is not possible, and u_ℓ is no longer the appropriate parameter to express the burning rate.

A homogeneous flammable mixture maintained at temperature T and pressure p will autoignite after a time, τ_i, the autoignition delay time. As with u_ℓ, τ_i can be computed from detailed chemical kinetics. The major problem is to quantify the details of

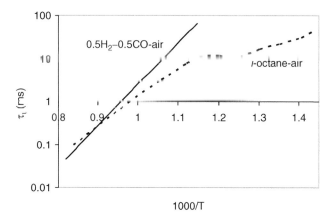

Figure 2.13   Variations of τ_i with 1000/T for two mixtures with $\phi = 0.50$ and $p = 4.0\,MPa$.

hydrocarbon breakdown and reactions of intermediates at lower temperatures (Pilling and Hancock, 1997). Under atmospheric conditions for most fuel–air mixtures, the value of τ_i is very large. Consequently, there is an extremely low probability of autoignition.

However, a different situation arises at high temperature. Variations of τ_i with $1000/T$ at 4 MPa and $\phi = 0.50$ are shown in Figure 2.13 for $0.5H_2 - 0.5CO$-air and for i-octane-air. The former comprises the modeled values of Bradley *et al.* (2002). The plotted relationship is close to a straight line of an Arrhenius rate law, the gradient of which yields a value of *E/R*, a global activation temperature.

For i-octane-air, the broken line relationship is comprised of the shock tube data of Fieweger *et al.* (1997), and at the lower temperatures, the chemical kinetically modeled values of Curran *et al.* (2002). At the higher temperatures, the plot is again close to a straight line, but as the temperature is reduced, so also is the gradient of the line – to such an extent that it even becomes slightly negative in the range 910 to 820 K, thereafter increasing sharply below 750 K. This tendency towards a zero, and even a negative temperature gradient becomes even more pronounced for the lower *ON*s of primary reference fuels, PRF (i-octane and n-heptane) (Fieweger *et al.*, 1997). *ON* is the percentage of i-octane in the PRF. At a given temperature and pressure, τ_i also decreases with the *ON* of a PRF. At 4 MPa and 900 K, τ_i for i-octane-air is 2.3 ms and for n-heptane-air it is 0.52 ms. Recirculation of both heat and hot gas can increase the temperature of reactants to give sufficiently low values of τ_i for autoignitive, flameless combustion. This can occur under conditions when flame propagation is not possible and is discussed further in Section 2.4.

2.3.2. AUTOIGNITION INSTABILITIES

Chemically driven instabilities can arise in flameless combustion. These take the form of low temperature oscillations in the regime of negative temperature coefficients for τ_i. For example, the reaction of an alkane radical with OH can create an alkyl radical. This reacts with O_2 to form a peroxy radical, which reacts to give degenerate branching agents, thereby accelerating the heat release rate and increasing the temperature. However, in this temperature range, this reverses the formation of peroxy radicals and the temperature falls (Proudler *et al.*, 1991). In a well-stirred reactor, these competing phenomena create temperature fluctuations and oscillatory reactions.

In practice, it is impossible to generate a perfectly homogeneous mixture. Within the mixture, small, yet significant localized gradients of temperature and concentration are inevitable. As a consequence, any autoignition is initiated at several separated locations, rather than instantaneously and uniformly throughout the mixture in a thermal explosion. Three-dimensional laser induced fluorescence in an HCCI engine has confirmed the occurrence of multiple isolated ignition spots, with ensuing combustion at distributed sites, rather than through flame propagation in a continuous flame sheet (Nygren *et al.*, 2002).

Autoignition at several such "hot spots" can create a high rate of change of the heat release rate. This, in turn, can generate strong pressure pulses, as described in Section 2.2.2. In spark ignition engines, these pulses can create Rayleigh–Taylor instabilities at the main flame front that further increase the burn rate by additional wrinkling. High speed photographs at 63 000 frames/s of autoignition and knock in a spark ignition engine suggest the presence of such instabilities (Kawahara *et al.*, 2007). Knock is less severe with lean mixtures.

2.3.3. AUTOIGNITION IN ENGINES

In practice, engine fuels are not characterized by a range of autoignition delay times, but by their cetane number (*CN*) for diesel fuels and by their research octane number (*RON*) and motor octane number (MON) for gasolines (Heywood, 1989). *ON*s are a guide as to whether gasolines will "knock." This might be taken as an indication of autoignition, though not of the details of it. Values of *RON* and *MON* are measured under closely defined conditions in a Co-operative Fuel Research (CFR) engine (American Society for Testing Materials, 2001a, b). These two test conditions originally were chosen as representative of the automotive operational regimes at the time. The *ON* is the percentage by volume of *i*-octane mixed with *n*-heptane in a PRF, that in the test just gives rise to engine knock under the same conditions as the actual fuel. For a given *RON* or *MON* test, there are corresponding loci of pressure and temperature in time, up to maximum values. Stoichiometry is varied in the tests to give the most knock-prone mixture (usually close to stoichiometric).

Historically, the operational regimes of spark ignition engines have increasingly departed from these test conditions. As a consequence, *OI* defined like an *ON* but under the particular engine operational conditions, has increasingly departed from the *RON* and *MON* values of the fuel. A linear weighting factor, *K*, has been used empirically to express *OI* in terms of *RON* and *MON*:

$$OI = (1 - K)RON + K \times MON. \tag{2.11}$$

(*RON–MON*) is termed the fuel sensitivity and is an important fuel characteristic.

With the passage of time, *K*, which was positive when the tests were devised, has tended to become negative and, with *RON>MON*, the *OI* has correspondingly increased to become greater than *RON* (Kalghatgi, 2005). Reasons for such "better than *RON*" ratings lie in the progressively decreasing inlet temperatures, allied to increasing cylinder pressures. Lower temperatures increase fuel differentiation in terms of *ON*.

Figure 2.13 demonstrates the superior resistance to autoignition of the $0.5H_2-0.5CO$ mixture over i-octane-air as the temperature decreases.

The influence of pressure upon τ_i is expressed as being proportional to p^{-n}. For a PRF, n is close to 1.7 (Douaud and Eyzat, 1978; Gauthier et al., 2004). In contrast, n for gasoline type fuels is somewhat less than this, ranging from unity, or even less, to 1.3 (Hirst and Kirsch, 1980; Gauthier et al., 2004). As a result, values of τ_i for a PRF decrease more than those for a gasoline as the pressure increases. Hence, what were originally near-identical values of τ_i for a gasoline and its corresponding PRF on the *RON* test will not match at a higher pressure. To achieve a match, the value of τ_i for the PRF must be increased to that for the gasoline by increasing the *ON* of the PRF. As a result, the same fuel in this operational regime at higher pressure has a higher *OI* than the *RON* (Bradley et al., 2004; Bradley and Head, 2006). This is particularly relevant to fuels for HCCI engines, which tend to operate at low ϕ, higher p, and lower T.

The *CN* is found in an analogous way to the *ON*, but it is based upon the percentage of cetane, n-hexadecane (low τ_i) in a mixture with heptamethylnonane (high τ_i). Alone, the latter has a *CN* of 15 (Heywood, 1989). As τ_i increases, *CN* decreases and *ON* increases. The *CN* of a diesel fuel is greater than about 30, whilst that of a gasoline is less than about 30 (with *RON* greater than about 60).

On the basis of laser-sheet imaging, Dec (1997) has presented a conceptual model of direct injection diesel combustion. Autoignition occurs at multiple points in a region where the fuel is vaporized and ϕ lies between 2 and 4, where τ_i might be anticipated to be minimum. This is followed by premixed rich combustion and soot formation. Autoignition is facilitated by using diesel fuels with low values of τ_i, smaller than those of gasolines, and closer to those of n-heptane. As a result, after injection near top dead center, autoignition occurs shortly afterwards, before the injected fuel and air are well mixed.

Consequently, there is a wide distribution of equivalence ratios in the mixture, the mean of which increases with load. Kalghatgi et al. (2006) showed that this wide distribution led to high emissions of soot and/or NO_x. In contrast, when they injected gasoline fuel into the same diesel engine, the greater values of τ_i and of fuel volatility resulted in enhanced mixing before autoignition occurred. This created a leaner mixture, which further increased τ_i. Both these factors reduced emissions of soot and NO_x.

To enhance mixing before autoignition occurs in HCCI engines, the fuel, usually a gasoline, is injected earlier to create a more uniformly homogeneous mixture. If the mixture were to be near-stoichiometric, the rate of change of heat release rate upon autoignition would cause excessive engine knock, greater than that in knocking combustion in a spark ignition engine. This is avoided by the use of lean mixtures, usually with some recirculation of burned gas, which also gives improved efficiencies (see Figure 2.1). Furthermore, in turbulent mixtures with a high value of K, lean turbulent flame propagation might not be possible and combustion would have to be entirely in the autoignitive, flameless mode.

2.3.4. AUTOIGNITIVE FRONT PROPAGATING VELOCITIES AND DETONATION

Here we consider how a reaction front can originate and spread through a gradient of reactivity within a single hot spot. If there is a spatial gradient in, say T, of a given mixture, there also will be an associated spatial gradient in τ_i. If τ_i increases with the radius, r, from the center of the hot spot, then autoignition will occur at increasing times at increasing distances. This will create a localized reaction front propagating at a velocity relative to the unburned gas of u_a, equal to $\partial r/\partial \tau_i$, or

$$u_a = \left(\frac{\partial r}{\partial T}\right)\left(\frac{\partial T}{\partial \tau_i}\right). \tag{2.12}$$

For illustrative purposes, it is assumed τ_i can be expressed locally, at constant pressure, in Arrhenius form by

$$\tau_i = A \exp(E/RT), \tag{2.13}$$

where A is a numerical constant and E is a global activation energy. For a gasoline, the greater the fuel sensitivity, the greater the value of E becomes (Bradley and Head, 2006). From Equations (2.12) and (2.13):

$$u_a = (T^2/(\tau_i E/R))(\partial T/\partial r)^{-1}. \tag{2.14}$$

This type of autoignitive reaction front propagation, in which the propagation velocity is inversely proportional to the product of τ_i and the temperature gradient, is very different from that in a laminar or turbulent flame. In those regimes where flame propagation also is possible, the two modes of burning can co-exist. This has been demonstrated for lean H_2–air mixtures in the direct numerical simulations of Chen *et al.* (2006).

Values of u_a at 4 MPa with $\phi = 0.5$, given by Equation (2.14) and corresponding to the data for the two mixtures in Figure 2.13, are given in Figure 2.14 at the different temperatures. A hot spot temperature gradient of 1 K/mm is assumed. As the temperatures are increased above 1000 K, for both mixtures, u_a becomes appreciable and eventually supersonic. Clearly, from Equation (2.14), an increase in $\partial T/\partial r$ decreases u_a. As $\partial T/\partial r \to 0$, $u_a \to \infty$ and, in the limit, a *thermal explosion* occurs throughout the mixture.

A special condition arises when the autoignition front moves into the unburned mixture at approximately the acoustic velocity, a (Zeldovich, 1980; Makhviladze and Rogatykh, 1991). The critical temperature gradient for this condition, from Equation (2.12), is

$$\left(\frac{\partial T}{\partial r}\right)_c = \frac{1}{a(\partial \tau_i/\partial T)}. \tag{2.15}$$

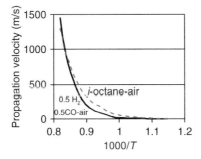

Figure 2.14 Variations of propagation velocity, u_a, with $1000/T$ at 4 MPa for the two mixtures of Figure 2.13, for $\partial T/\partial r = 1$ K/mm.

This gradient, which is a function of the reactants, couples the leading edge of the pressure wave generated by heat release with the autoignition front. The fronts are mutually reinforced and united to create a damaging pressure spike propagating at high velocity in a *developing detonation*. It is convenient to normalize temperature gradients by this critical value, to define a parameter ξ

$$\xi = \left(\frac{\partial T}{\partial r}\right)\left(\frac{\partial T}{\partial r}\right)_c^{-1}. \tag{2.16}$$

Values of ξ are only known for the initial boundary condition. In practice, heat conduction, species diffusion, and some reaction modify the initial boundary before autoignition occurs. From Equations (2.12), (2.15), and (2.16),

$$\xi = a/u_a. \tag{2.17}$$

The probability of a detonation developing at the outer radius of a hot spot, r_o, also depends upon the chemical energy that can be unloaded into the developing acoustic wave during the time r_o/a, it takes to propagate down the temperature gradient. The chemical time for energy release defines an excitation time, τ_e, which can be obtained from chemical kinetic modeling (Lutz *et al.*, 1988). Here, it is taken to be the time interval from 5% to the maximum chemical power. The ratio of the two times yields (Gu *et al.*, 2003)

$$\varepsilon = (r_o/a)/\tau_e. \tag{2.18}$$

This is an approximate indicator of the energy fed into the acoustic wave. It is similar to an inverse Karlovitz stretch factor (or Damköhler number), in that it is an aerodynamic time divided by a chemical time.

Shown in Figure 2.15, from Bradley *et al.* (2002), is the temporal development of a detonation, $\xi = 1$, at a hot spot of radius of 3 mm for $0.5H_2-0.5CO$-air, $\phi = 1.0$, initial $T = 1200$ K, and $p = 5.066$ MPa. The initial hot spot maximum temperature elevation is 7.28 K. The radial profiles of temperature and pressure, some of which are numbered, are given for different time intervals. The times for the numbered profiles are given in the figure caption. The pressure and reaction fronts are soon coupled, with a steep increase in the temperature, immediately after the sudden increase in pressure.

The detonation wave becomes fully developed and travels with a speed of 1600 m/s (see Figure 2.15C). This is close to the Chapman–Jouguet velocity. Shortly afterwards, autoignition occurs throughout the remaining mixture in a thermal explosion, indicated by (i), at $\tau_i = 39.16\,\mu s$. At this time, the detonation wave has a radius of 4.6 mm. In practice, most *developing detonations* are confined within the hot spot. However, with the more energetic mixtures at the highest temperatures, such as in the present case, they can propagate outside it. This creates the potential for damaging consequences, particularly when a significant amount of the remaining charge is consumed in the detonation front.

Shown in Figure 2.16 is a plot of ξ against ε. This defines a peninsula, within which localized detonations can develop inside a hot spot. The points on the figure were generated computationally from chemical kinetic models of the development of autoignitive propagation for mixtures of $0.5H_2-0.5CO$-air, at equivalence ratios between 0.4 and 1.0 and for hot spots of different radii. These data were processed to identify the upper and lower limits for the *development of detonations*, ξ_u and ξ_l, and these define

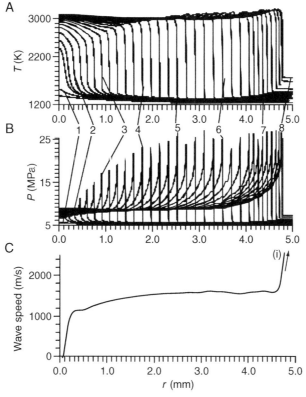

Figure 2.15 History of a hot spot: $r_o = 3\,mm$, $\xi = 1$, stoichiometric $0.5H_2{-}0.5CO$-air, $T_o = 1200\,K$, $P_o = 5.066\,MPa$, and $\tau_i = 39.16\,\mu s$. Time sequence (μs) $1 - 35.81$, $2 - 36.16$, $3 - 36.64$, $4 - 37.43$, $5 - 37.72$, $6 - 38.32$, $7 - 38.86$, $8 - 39.13$. Plots show (A) temperature, (B) pressure, and (C) combustion wave speed (Bradley *et al.*, 2002).

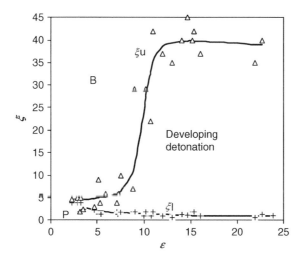

Figure 2.16 Developing detonation peninsula on plot of ξ against ε (Bradley *et al.*, 2002). Supersonic and subsonic autoignitive fronts propagate in regions P and B, respectively.

the limits of the peninsula shown in the figure. It was found that at sufficiently high values of ε, greater than about 80, the detonations could propagate outside the hot spot with the Chapman–Jouguet detonation velocity.

A variety of propagation modes of the reaction front are possible and these are discussed in more detail by Bradley *et al.* (2002) and Gu *et al.* (2003). *Thermal explosions* occur when $\xi = 0$. At slightly higher values of ξ, but less than ξ_1 (regime P on the figure), *supersonic autoignitive fronts* occur. Above ξ_u, in regime B, the fronts propagate *subsonically*. As ϕ decreases, τ_e increases, ε decreases and, as can be seen from Figure 2.16, the detonation regime narrows appreciably. Conversely, as ϕ increases from such low lean values, the detonation peninsula broadens appreciably. This corresponds to increases in ϕ above about 0.6.

2.4. RECIRCULATION OF HEAT FROM BURNING AND BURNED GAS

Whether combustion is in the flame propagation or autoignitive mode, its rate is usually enhanced by an increase in temperature. This can be achieved by recirculating heat from exhaust gases through a heat exchanger to the incoming reactants and also by recirculating hot gas itself, either externally or internally (Wünning and Wünning, 1997; Katsuki and Hasegawa, 1998). With lean mixtures, the increase above the temperature of the reactants arising from recirculated heat can be greater than that arising from combustion. Consequently, parasitic heat loss and pumping power can be a significant proportion of the heat release, thereby reducing the efficiency of the process. Three modes of heat recirculation are indicated schematically in Figure 2.17. These involve (i) recuperative heat exchange and recirculation of burned gas, (ii) spouted beds, and (iii) filtrational combustion.

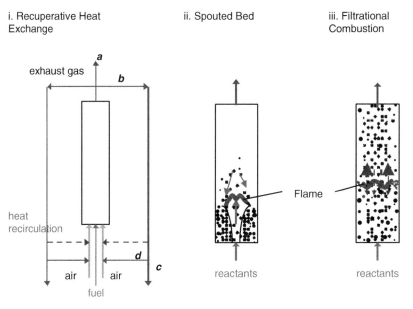

Figure 2.17 Different modes of heat recirculation. (i) For $b = c = d = 0$: conventional burner. For $b \leq a$, $d = 0$: recuperative heat recirculation, indicated by dashed line. For $b \leq a$, $d = b$, $c = 0$: both recuperator heat and burned gas recirculation; (ii) spouted bed with fountain of particles; and (iii) filtrational combustion with solid porous medium. (See color insert.)

Referring to Figure 2.17(i), in a simple burner without any recirculation, the flow b is zero. With recuperative heat recirculation solely via a heat exchanger, as indicated by the dashed lines, $b < a$ and $d = 0$. A compact heat exchanger working on this principle is the double spiral "Swiss roll" burner of Weinberg (1986). In this, the reactants flowed inwards through an outer channel of the heat exchanger to the combustion chamber, which was at the center of the spiral. Thereafter, the combustion products spiraled outwards and transferred heat to the reactants in the counter-flow heat exchanger. The increased temperature of the reactants enabled leaner mixtures to be burned. A flow that was too high led to flame blow-off, while one that was too low led to flash-back. Such an atmospheric pressure burner was able to burn CH_4-air in a stable flame with ϕ as low as 0.16. This was lower than the flammability limit of $\phi = 0.53$ at atmospheric temperature and pressure.

A combination of both heat and hot gas recirculation appears to be necessary to achieve sufficiently low values of τ_i for autoignitive, flameless combustion in very lean mixtures. This corresponds to the conditions $b < a$, $d = b$, and $c = 0$ in Figure 2.17(i). The burner described by Plessing *et al.* (1998) operated in this mode with hot gas recirculation within the combustion chamber. At start-up, CH_4 and air were first fed separately to the burner, with heat transfer from exhaust gas to the incoming air, as well as internal hot gas recirculation. This enabled stable, lean flames to be established, with up to 30% of the fuel and air flow recirculating. This regime is indicated by S in Figure 2.18. Transition to the flameless mode involved switching the fuel flow to a central nozzle and increasing the internal recirculation. Eventually, the increased dilution quenched the turbulent flame, in the regime indicated by Q, as the recirculation, d, increased. At temperatures above 1000 K, the mixture re-ignited in the stable "flameless oxidation" mode, regime F in Figure 2.18, indicated by the shading. For the idealized conditions of a perfectly stirred reactor, a necessary condition for combustion would be that the residence time in the reactor must be greater than the autoignition delay time.

The nature of the flameless oxidation in the burner of Plessing *et al.* was diagnosed with OH laser-induced pre-dissociative fluorescence (LIPF) and Rayleigh thermometry. Images of the reaction zone are compared with those of a highly turbulent flame in Figure 2.19. Color-coded temperatures are on the left and OH intensities are on the right. The upper, highly turbulent flame images indicate a thin, convoluted flame reaction sheet. The lower images of flameless oxidation indicate a very different spotty structure, with reaction at small disconnected hot spots and with smaller increases in temperature. The OH concentrations are fairly uniform in both cases. For the highly

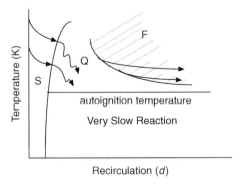

Figure 2.18 Burner of Plessing *et al.* (1998). Transition from flame to autoignitive combustion. S is regime of stabilized flame, Q that of flame quenching, and F that of stable autoignitive burning.

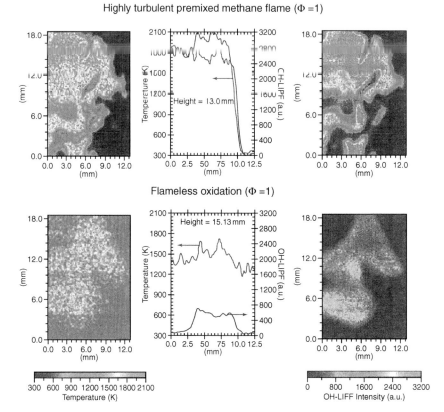

Figure 2.19 Simultaneous temperature and OH-LIPF images of premixed highly turbulent flame and flameless oxidation (Plessing *et al.*, 1998, p. 3202). (See color insert.)

turbulent flame, this was because of the turbulence. For the less turbulent, flameless oxidation, it was because of the more uniformly distributed reaction. Maximum temperatures in the flameless oxidation were about 1650 K, with a temperature rise due to reaction of about 200–400 K and a NO_x concentration of about 10 ppm. The latter value was below the 30 ppm found in the stabilized flame. Chemical times were greater than turbulent times and reaction was likened to that in a well-stirred reactor.

Recirculation of hot gas also is important in explosive combustion, as in HCCI engines. Recirculation both dilutes the mixture further and increases its temperature. Dilution with inert gases reduces the rate of burning. On the other hand, the increased temperature and any recirculated active radicals increase the rate of burning. This increase tends to predominate over the dilution decrease much more in the autoignitive mode, than in the flame propagation mode. It is one reason why the autoignitive mode is so often the only means of achieving very lean burn.

Weinberg (1986) has described a variety of heat-recirculating burners, including the spouted bed burner shown in Figure 2.17(ii). In this, a central jet of gaseous reactants causes chemically inert particles in a bed at the bottom of the reactor to rise in a fountain. A conical flame stabilizes where the fountain emerges from the top of the bed. The particles are heated as they pass through the flame, and the particle bed reaches a temperature in the region of 1000 K. The particles recirculate due to gravity and transfer

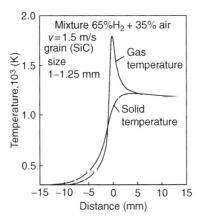

Figure 2.20 Gas and porous medium axial temperature distributions during filtrational combustion in low velocity regime (Babkin, 1993). Flow from left to right.

heat to the incoming gaseous reactants. Stable, lean limit combustion of CH_4–air mixtures was observed down to $\phi = 0.28$.

With regard to the filtrational combustion of Figure 2.17(iii), combustion within porous inert media has been reviewed by Howell *et al.* (1996). Different regimes of flame propagation have been identified by Babkin (1993). A flame established in a tube filled with an inert, porous medium, through which a homogeneous mixture flows, can move either with, or against, the direction of flow. Figure 2.20 shows temperature profiles in the gas and the solid porous medium in a *low velocity regime*. Heat conducts through the porous medium in the direction upstream from the flame and this heats the cooler flowing mixture. Sufficient heat is transferred to initiate reaction in the gas. As a result, this becomes hotter than the solid, but cools as it transfers heat to it. The high thermal conductivity of the solid effectively conducts sufficient heat from hot gases to cold reactants. The porous medium acts as a recuperator, transferring heat from hot gas to cold reactants, in a not-dissimilar fashion to what occurs in the Swiss roll and the spouted bed burners. Hsu *et al.* (1993) were able to burn CH_4-air down to $\phi = 0.41$ in a porous ceramic burner.

With increasing pore size there is a transition to a *high velocity regime* of the reaction front, with reduced thermal coupling of gas and solid in the reaction zone and no heat recuperation. There is an increasing dominance of aerodynamic and turbulence effects. A pressure wave builds up in front of the combustion zone and a *sonic velocity regime* is entered, with a sudden jump in the reaction wave velocity. Finally, there is a transition to a *low velocity, subsonic detonation*, which evolves spontaneously from deflagrative combustion (Brailovsky and Sivashinsky, 2000).

2.5. FLAME STABILIZATION

In steady state combustion, the flame should neither flash-back nor blow-off, despite some inevitable fluctuations in ϕ, velocity, pressure, and temperature. As mentioned in Section 2.1, the possibility of flash-back down an entry duct only arises when there is premixing ahead of the combustion chamber. Minimization of NO_x formation might proscribe near-stoichiometric burning, but flash-back is still possible with lean premixtures. In laminar flow, flash-back occurs when the burning velocity is greater than the low flow velocity adjacent to the wall of the tube, along which the reactants flow. Prevention requires that the laminar burning velocity be reduced at the wall.

This is achieved by a combination of heat loss to the tube and, for positive Markstein numbers, flame stretch at the wall. Close to the wall, velocity gradients are in the region of $10^2 - 10^3$ s^{-1} (Berlad and Potter, 1957; Lewis and von Elbe, 1987). These values are comparable to the values of flame extinction stretch rates in Figure 2.2. For high velocity turbulent flow of a flammable mixture along a flat surface, flash-back of a turbulent flame is improbable in the outer region of the flow, as the flow velocity will always exceed the burning velocity. Close to the surface, within the viscous sub-layer, velocity gradients are generally sufficiently high to quench the flame and prevent flash-back (Bradley, 2004).

The stability and overall high volumetric intensity of turbulent combustion in gas turbine and furnace combustion chambers depend upon the patterns of gas recirculation that mix burning and burned gas with the incoming reactants. Computational fluid dynamics is clearly a powerful design tool in this context. Bradley *et al.* (1998b) have described a computational and experimental study in which a swirling, turbulent pre-mixture of CH_4-air flowed from a burner tube, with a stepped increase in internal diameter, into a cylindrical flame tube at atmospheric pressure. The step generated an outer recirculation zone at the corner, shown by the computed streamlines in Figure 2.21c. A matrix of small diameter hypodermic tubes housed within a cylindrical burner tube rotating at 54 000 rpm produced the swirl. This generated a radially outward cold gas flow and an inward hot gas flow to create a central recirculation zone. The mean axial entry velocity and the swirl number were maintained constant and the equivalence ratio was varied. Computed parameters were obtained from a Reynolds stress, stretched laminar flamelet model.

The computed concentration of NO in the gases leaving the 300 mm long, silica glass flame tube of 39.2 mm diameter, decreased from 20 ppm at $\phi = 0.75$ to 2 ppm at $\phi = 0.56$. At this lowest attainable value of ϕ, the NO was almost entirely created as "prompt" NO in the flame front, the temperature being too low along the tube for much production of thermal NO. Shown in Figures 2.21 and 2.22 underneath (a) self-luminescent photographs of the flames are computed contours within the flame tube of (b) mean volumetric heat release rate, (c) streamlines, (d) mean temperatures, and (e) Karlovitz stretch factor, K. For both sets of contours, the swirl number was 0.72 and the mean axial entry velocity to the flame tube was 10 m/s. Figure 2.21 shows conditions for $\phi = 0.59$ and Figure 2.22 for $\phi = 0.56$. In both cases, the outer, corner, doughnut-like recirculation zone was created by the sudden step, and the inner one, with reverse flow along the center line, was created by the swirl. The reverse flow recirculated hot gas into the incoming reactants.

In Figure 2.21, high values of K marginally inhibited combustion close to the corners. With a slightly higher value of ϕ, the value of K was halved in this region and a more compact flame was generated, with heat release close to the end face and tube walls. At this value of ϕ, small changes in velocity and ϕ caused experimental flame extinctions and re-ignitions. The corresponding fluctuations in heat release rate generated pressure oscillations. These could amplify due to resonances with the natural acoustic wavelengths of the tube and Rayleigh–Taylor instabilities. The combination of the streamline flow pattern and the contours of K, with flame extinctions at the higher

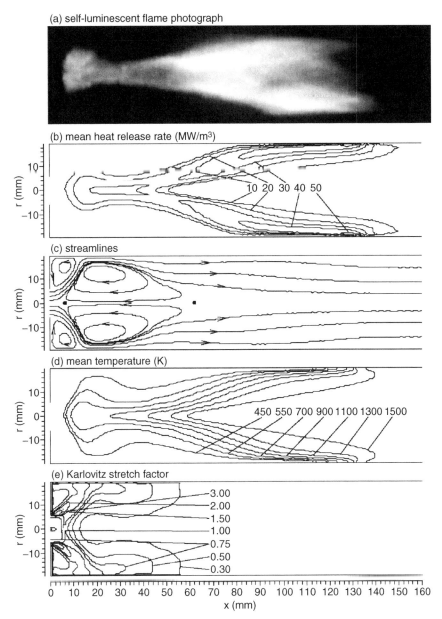

(a) self-luminescent flame photograph

(b) mean heat release rate (MW/m³)

10 20 30 40 50

(c) streamlines

(d) mean temperature (K)

450 550 700 900 1100 1300 1500

(e) Karlovitz stretch factor

3.00
2.00
1.50
1.00
0.75
0.50
0.30

x (mm)

Figure 2.21 Swirling, pre-mixed CH₄-air flame for $\phi = 0.59$: (a) flame photograph, (b) contours of heat release rate, (c) streamlines, (d) mean temperature, and (e) Karlovitz stretch factor (Bradley *et al.*, 1998).

values, determined the location of the flame. At the lower value of $\phi = 0.56$, Figure 2.22 shows how the higher upstream values of K caused the flame to move downstream, where a V-flame anchored precariously between the two stagnation-saddle points. As the mixture was leaned off further, the leading edge of the flame moved towards the downstream stagnation-saddle point. The flame was very unstable, with small changes in velocity or ϕ causing it to fluctuate with periodic extinctions and re-ignitions, and eventually blew off.

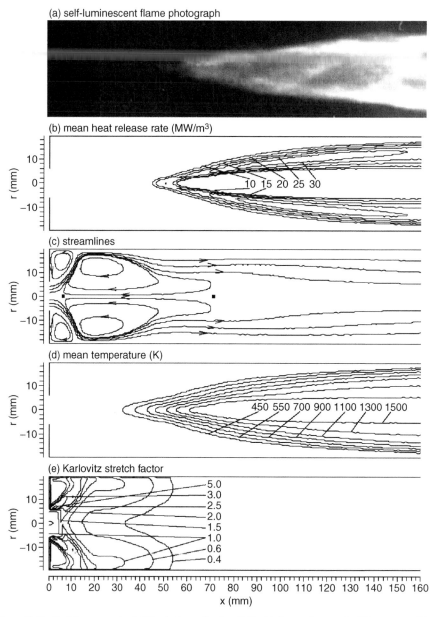

Figure 2.22 Swirling, pre-mixed CH$_4$-air flame for $\phi = 0.56$: (a) flame photograph, (b) contours of heat release rate, (c) streamlines, (d) mean temperature, and (e) Karlovitz stretch factor (Bradley *et al.*, 1998).

De Zilwa *et al.* (2002) also have shown how extinctions and oscillations are related to stabilization. Candel (1992) has reviewed combustion oscillations arising from the coupling of flame fluctuations with acoustic waves and described how actively perturbing combustion parameters can control and suppress instabilities. An example of this is the active control system of Evesque *et al.* (2002), in which fuel was injected unsteadily to alter the heat release rate in response to an input signal triggered by the oscillations.

Another technique is to inject fuel locally into regions of high K. Some of these issues are discussed in Chapter 7.

Upstream of a combustion chamber, if the residence time of a premixture at an elevated temperature is greater than τ_i, autoignition will occur. Temperatures and pressures, and hence τ_i, may not be constant, in which case autoignition will occur when the Livengood and Wu (1955) integral

$$\int_0^t \frac{dt}{\tau_i} \tag{2.19}$$

attains a value of unity. This becomes more probable when recirculation zones are created, within which the reactants might have an increased residence time.

With regard to the lift-off and stability of non-premixed fuel jet flames in still air, the underlying mechanism for the lift-off has been described in Section 2.1. If the jet velocity increases, the flame lifts further from the burner tube, until eventually it blows-off. Dimensionless correlations of lift-off heights and blow-off velocities have been developed that summarize the results of both computations and experiments (Bradley *et al.*, 1998a). Surrounding a fuel jet with hot recirculated gases with added air enables very lean mixtures to be burned. The high strain rate that otherwise inhibits combustion and the enhanced mixing at the base of the lifted flame contribute to flame stability. This system is characterized as a fuel jet in hot co-flow. With hot co-flows at 1300 K and as little as 3% oxygen by mass, Dally *et al.* (2002) measured temperature increases due to combustion, which were as low as 100 K. Concentrations of NO were less than 5 ppm and were fairly uniformly dispersed. Combustion was governed by low temperature chemical kinetics, rather than the high temperature kinetics associated with a propagating flame.

2.6. CONCLUSIONS

Analyses of different engine cycles have demonstrated how engine efficiencies and NO_x emissions are related to the combustion process. The advantages of lean combustion became apparent. This led to consideration of the different modes of burning and identification of the relevant parameters for lean and stable combustion. With turbulent premixed combustion, lean flames of gasoline fuels are quenched when $\phi < 0.7$. In the leaner regime, flame propagation can be achieved by the addition of H_2, which inhibits flame quenching. Another alternative, which enables hydrocarbons to be burned down to about $\phi = 0.25$ is to adopt autoignitive, or flameless, combustion.

In this regime, because practical mixtures are not homogeneous, different modes of propagation of reaction can occur at the distributed hot spots. These modes have propagation speeds that range from subsonic to potentially damaging *developing detonations*, which might propagate outside the hot spot. They are relevant to engine knock, but operational regimes in HCCI engines can be far removed from those of the *RON* and *MON* tests. Consequently, caution is necessary when using *RON* and *MON* values as guides, particularly for HCCI combustion. In ultra-lean, flameless combustion the temperature rise may be no more than a few hundred degrees. To achieve practical volumetric rates of combustion, operation of the device at higher pressure becomes necessary.

Flame instabilities comprise D–L, T-D, and Rayleigh–Taylor instabilities. These are important in both explosive and continuous combustion, particularly at low or negative values of Markstein numbers, and hence with lean H_2 and CH_4 mixtures. Their effects can be amplified by feedback mechanisms that can result in large increases in burning velocities, accompanied by dangerously large amplitudes of pressure oscillations. In explosions, increasing turbulence reduces these effects. The continuous, stable combustion of lean mixtures usually is achieved through the recirculation of heat from burned gases and/or the hot gases themselves. As the lean limit for turbulent burning is approached, combustion becomes intermittent, with flame extinctions and re-ignitions. These trigger severe pressure oscillations and instabilities, which can be ameliorated by appropriate active control systems. Non-premixed jet flames have much less, if any, recirculation and are stabilized in a regime bounded by lift-off and blow-off.

The low temperature, flameless mode of lean combustion also depends upon heat and mass recirculation. This is a quieter combustion, which avoids the pulsations arising from flame extinctions and re-ignitions. However, it can be susceptible to chemically driven oscillations. Filtrational combustion, like flameless combustion, also enables combustion to occur beyond the lean flammability limit and a variety of combustion regimes is possible.

ACKNOWLEDGMENTS

I am pleased to acknowledge the advice of Chris Sheppard, at Leeds; and Roger Cracknell, Gautam Kalghatgi, and Harold Walmsley, at Shell Global Solutions.

REFERENCES

Abdel-Gayed, R.G. and Bradley, D. (1985). Criteria for turbulent propagation limits of premixed flames. *Combust. Flame* **62**, 61–68.

Abdel-Gayed, R.G., Bradley, D., Hamid, M.N., and Lawes, M. (1985). Lewis number effects on turbulent burning velocity. *Proc. Combust. Inst.* **20**, 505–512.

Abdel-Gayed, R.G., Bradley, D., and Lung, F.K.-K. (1989). Combustion regimes and the straining of turbulent premixed flames. *Combust. Flame* **76**, 213–218.

Aceves, S.M. and Smith, J.R. (1997). Hybrid and Conventional Hydrogen Engine Vehicles that Meet EZEV Emissions, SAE Paper 970290.

Aldredge, R.C. and Killingsworth, N.J. (2004). Experimental evaluation of Markstein-number influence on thermo-acoustic instability. *Combust. Flame* **137**, 178–197.

Al-Shahrany, A.S., Bradley, D., Lawes, M., and Woolley, R. (2005). Measurement of unstable burning velocities of iso-octane-air mixtures at high pressure and the derivation of laminar burning velocities. *Proc. Combust. Inst.* **30**, 225–232.

Al-Shahrany, A.S., Bradley, D., Lawes, M., Liu, K., and Woolley, R. (2006). Darrieus–Landau and thermo-acoustic instabilities in closed vessel explosions. *Combust. Sci. Technol.* **178**, 1771–1802.

American Society for Testing Materials (2001a). Designation: D 2699 – 01a, Standard Test Method for Research Octane Number of Spark-Ignition Engine Fuel.

American Society for Testing Materials (2001b). Designation: D 2700 – 01a, Standard Test Method for Motor Octane Number of Spark-Ignition Engine Fuel.

Babkin, V.S. (1993). Filtrational combustion of gases: Present state of affairs and prospects. *Pure Appl. Chem.* **65**, 335–344.

Batley, G.A., McIntosh, A.C., and Brindley, J. (1996). Baroclinic distortion of laminar flames. *Proc. R. Soc. London* **A452**, 199–221.

Bechtold, J.K. and Matalon, M. (1987). Hydrodynamic and diffusion effects on the stability of spherically expanding flames. *Combust. Flame* **67**, 77–90.

Berlad, A.L. and Potter, A.E. (1957). Relation of boundary velocity gradient for flash-back to burning velocity and quenching distance. *Combust. Flame* **1**, 127–128.

Borghi, R. (1985). On the structure and morphology of turbulent premixed flames. In: *Recent Advances in the Aerospace Science*. Plenum, New York, pp. 117–138.

Bradley, D. (1999). Instabilities and flame speeds in large-scale premixed gaseous explosions. *Phil. Trans. R. Soc. London* **A357**, 3567–3581.

Bradley, D. (2002). Problems of predicting turbulent burning rates. *Combust. Theory Model.* **6**, 361–382.

Bradley, D. (2004). Flame propagation along flat plate turbulent boundary layers. In: Bradley, D., Drysdale, D., and Molkov, V. (Eds.). Proceedings of the Fourth International Seminar on Fire and Explosion Hazards, FireSERT – University of Ulster, Londonderry, Northern Ireland, pp. 201–208.

Bradley, D. and Harper, C.M. (1994). The development of instabilities in laminar explosion flames. *Combust. Flame* **99**, 562–572.

Bradley, D. and Head, R.A. (2006). Engine autoignition: The relationship between octane numbers and autoignition delay times. *Combust. Flame* **147**, 171–184.

Bradley, D., El-Din Habik, S., and El-Sherif, S.A. (1991). A generalisation of laminar burning velocities and volumetric heat release rates. *Combust. Flame* **87**, 336–346.

Bradley, D., Lau, A.K.C., and Lawes, M. (1992). Flame stretch rate as a determinant of turbulent burning velocity. *Phil. Trans. R. Soc. London* **A338**, 357–387.

Bradley, D., Gaskell, P.H., and Gu, X.J. (1996). Burning velocities, Markstein lengths, and flame quenching for spherical methane-air flames: A computational study. *Combust. Flame* **104**, 176–198.

Bradley, D., Gaskell, P.H., and Gu, X.J. (1998a). The mathematical modeling of liftoff and blowoff of turbulent non-premixed methane jet flames at high strain rates. *Proc. Combust. Inst.* **27**, 1199–1206.

Bradley, D., Gaskell, P.H., Gu, X.J., Lawes, M., and Scott, M.J. (1998b). Premixed turbulent flame instability and NO formation in a lean-burn swirl burner. *Combust. Flame* **115**, 515–538.

Bradley, D., Sheppard, C.G.W., Woolley, R., Greenhalgh, D.A., and Lockett, R.D. (2000). The development and structure of flame instabilities and cellularity at low Markstein numbers in explosions. *Combust. Flame* **122**, 195–209.

Bradley, D., Morley, C., Gu, X.J., and Emerson, D.R. (2002). Amplified Pressure Waves during Autoignition: Relevance to CAI Engines, SAE Paper 2002–01–2868.

Bradley, D., Gaskell, P.H., Gu, X.J., and Sedaghat, A. (2003). Generation of PDFS for flame curvature and for flame stretch rate in premixed turbulent combustion. *Combust. Flame* **135**, 503–523.

Bradley, D., Morley, C., and Walmsley, H.L. (2004). Relevance of Research and Motor Octane Numbers to the Prediction of Engine Autoignition, SAE Paper 2004–01–1970.

Bradley, D., Gaskell, P.H., Gu, X.J., and Sedaghat, A. (2005). Premixed flamelet modelling: Factors influencing the turbulent heat release rate source term and the turbulent burning velocity. *Combust. Flame* **143**, 227–245.

Bradley, D., Lawes, M., Liu, K., and Woolley, R. (2007a). The quenching of premixed turbulent flames of iso-octane, methane and hydrogen at high pressures. *Proc. Combust. Inst.* **31**, 1393–1400.

Bradley, D., Lawes, M., Liu, K., Verhelst, S., and Woolley, R. (2007b). Laminar burning velocities of lean hydrogen-air mixtures at pressures up to 1.0 MPa. *Combust. Flame* **149**, 162–172.

Brailovsky, I. and Sivashinsky, G. (2000). Hydraulic resistance as a mechanism for deflagration-to-detonation transition. *Combust. Flame* **122**, 492–499.

Burke, S.P. and Schumann, T.E.W. (1928). Diffusion flames. *Proc. Combust. Inst.* **1**, 1–11.

Cambray, P. and Joulin, G. (1992). On moderately-forced premixed flames. *Proc. Combust. Inst.* **24**, 61–67.

Cambray, P. and Joulin, G. (1994). Length scales of wrinkling of weakly-forced, unstable premixed flames. *Combust. Sci. Technol.* **97**, 405–428.

Candel, S.M. (1992). Combustion instabilities coupled by pressure waves and their active control. *Proc. Combust. Inst.* **24**, 1277–1296.

Chen, J.H., Hawkes, E.R., Sankaran, R., Mason, S.D., and Im, H.G. (2006). Direct numerical simulation of ignition front propagation in a constant volume with temperature inhomogeneities: Fundamental analysis and diagnostics. *Combust. Flame* **145**, 128–144.

Clavin, P. (1985). Dynamic behaviour of premixed flame fronts in laminar and turbulent flows. *Prog. Energy Combust. Sci.* **11**, 1–59.

Curran, H.J., Gaffuri, P., Pitz, W.J., and Westbrook, C.K. (2002). A comprehensive modeling study of iso-octane oxidation. *Combust. Flame* **129**, 253–280.

Dally, B.B., Karpetis, A.N., and Barlow, R.S. (2002). Structure of turbulent non-premixed jet flames in a diluted hot coflow. *Proc. Combust. Inst.* **29**, 1147–1154.

De Zilwa, S.R.N., Emiris, I., Uhm, J.H., and Whitelaw, J.H. (2002). Combustion of premixed methane and air in ducts. *Proc. R. Soc. London* **A457**, 1915–1949.

Dec, J. (1997). A Conceptual Model of DI Diesel Combustion Based on Laser-sheet Imaging, SAE Paper 970873.

Donbar, J.M., Driscoll, J.F., and Carter, C.D. (2001). Strain rates measured along the wrinkled flame contour within turbulent non premixed jet flames. *Combust. Flame* **126**, 1239–1257.

Dong, Y., Holley, A.T., Andac, M.G., Egolfopoulos, F.N., Davis, S.G., Middha, P., and Wang, H. (2005). Extinction of premixed H_2/air flames: Chemical kinetics and molecular diffusion effects. *Combust. Flame* **142**, 374–387.

Douaud, A.M. and Eyzat, P. (1978). Four-Octane-Number Method for Predicting the Anti-knock Behaviour of Fuels and Engines. *SAE Trans.* **87**, SAE Paper 780080.

Evesque, S., Dowling, A.P., and Annaswamy, A.M. (2002). Self-tuning regulators for combustion oscillations. *Proc. R. Soc. London* **A459**, 1709–1749.

Farrell, J.T., Johnston, R.J., and Androulakis, I.P. (2004). Molecular Structure Effects on Laminar Burning Velocities at Elevated Temperature and Pressure, SAE Paper 2004–01–2936.

Fieweger, K., Blumenthal, R., and Adomeit, G. (1997). Self-ignition of S.I. engine model fuels: A shock tube investigation at high pressure. *Combust. Flame* **109**, 599–619.

Gauthier, B.M., Davidson, D.F., and Hanson, R.K. (2004). Shock tube determination of ignition delay times in full-blend and surrogate fuels. *Combust. Flame* **139**, 300–311.

Gu, X.J., Emerson, D.R., and Bradley, D. (2003). Modes of reaction front propagation from hot spots. *Combust. Flame* **133**, 63–74.

Heywood, J.B. (1989). *Internal Combustion Engine Fundamentals*. McGraw-Hill, New York.

Hirst, S.L. and Kirsch, L.J. (1980). The application of a hydrocarbon autoignition model in simulating knock and other engine combustion phenomena. In: *Combustion Modeling in Reciprocating Engines*. Plenum Publishing, New York.

Holley, A.T., Dong, Y., Andac, M.G., and Egolfopoulos, F.N. (2006). Extinction of premixed flames of practical liquid fuels: Experiments and simulations. *Combust. Flame* **144**, 448–460.

Howell, J.R., Hall, M.J., and Ellzey, J.L. (1996). Combustion of hydrocarbon fuels within porous inert media. *Prog. Energy Combust. Sci.* **22**, 121–145.

Hsu, P.-F., Evans, W.D., and Howell, J.R. (1993). Experimental and numerical study of premixed combustion within nonhomogeneous porous ceramics. *Combust. Sci. Technol.* **90**, 149–172.

Jamal, Y. and Wyszyński, M.L. (1994). Onboard generation of hydrogen-rich gaseous fuels – a review. *Int. J. Hydrogen Energy* **19**, 557–572.

Kalghatgi, G.T. (2005). Auto-ignition Quality of Practical Fuels and Implications for Fuel Requirements of Future SI and HCCI Engines, SAE Paper 2005–01–0239.

Kalghatgi, G.T., Risberg, P., and Angström, H.-E. (2006). Advantages of Fuels with High Resistance to Auto-ignition in Late-injection, Low-temperature, Compression Ignition Combustion, SAE Paper 2006–01–3385.

Katsuki, M. and Hasegawa, T. (1998). The science and technology of combustion in highly preheated air. *Proc. Combust. Inst.* **27**, 3135–3146.

Kawahara, N., Tomita, E., and Sakata, Y. (2007). Auto-ignited kernels during knocking combustion in a spark-ignition engine. *Proc. Combust. Inst.* **31**, 2999–3006.

Kobayashi, H., Tamura, T., Maruta, K., Niioka, T., and Williams, F. (1996). Burning velocity of turbulent premixed flames in a high-pressure environment. *Proc. Combust. Inst.* **26**, 389–396.

Kobayashi, H., Kawabata, Y., and Maruta, K. (1998). Experimental study on general correlation of turbulent burning velocity at high pressure. *Proc. Combust. Inst.* **27**, 941–948.

Kobayashi, H., Seyama, K., Hagiwara, H., and Ogami, Y. (2005). Burning velocity correlation of methane/air turbulent premixed flames at high pressure and high temperature. *Proc. Combust. Inst.* **30**, 827–834.

Law, C.K., Zhu, D.L., and Yu, G. (1986). Propagation and extinction of stretched premixed flames. *Proc. Combust. Inst.* **21**, 1419–1426.

Lawes, M., Ormsby, M.P., Sheppard, C.G.W., and Woolley, R. (2005). Variations of turbulent burning rate of methane, methanol and iso-octane air mixtures with equivalence ratio at elevated pressure. *Combust. Sci. Technol.* **177**, 1273–1289.

Lewis, B. and von Elbe, G. (1987). *Combustion, Flames and Explosions of Gases*, 3rd edition. Academic Press, New York.

Livengood, J.C. and Wu, P.C. (1955). Correlation of autoignition phenomena in internal combustion engines and rapid compression machines. *Proc. Combust. Inst.* **5**, 347–356.

Lutz, A.E., Kee, R.J., Miller, J.A., Dwyer, H.A., and Oppenheim, A.K. (1988). Dynamic effects of autoignition centers for hydrogen and $C_{1,2}$-hydrocarbon fuels. *Proc. Combust. Inst.* **22**, 1683–1693.

Makhviladze, G.M. and Rogatykh, D.I. (1991). Non-uniformities in initial temperature and concentration as a cause of explosive chemical reactions in combustible gases. *Combust. Flame* **87**, 347–356.

Mandilas, C., Ormsby, M.P., Sheppard, C.G.W., and Woolley, R. (2007). Effects of hydrogen addition on laminar and turbulent premixed methane and iso-octane-air flames. *Proc. Combust. Inst.* **31**, 1443–1450.

Markstein, G.H. (1964). *Non-steady Flame Propagation*. AGARDograph 75. Pergamon Press, Oxford.

Miller, C.D. (1947). Roles of detonation waves and autoignition in SI engines. *SAE Q. Trans.* **1**, January.

Morley, C. (2007). Gaseq Chemical Equilibrium Program, c.morley@ukgateway.net.

Mueller, C.J., Driscoll, J.F., Reuss, D.L., and Drake, M.C. (1996). Effects of unsteady stretch on the strength of a freely propagating flame wrinkled by a vortex. *Proc. Combust. Inst.* **26**, 347–355.

Nygren, J., Hult, J., Richter, M., Aldén, M., Christensen, M., Hultqvist, A., and Johansson, B. (2002). Three-dimensional laser induced fluorescence of fuel distributions in an HCCI engine. *Proc. Combust. Inst.* **29**, 679–685.

Ó Conaire, M., Curran, H.J., Simmie, J.M., Pitz, W.J., and Westbrook, C.K. (2004). A comprehensive modeling study of hydrogen oxidation. *Int. J. Chem. Kinet.* **36**, 603–622.

Peters, N. (2000). *Turbulent Combustion*. Cambridge University Press, Cambridge.

Pickett, L.M. and Siebers, D.L. (2002). An investigation of diesel soot formation processes using micro-orifices. *Proc. Combust. Inst.* **29**, 655–662.

Pilling, M.J. and Hancock, G. (Eds.) (1997). *Low Temperature Combustion and Autoignition*. Elsevier, Amsterdam.

Plessing, T., Peters, N., and Wünning, J.G. (1998). Laseroptical investigation of highly preheated combustion with strong exhaust gas recirculation. *Proc. Combust. Inst.* **27**, 3197–3204.

Proudler, V.K., Cederbalk, P., Horowitz, A., Hughes, K.J., and Pilling, M.J. (1991). Oscillatory ignitions and cool flames in the oxidation of butane in a jet-stirred reactor. In: *Chemical Instabilities, Oscillations and Travelling Waves. Phil. Trans. R. Soc. London* **A337**, 195–306.

Rayleigh, J. (1878). The explanation of certain acoustical phenomena. *Nature* **18**, 319–321.

Searby, G. (1992). Acoustic instability in premixed flames. *Combust. Sci. Technol.* **81**, 221–231.

Singh, K.K., Zhang, C., Gore, J.P., Mongeau, L., and Frankel, S.H. (2005). An experimental study of partially premixed flame sound. *Proc. Combust. Inst.* **30**, 1707–1715.

Stokes, J., Lake, T.H., and Osborne, R.J. (2000). A Gasoline Engine Concept for Improved fuel Economy – the Lean Boost System, SAE Paper 2000–01–2902.

Van Blarigan, P. (1996). Development of a Hydrogen Fueled Internal Combustion Engine Designed for Single Speed/Power Operation, SAE Paper 961690.

Verhelst, S., Woolley, R., Lawes, M., and Sierens, R. (2005). Laminar and unstable burning velocities and Markstein lengths of hydrogen–air mixtures at engine-like conditions. *Proc. Combust. Inst.* **30**, 209–216.

Warnatz, J., Maas, U., and Dibble, R.W. (1996). *Combustion*. Springer-Verlag, New York.

Weinberg, F.J. (1986). Combustion in heat-recirculating burners. In: Weinberg, F.J. (Ed.) *Advanced Combustion Methods*. Academic Press, London, pp. 183–236.

Wünning, J.A. and Wünning, J.G. (1997). Flameless oxidation to reduce thermal NO-formation. *Prog. Energy Combust. Sci.* **23**, 81–94.

Zeldovich, Y.B. (1980). Regime classification of an exothermic reaction with nonuniform initial conditions. *Combust. Flame* **39**, 211–214.

Chapter 3

Highly Preheated Lean Combustion

Antonio Cavaliere, Mariarosaria de Joannon, and Raffaele Ragucci

Nomenclature

c_p	Specific heat
CI	Compression ignition
DI	Diffusion ignition
HBBI	Homogeneous burnt back-mixing ignition
HC	Homogeneous charge
HCCI	Homogeneous charge compression ignition
HCDI	Homogeneous charge diffusion ignition
HDDI	Hot diluted diffusion ignition
HFDF	Fuel is hot and diluted
HFDO	Fuel is hot and the oxidant is diluted
HFFI	Homogeneous flow fluent ignition
HiTAC	High temperature air combustion
HiTeCo	High temperature combustion
HODF	Oxidant is hot and the fuel is diluted
HODO	Oxidant is hot and diluted
k	Asymptotic strain rates, $2v_0/L$
L	Length
MILD	Moderate and intense low oxygen dilution
$\dot{m}_{dil}$	Mass flow rate of diluent
$\dot{m}_{fuel}$	Mass flow rate of fuel
$\dot{m}_i^{inlet}$	Mass flow rate at an inlet
$\dot{m}_{ox}$	Mass flow rate of oxidant
$p^{reactor}$	Reactor pressure
T_3	Turbine inlet temperature
T^0	Temperature in standard conditions
T_{dil}	Temperature of diluent
$T_{eq}(Z)$	Equilibrium mixture temperature
$T_{froz}(Z)$	Frozen mixture temperature
T_{fuel}	Temperature of fuel

T_i^inlet	Temperature at an inlet
T_ign'	Autoignition temperature for a stoichiometric fuel air mixture
$T_\text{ign}(Z)$	Mixture autoignition temperature
T_IGN	Autoignition temperature
T_in	Inlet temperature
T_inlet	Average of all inlet temperatures
$T_\text{max}(Z)$	Maximum temperature
T_ox	Temperature of oxidant
T_s	Stoichiometric temperature
T_si	Autoignition temperature of a well-stirred reactor
T_x	Brayton cycle temperature
T_WSR	Working temperature of a Well Stirred Reactor
X_{O_2}	Oxygen molar fraction
Y_dil	Local mass fraction of diluent
Y_{f_1}	Fuel mass fraction in the fuel stream
Y_f	Local mass fraction of fuel
Y_ox	Local mass fraction of oxidant
Y_{ox_2}	Oxidant mass fraction in the oxidant stream
Z	Mixture fraction
Z_s	Stoichiometric mixture fraction
α	Equivalence ratio
α_max	Maximum allowable equivalence ratio
ΔH_c	Heat of combustion
ΔT	Temperature increase
ΔT_IGN	Maximum ignition–inlet temperature differential
$\Delta T_\text{IGN}'$	Standard autoignition temperature difference
ΔT_MAX	Maximum temperature increase
ΔT_INLET	Differential ignition–inlet temperature
ϕ	Inlet conditions variables vector
η_c	Compressor efficiency
η_iso	Isentropic efficiency
η_r	Regenerator efficiency
η_t	Turbine efficiency
v_s	Stoichiometric mass ratio
π	Compression ratio
$\pi^{(\gamma-1)/\gamma}$	Isentropic compression relation
τ	Temperature ratio
τ^reactor	Average residence time

3.1. INTRODUCTION

The terms "high temperature air combustion (HiTAC)" and "moderate and intense low-oxygen dilution (MILD)" combustion have only recently been defined. First published in the Combustion Institute Proceedings in 1998 (Katsuki and Hasegawa, 1998) and then in the Mediterranean Combustion Symposium Proceedings in 2000 (de Joannon *et al.*, 2000), the definitions of these terms arose from a series of publications which have stressed features of interest pertaining to energy conversion and pollutant formation reduction. Proceedings papers concerning HiTAC (Crest, 2000; HiTAC, 2000, 2002) have added information on possible variants of the process. The review paper of Cavaliere and de Joannon (2004) and the monographic papers of de Joannon *et al.* (2000, 2007) and Peters (2001) have discussed the possibility of extending the application of MILD combustion by studying the process in a simple well-stirred reactor. Although classifying a process as HiTAC or MILD combustion is straightforward when the process is evaluated based on its beneficial effects, it is not always clear to which elementary process these effects should be attributed.

In this respect, possible applications extend beyond the limits identified in the previously referenced papers. Several works (Robertson and Newby, 2001; Yamada *et al.*, 1999; www.floxcom.ippt.gov.pl; Yasuda, 1999; Masunaga, 2001) have reported practical experiences that can be directly or indirectly related to HiTAC and MILD combustion. Other applications have also been identified and given specific names such as superadiabatic combustion (Howell *et al.*, 1996), indirect combustion, and flameless combustion (Wünning and Wünning, 1997; Milani and Wünning, 2002), which are strongly related to the processes presented here.

The potential of these processes has not yet been shown because the basic correlation between processes and performance, are repeatability and reliability, which are needed to tailor the processes to specific engineering design requirements, is not yet completely understood. Consequently, this chapter aims to clarify the fundamental aspects of these high preheat combustion processes, giving rigorous definitions and identifying precise elementary features. These can be used as a reference guide in understanding the observed behavior of real systems. In addition, this chapter suggests possible simple experiments that would result in the systemization of general knowledge concerning high temperature combustion (HiTeCo) and MILD combustion. Furthermore, this chapter identifies a range of practical applications that, by taking advantage of these unusual combustion modes, could enhance combustion performance in terms of energy conversion efficiency and emission reduction. Some preliminary practical experiences and basic concepts related to gas turbine applications are presented as leavening yeast for further enhancements in this field.

3.2. MILD COMBUSTION

An ensemble of inlet variables and boundary conditions that determine which single or multiple processes exist within a reactor can be used to characterize the reactor and its operation. These variables can be expressed in terms of a vector ϕ:

$$\phi = \phi(\dot{m}_i^{\text{inlet}}, T_i^{\text{inlet}}, p^{\text{reactor}}, \tau^{\text{reactor}}), \qquad (3.1)$$

where $\dot{m}_i^{\text{inlet}}$ and T_i^{inlet} are the mass flow rate and temperature of the reactants and diluents, respectively, p^{reactor} is the pressure inside the reactor (under the assumption of quasi-isobaricity), and τ^{reactor} is the average residence time. Note that this ϕ is not the equivalence ratio, as it denotes in other chapters of this book.

Figure 3.1 presents a schematic overview of a reactor. Oxidants, fuels, and diluents flow into the reactor with mass flow rates $\dot{m}_{\text{ox}}$, $\dot{m}_{\text{fuel}}$, and $\dot{m}_{\text{dil}}$, respectively, and with temperatures T_{ox}, T_{fuel}, and T_{dil}. It should be stressed that the diluent flow can be either

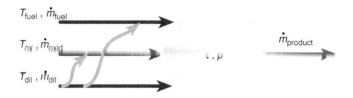

Figure 3.1 Schematic overview of a reactor.

fed to the system as a separate stream or premixed with the other flows. In the latter case, oxygen and fuel molar fractions in the oxidant and fuel streams, respectively, are lower than unity.

When the variable vector is known, it is possible to define a number of synthetic quantities that can be used to classify processes taking place in the reactor. The average inlet temperature, T_{inlet}, is the average of all the inlet temperatures weighted by their corresponding mass flows and heat capacities. Another way to identify the system status is by means of its frozen mixture temperature, $T_{froz}(Z)$, which represents the temperatures of all possible frozen mixtures consistent with the inlet conditions. In premixed systems, the two quantities coincide because only one inlet temperature can be fixed, whereas in reactors in which fuels, oxidants, and sometimes diluents are fed separately, the two temperatures can be different. The average inlet temperature is easy to calculate even in the presence of multiple flows. The frozen mixture temperatures depend on the mixture fractions, which depend in turn on the number of inlet flows. In practice, a great number of processes utilize flows of fuel and oxidant in which one or both of these streams are diluted. In this case, it is possible to define a simple mixture fraction that varies between 0 and 1 (Cavaliere and Ragucci, 2001). The temperature associated with this mixture is the average temperature weighted by the composition of the mixture, and is therefore a function of the composition (i.e., of the mixture fraction).

Other quantities of interest are the autoignition temperature of the system, T_{IGN}, and the mixture autoignition temperature, $T_{ign}(Z)$. These are the temperatures at which minimum conversion occurs starting either from the average composition, or from a frozen mixture consistent with the inlet flows for a period shorter than the residence time of the reactor. Finally, the maximum temperature, $T_{max}(Z)$, is defined as the temperature achievable for the maximum oxidative or pyrolytic conversion of a mixture, Z, consistent with the reactor conditions.

It is important to understand the meaning and the usefulness of these temperatures. Combining the temperatures forms a maximum temperature increase, ΔT_{MAX}, and a maximum ignition–inlet differential, ΔT_{IGN}. The maximum temperature increase is the greater of two maxima evaluated by means of Equations (3.2) and (3.3):

$$\Delta T_{MAX} = \max \left[T_{max}(Z) - T_{froz}(Z) \right] \text{ for } T_{eq}(Z) > T_{froz}(Z), \tag{3.2}$$

$$\Delta T_{MAX} = \max \left[T_{froz}(Z) - T_{eq}(Z) \right] \text{ for } T_{eq}(Z) > T_{froz}(Z). \tag{3.3}$$

The first maximum is defined by the greatest difference between maxima for $T_{eq}(Z) > T_{froz}(Z)$, and is between the frozen mixture maximum temperature and the inlet frozen mixture temperature according to Equation (3.2). The second maximum is for the difference between the mixture frozen temperature and the mixture equilibrium temperature $T_{eq}(Z)$ according to Equation (3.3) that is generally referred to pyrolytic conditions.

The maximum temperature increase is a measure of the possible incremental temperature increase or decrease that the process can experience, and is related to exothermicity and endothermicity as well as to the mass of the inert in which the reaction products are mixed. Nearly isothermal processes develop when the maximum temperature increase is around zero, whereas oxidation and pyrolysis take place when this quantity is positive or negative, respectively.

The differential inlet–ignition temperature is the maximum difference between the inlet frozen mixture temperature and the autoignition frozen mixture temperature according to Equation (3.4):

$$\Delta T_{INLET} = \max \left[T_{froz}(\Sigma) \quad T_{ign}(\Sigma) \right]. \tag{3.4}$$

In this case, the difference represents the tendency of the system to undergo explosion or autoignition. That is, the reactants introduced in the reactor can evolve along a reaction path without additional heating from an external source.

The combination of the two difference values identifies regimes which are well known as large macro-area processes. They are reported in Figures 3.2 and 3.3. The first figure identifies four broad categories on a diagram in which the maximum temperature difference and the inlet–ignition differential are reported on the ordinate and abscissa axes, respectively. The upper left quadrant is the region of assisted-ignited combustion for which the maximum temperature difference is positive compared to the inlet–ignition differential, which is negative. This is the classical combustion condition in which ignition of the mixture is ensured by heat introduced externally to the reactor, because the reactants do not carry enough sensible enthalpy to initiate the process. Processes located in the upper right quadrant of the figure are classified as autoignited combustion. In this case, the inlet–ignition differential temperature is positive, so ignition occurs spontaneously. Beneath the abscissa, the processes are characterized by a negative maximum temperature increase. These processes are referred to as pyrolysis because they are endothermic in nature. Pyrolysis is subdivided into two categories: autoincepted pyrolysis and assisted-incepted pyrolysis. Autoincepted pyrolysis does not need an external source to make the processes effective. It is defined by the lower right quadrant of Figure 3.2. Assisted-incepted pyrolysis needs external heating or a catalytic device in order to develop. It is defined by the lower left quadrant of the figure where the inlet temperature is lower than the ignition temperature. In the case of pyrolysis, this inlet temperature should be named the inception temperature.

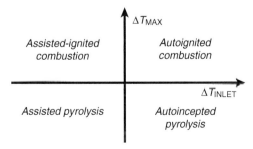

Figure 3.2 Combustion regimes on a temperature plot. The maximum temperature increase of a process is reported on the ordinate, and the inlet–ignition temperature difference is reported on the abscissa.

This broad classification can be refined by means of comparison between two differentials, that is the maximum temperature increase ΔT_{MAX} of the equation (3.2) and the standard autoignition temperature difference ΔT_{IGN}^{o}, which is the difference between the autoignition temperature at the stoichiometric air–fuel mixture fraction, and the temperature in standard conditions, T^{o}, according to Equation (3.5):

$$\Delta T'_{IGN} = T'_{ign} - T^{o}. \tag{3.5}$$

Figure 3.3 reports the possible subregimes which can be identified when the maximum temperature increase is higher or lower than the standard autoignition temperature differential. Specifically, the region under the horizontal line in the assisted-ignited combustion quadrant on the left side identifies conditions in which no combustion process should occur. This is the region in which the maximum temperature increase is still lower than the temperature increment needed to activate the processes even in a hot diluted system. Therefore, reaction is not assured to take place in a diluted system. The only possible processes under these conditions are those that develop when an external aid, which lowers or overcomes the activation energy of the processes, is added (i.e., catalytic assistance or external heat addition). It is appropriate that this category be named pilot combustion. In the upper part of the quadrant, all the processes recognized in classical combustion (Williams, 1985; Kuo, 1986; Griffiths and Barnard, 1995; Peters, 2000) as deflagration, detonation, or diffusion flames are reported. They are called feedback combustion processes because the enthalpy of the combustion products can heat the reactants up to a temperature that allows oxidation to occur.

Finally, the right quadrant of Figure 3.3 is subdivided into fields named HiTeCo or HiTAC, and MILD combustion. All of these are characterized by autoignition of the mixture by means of the sensible enthalpy carried by the reactants themselves, but they differ since the maximum temperature increase can be higher or lower than the standard ignition temperature differential.

Understanding the subdivision between HiTeCo and MILD combustion is useful in order to stress that the oxidation process is under the control of either the temperature of the reactant flows, or of the temperature which develops during combustion. This distinction is easy to understand in the limiting cases where the maximum temperature increase is much higher or lower than the temperature for autoignition; in this case the ratio of maximum to autoignition temperatures then tends to infinity or zero. In the particular case where the ratio is around unity, however, the reactive structure is located in the same region for HiTeCo and MILD combustion. Here, oxidation is driven by the

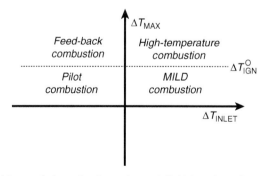

Figure 3.3 Non-pyrolytic combustion regimes subdivided on the ordinate of Figure 3.2.

enthalpy content of the reactants or by the enthalpy released during the reaction. This is not a general rule, and the transition may occur at different values of the temperature ratio, as can occur when the reactive structure is positioned in different regions and is under the control of different factors.

3.3. SIMPLE PROCESSES IN MILD COMBUSTION

MILD combustion is controlled by different factors according to the simple processes it undergoes. Simple processes are those that evolve in simple reactors like a well-stirred reactor, batch reactor, plug flow reactor, or counter-diffusion reactor. In each of these simple processes, elementary phenomena like convection, diffusion, and reactions develop in a specific way. The general feature is that the heat released by the MILD combustion is comparable to or less than the heat content of the reactants in each of these simple processes, characterizing as well as differentiating them from standard feedback combustion and HiTeCo processes. It is anticipated that the main difference consists of the fact that in standard combustion, there is always an internal flame structure that has relatively small scales with respect to the fluid dynamic scales in which the reaction heat is released. This relatively rapid heat release in a thin zone increases the local temperature to such a high level that enthalpy diffusion out of this flame zone preheats the reactants, making them more reactive and prone to ignite.

Initiation and stabilization of these thin flames requires an external source of heat, which can be removed when the flame is stabilized. This calls for at least two stable solutions of the equation describing feedback combustion, namely the frozen and reactive solutions. In contrast with HiTeCo, it is not possible to stabilize a frozen condition for thin reaction zones. Hence, extinction of the whole process is not possible, in principle. Nevertheless, if the maximum temperature increase in the reaction layer is relatively high with respect to the inlet–ignition temperature differential, then the reactive solution retains similarity to the feedback solution (essentially, the reaction provides its own thermal feedback). In fact, the oxidative reaction rate is very high in the same mixture fraction region as in the feedback combustion region, since the temperature is much greater than it is in the mixture fraction region far from stoichiometric. On the other hand, when the maximum temperature increase is less than the inlet–ignition temperature differential, the temperature in the stoichiometric mixture fraction region is not much higher than the temperature in the mixture fraction region far from stoichiometric conditions. In this case, it follows that the reactivity of any mixture fraction becomes comparable, and so reactions can be distributed more homogeneously.

In limited cases, the maximum temperature increase is small enough that the temperature at the stoichiometric condition is lower than temperatures at other mixture fractions. This is anticipated to occur when highly diluted fuel is mixed with very hot oxidant. In this situation, the reactivity can be even higher in mixture fraction regions away from stoichiometric. This possibility is noteworthy because it demonstrates an asymptotic heat release limit beyond which the traditional flame structure cannot occur.

In comparison, HiTeCo regimes contain multiple oxidative structures that cannot be neglected, while in strict MILD combustion regimes, distributed reactions are expected to play a primary role. Essentially, hysteresis effects are still present in HiTeCo, whereas they are not consistent with MILD combustion. Some authors use this feature as the definition of MILD combustion, even though it is limited to well-stirred reactor conditions. The meaning of multiplicity and how it is depressed in MILD combustion in well-stirred reactors as well as in other simple reactors will be explored further. This allows for investigation of the structure of the oxidation region in which MILD combustion takes place, and refines the definitions of HiTeCo and MILD combustion given in the previous section. Finally, analysis of the aforementioned simple processes requires study since they can be combined into more complex processes in which turbulent mixing is also included.

Simple processes in simple reactors can be classified according to Table 3.1. Homogeneous burnt back-mixing ignition (HBBI) and homogeneous charge compression ignition (HCCI) are zero-dimensional processes in space. Therefore, their analysis is relatively simple compared to the other three, which are one-dimensional in space. Autoignition is ensured in various processes in different ways, and this will be discussed in the following subsection. In the last processes, the homogeneous mixture evolves from unreacted to reacted conditions continuously in a similar way. The only difference is that the charge is quiescent in the HCCI process and is transported in the homogeneous flow fluent ignition (HFFI) process along one direction. This means that the temporal evolution of the mixture in HCCI is the same as that along the spatial coordinate in HFFI.

In order to understand the differences among these processes as compared to "standard" feedback combustion, the behavior of temperatures involved in defining MILD combustion, that is the "maximum allowable," frozen and autoignition temperatures, should be described in terms of the mixture fraction of frozen mixture for feedback combustion. A plot of this comparison is provided in Figure 3.4. The equilibrium

Table 3.1 Simple ignition processes in simple reactors

Process Acronym	Ignition Type	Reactor Type
HBBI	Homogeneous burnt back-mixing	Well-stirred
HCCI	Homogeneous charge compression	Batch
HCDI	Homogeneous charge diffusion	Counter-diffusion flow
HDDI	Hot diluted diffusion	Counter-diffusion flow
HFFI	Homogeneous flow fluent	Plug flow

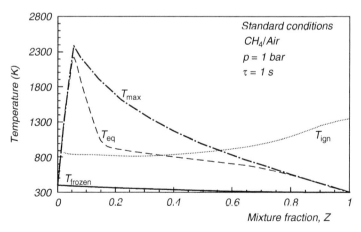

Figure 3.4 Frozen, maximum, equilibrium, and autoignition temperatures as a function of Z.

temperature of the system is also reported in the figure in order to evaluate the effectiveness of these temperatures on the reaction propensity. These behaviors will later be judged against similar plots obtained for inlet conditions related to the different processes listed previously. The same reference conditions are utilized in these plots in order to make the examples easier to compare.

In particular, an oxygen–methane system diluted with nitrogen is considered first. Reactants are injected in an atmospheric pressure reactor for a residence time of 1 second. Details of the computation procedures are described in a previous review paper (Cavaliere and de Joannon, 2004). Figure 3.4 refers to an undiluted air–methane system in which the air is at 400 K and the methane is at 300 K. The solid line indicates the frozen temperature, which ranges between the two inlet temperatures, whereas the dashed dotted line is the maximum achievable temperature for each mixture, reaching an absolute maximum value around 2400 K at the stoichiometric mixture fraction condition. The maximum temperature decreases steeply and gently on the lean and rich sides, respectively. The autoignition temperature is reported with a dotted line; it is nearly constant at 900 K for mixture fractions lower than 0.5, and then increases up to an autoignition temperature of 1400 K for very rich conditions. At the richest conditions, the concept of ignition may not be particularly relevant in any case. For mixture fractions higher than 0.7, the equilibrium temperature (reported by the dashed line) begins to track the maximum temperature, as it does on the lean side nearly up to the stoichiometric value. It decreases steeply on the rich side up to a mixture fraction value of 0.15 and is nearly constant between 0.15 and 0.6.

The full meaning of these plots will be clarified in the description of MILD combustion processes later in this chapter, but some initial characteristics are evident from the examples just provided. Differences among temperatures at fixed mixture fractions are relatively large for a wide range of compositions, and partial combustion or mixing between reacted and unreacted flows can yield, in principle, a mixture with temperatures between plotted pairs.

3.3.1. HBBI in Well-Stirred Reactors

Homogeneity of inlet composition simplifies the definition of MILD combustion with respect to the determination of the inlet–ignition temperature differential. In fact, in this case, both the inlet and ignition temperatures are explicitly determined since they refer to only one inlet flow, which is homogeneous in composition. In contrast, evaluation of the maximum temperature is complicated, albeit not difficult. Here, the residence time and reactor volume also determine the level of additional oxidized species present in the reactor. This quantity is mixed with reacting species and determines a mass averaged temperature, which is intermediate between the inlet and oxidized species temperatures. For the particular case where the maximum oxidation level is relatively low due to inlet dilution, and when the requirements for MILD combustion are satisfied, all possible conditions with varying residence times and reactor volumes can be included in this combustion category.

In Figure 3.5, working temperature (T_{WSR}) is computed for adiabatic conditions for three values of oxygen molar fraction (X_{O_2}), and are reported as a function of inlet temperature (T_{in}). Results of calculations at $X_{O_2} = 0.2$, represented in Figure 3.5 by a dashed line, show the typical trend expected for such a system. This anticipated

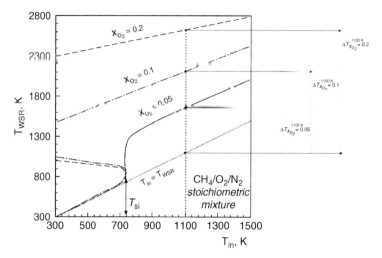

Figure 3.5 Working temperature of a well-stirred reactor as a function of inlet temperature for a stoichiometric mixture of methane, oxygen, and nitrogen (Cavaliere and de Joannon, 2004).

S-shaped curve is partially visible in the figure. According to the classical literature on reactor behavior (Aris, 1969; Fogler, 1986), the autoignition temperature (T_{si}) of a well-stirred reactor is the inlet temperature at which any differential temperature increase makes the system reach the higher branch of the S-shaped curve, at which point the chemical process self-sustains.

Thus, for T_{in} equal to or higher than T_{si}, the mixture ignites and burns, thus increasing the temperature of the system. The maximum temperature increase (ΔT) is the difference between the maximum temperature which occurs in the reactor, and the temperature of the inlet reactants T_{in}. In this case, for $T_{in} = 1100\,K$, the resulting ΔT is about 1600 K with much of the hydrocarbon fuel being consumed during the residence time under consideration. If the operating conditions are then changed by increasing the dilution level with a constant methane/oxygen ratio, the system responds by reducing the temperature increase during oxidation. With $X_{O_2} = 0.1$ and $T_{in} = 1100\,K$, ΔT is about 1000 K and becomes 550 K for $X_{O_2} = 0.05$. This is a relatively small temperature increase compared to those that occur in conventional combustion processes. The latter condition can be considered to belong to the MILD combustion category. This means that the process evolves in a rather narrow temperature range, which could be in an intermediate region between the very fast kinetics of oxidative non-diluted conditions and the relatively slow kinetics linked to low temperature self-ignition regimes.

Of course, real behavior is more complex since the yield of oxidized products is lower than unity in some conditions. In addition, the reactor can be fed rich mixtures, resulting in partial oxidation of the fuel, which can be analyzed according to the pyrolytic route for very limited conditions. These conditions have been analyzed in greater detail by de Joannon *et al.* (2000).

3.3.2. HCCI IN BATCH REACTORS

The inclusion of the HCCI process in the MILD combustion category is very straightforward since the inlet and outlet temperatures are easily determined.

Analogously to HBBI, this process is homogeneous and the temperature reached during compression is easily measured experimentally or evaluated through the compression ratio. The fuel–oxidant mixture progresses along the oxidation path toward the temperature that can be reached according to the residence time in the batch reactor. Also in parallel to the HBBI condition, the maximum possible level of oxidation can determine the maximum temperature attainable in the charge. This temperature is kept at a lower level than that of spark-ignited processes, because the charge is very lean and homogeneously distributed. The case in which this temperature is lower than suitable for feedback combustion allows for all possible processes to be considered to occur in MILD combustion conditions.

The HCCI process is suitable for applications mimicking premixed charge engines. The transition from prototype engines to products of widespread use is the subject of many studies (Kraft *et al.*, 2000; Flowers *et al.*, 2001, 2002). The process has advantages in terms of low nitrogen oxide emission and soot depression (Kraft *et al.*, 2000; Flowers *et al.*, 2002), but it has not been fully evaluated in terms of the possible increase in partial oxidation products and carbonaceous species different from soot. These conflicting aspects that increase some pollutants while decreasing others are related to the relatively low temperatures and high oxygen concentrations produced at the end of the combustion process. It is known, for example, that thermal NO_x is affected by both temperature and oxygen concentration, but temperature has a greater effect on the process (Beér, 2000). Also, soot formation is reduced when both the temperature and the carbon/oxygen ratio is very low (Böhm *et al.*, 1988; Wang and Cadman, 1998; Douce *et al.*, 2000). At the same time, partial oxidation of the fuel is possible due to low temperatures particularly because the part of the engine that can be further cooled by the chamber wall may be relatively extensive in spark-ignited engines, which normally reach higher charge temperatures.

Furthermore, the HCCI process is difficult to adjust according to the required mechanical power because ignition delay varies with air–fuel ratio, making it difficult to complete oxidation across all engine speeds. This disadvantage can be overcome by using flue gas recirculation in order to dilute the air–fuel charge without changing the beneficial effect of reducing the final temperature (Smith *et al.*, 1997; Chen and Milovanovic, 2002; Ng and Thomson, 2004; Xing-Cai *et al.*, 2005). Except to note that HCCI can be classified as a highly preheated lean combustion process, further details regarding the HCCI process are not presented here but they can be found in the references listed.

3.3.3. HCDI in Counter-Diffusion Flow Reactor

A fuel lean mixture outside the flammability (or inflammability) limit is defined as a mixture in which deflagration cannot be stabilized. Nevertheless, oxidation may occur when the mixture is heated by an external source by means of enthalpy and/or mass diffusion through the mixture. Several experimental studies have shown the occurrence of this process when the source is a heated wall or a high temperature inert flow in a laminar (Daribha *et al.*, 1980; Smooke *et al.*, 1990; Zheng *et al.*, 2002) or turbulent (Blouch *et al.*, 1998; Mastorakos *et al.*, 1995) counter-diffusion reactor. These works were more focused on the ignition and extinction limit of the system rather than on their oxidation structure, even though they report many significant features of the

process. In particular, the peculiarity of oxidation under these conditions is not empha-
sized. In the taxonomy of highly preheated combustion, however, it deserves a specific
name in order to refer to it in an unambiguous way. An analogy to the HCCI process is
suitable for this purpose. The first part of the acronym, HC, refers to the homogeneous
charge of air and fuel and should be kept, while the second part of the acronym, CI
(compression ignition), can be replaced with DI, which stands for diffusion ignition,
to underline that they are similar in the autoignition of the mixture but differ in the cause
of this process. In the first case, autoignition is due to compression heating, whereas
in the second case it is due to enthalpy/mass diffusion, which mixes the HC with a
high temperature inert flow. Hence, the acronym for the process is HCDI, that is
Homogeneous charge diffusion ignition.

The main feature expected of this type of process consists of the whole oxidation
structure which is placed inside the diffusive controlled mixture layer. Providing
examples of this structure requires reference to the counter-flow configuration shown
in Figure 3.6 for an oxygen, a methane, and a nitrogen system. As illustrated by the
figure, two flows proceed along the x direction, which is the coordinate axis with the
origin at the stagnation point. The flows have defined velocities at infinite x on both
the positive and negative sides and are characterized by a linear decrease of the velocity
component on the symmetry axis coupled with a linear increase of the orthogonal
component of the velocity. Therefore, the fluid dynamic pattern is characterized by a
single parameter, that is the velocity gradient of the two velocity components.

In practice, this pattern is created around the stagnation point of two opposed flat
(uniform velocity) jets, which flow through orifices placed at a fixed distance from each
other. A mixing layer occurs around the stagnation point, in which the species of both
jets are diffused into each other, yielding a continuous variation of the mixture fraction.
Reactions can also take place inside the mixing layer where suitable composition and
temperature conditions are created. Figure 3.6 shows schematically the location of a
reactive region by the shaded area. The region is deliberately shown away from the
stagnation point in the symmetric plane in order to stress that the proper composition
and temperature for reaction may be placed at any location.

The use of this counter-flow configuration allows for the illustration of a wide variety
of HCDI structures as well as more invariant deflagration and diffusion flame structures
by comparing a spatial or mixture fraction coordinate to different parameters involved
in the controlling process. For instance, the role of the inert flow temperature is shown

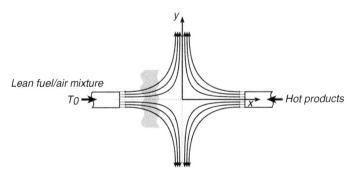

Figure 3.6   Schematic of counter-flow configuration for flame stabilization.

in Figure 3.7. The plot reports three ensembles of spatial distributions of enthalpy production per unit mass. The three frames correspond to inert flow temperatures of 1400, 1600, and 1800 K, respectively, and the three curves in each frame correspond to three asymptotic strain rates, k (obtained as $k = 2v_0/L$), of 20, 55, and 80 s^{-1}.

The heat release values in Figure 3.7 for $k = 20$ s^{-1} at 1400 K are on the negative side, quite far from the stagnation point. This means that the reaction zone is placed in the homogeneous air–fuel mixture. The other two curves lie on the positive side with respect to the stagnation point. They are centered furthest from the stagnation point for the highest strain rate and are shifted toward the stagnation point for the intermediate strain rate. The bell-shaped curves are quite broad and extend for a large part of the diffusive layer.

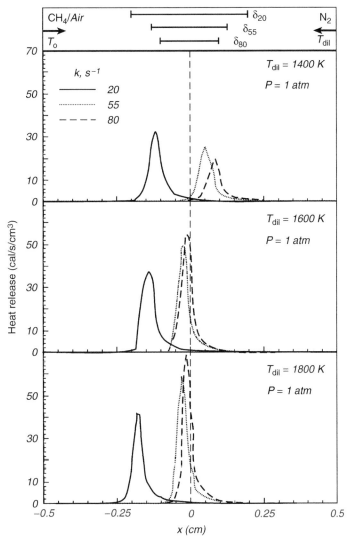

Figure 3.7 Spatial distribution of heat release in a counter-flow configuration at 1400, 1600, and 1800 K.

The enthalpy production at 1600 K for $k = 55\,s^{-1}$ and $80\,s^{-1}$ is positioned on the opposite side of the stagnation point with respect to the position at 1400 K, which is the side from which the air–fuel mixture enters. These two curves are different from those in the previous case as they are quite thin, and are shifted closer to the stagnation point. The same trend is also confirmed at the higher temperature of 1800 K, even though the profiles are nearer the stagnation point. Notably, the peak of the enthalpy production is comparable in amplitude for the last two cases, whereas it is quite low for the process at 1400 K.

On top of the figure, the mixing layer thicknesses are reported for the three strain rates in the form of dark bars which extend in the range of the spatial axis, shown as the abscissa of the plots. Note that the heat release profiles are approximately inside the corresponding mixing layer thicknesses.

In summary, the figure demonstrates that a great variety of oxidation structures may be stabilized in HCDI processes, but there is an unequivocal localization of the structure inside the mixing layer. In other words, the presence of the oxidation process is related to the diffusion between the inert and the air/fuel charge. Therefore, the acronym seems appropriate, but the structures do not share any characteristics with diffusion flames, neither in the provenance of the reactants nor in the invariance properties which are known for classical diffusion flames.

In order to give a reference framework for the analysis of these processes, two ensembles of plots, which refer to examples of two regimes identified in the literature

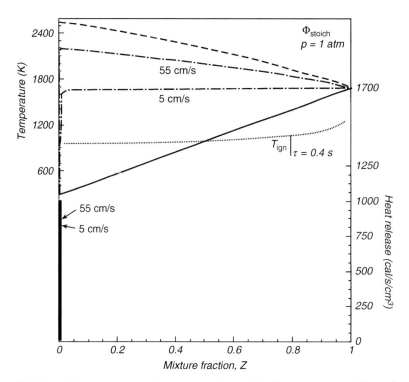

Figure 3.8   Frozen inlet temperature, maximum temperature, autoignition temperature, and heat release rate as function of the mixture fraction Z for different strain rates for a subadiabatic case.

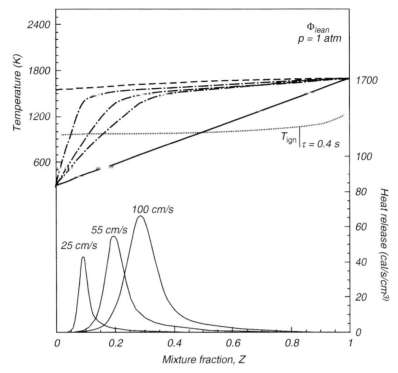

Figure 3.9 Frozen inlet temperature, maximum temperature, autoignition temperature, and heat release rate as function of the mixture fraction Z for different strain rates for a superadiabatic case.

(Libby and Williams, 1983; Puri and Seshadri, 1987) are reported in Figures 3.8 and 3.9. In Figure 3.8, the frozen inlet temperature, maximum temperature, and autoignition temperature are reported as a function of the mixture fraction with solid, dashed, and dotted lines, respectively. The mixture fraction in this case is the mass fraction of the inert species, and is thus zero for the air–fuel mixture and unity for the diluents, so the fractional values are percentages based upon the presence of the diluents in the mixture. The example refers to an adiabatic flame temperature of 2004 K and an inert temperature of 1700 K. Consequently, the maximum temperature is a decreasing linear function of the mixture fraction, as it is obtained from the weighted average of the enthalpies of the undiluted, reacted charge and the inert flow. This condition has been classified in the literature as the subadiabatic regime, since the enthalpy of the inert gases is lower than the enthalpy of the reactants (Libby and Williams, 1983). It refers to conditions which are used in some experimental works (Puri and Seshadri, 1987; Smooke *et al.*, 1990) and in some theoretical studies (Libby and Williams, 1982, 1983; Seshadri, 1983) in order to analyze the behavior of a counter-flow reactor with homogeneous air–fuel flow impinging on a hot inert flow.

The most important result of these papers is that they show the occurrence of premixed flame extinction. Furthermore, an abrupt displacement of the reaction zone from the reactant side to the inert side is also related to this transition (Libby and Williams, 1982), which separates the subcritical and supercritical regimes. In Figure 3.8, the vertical black bar in the lower part of the figure (heat release) and the dash-dot lines in the upper part of the figure (temperature) show a steep variation at very low values of the mixture fraction. This means that oxidation takes place on the reactant side because the zero value is the mixture without any dilution. Figure 3.9 shows the same quantities that are reported in Figure 3.8 for a superadiabatic condition, in which, as shown by the dashed line, the adiabatic flame temperature of the fresh air–fuel mixture is 1500 K and the temperature of the inert gases is 1700 K. The adiabatic flame temperature corresponds to an air–fuel ratio slightly lower than the lean flammability limit, and the inert gases are close to the maximum temperature at which a metallic material can survive. In other words, the example is of some interest from the perspective of practical applications.

In the same plot, the autoignition temperature is also reported as a function of the mixture fraction for minimum oxidation in a residence time of one second. This temperature is higher than the frozen mixture temperature, so the example pertains to an autoignited combustion regime. Furthermore, the maximum temperature increase is 1200 K, which is lower than the ignition temperature differential since the equivalence ratio of the mixture is above the flammability limit. Of course, it is possible to create superadiabatic combustion processes which are not MILD, but such processes are always at HiTeCo conditions because the inert gas temperature is higher than the adiabatic flame temperature, which is in turn higher than the autoignition temperature. Under these conditions, it is possible to stabilize two kinds of processes: the development of a deflagration at $Z = 0$, or an autoignition process in the range where the autoignition temperature is higher than the frozen mixture temperature.

In contrast, in the example of Figure 3.9, it is not possible to have a deflagrative premixed flame because the mixture is outside the flammability limit and only the autoignition structure can be stabilized. Thus, the temperature profiles reported with dash-dot lines and the enthalpy production reported with solid lines for three different stretch rates belong to the only possible regime, and the distinction between sub/supercritical conditions is meaningless.

3.3.4. HDDI in Counter-Diffusion Flow Reactor

Combustion processes in which the reactants are separated in feedback combustion yield diffusion flames. Differing from the case of premixed flames, diffusion flames do not have a specific designation to characterize the process generating their structure. For instance, from the discussion in the preceding section, deflagration is the process that represents the premixed flame, whereas the process that generates the diffusion flame has not been formally named. Thus, it is not possible to depict a perfect correspondence between the diffusion controlled processes in feedback combustion and those in MILD combustion, at least in terms of a glossary. On the contrary, it is possible to infer analogies when analyzing the simple reactor in which the processes are stabilized. In both cases, for example, the counter-diffusion reactor is fed with oxidant and fuel in opposite directions along the same axis. The main difference is that in MILD combustion, the oxidant and the fuel can be diluted with inert species and can be heated to such high temperatures that the frozen temperature is higher than the ignition temperature. Therefore, it is possible to envisage the four configurations listed in Table 3.2, each with a simple acronym of self-explaining terms. Specifically, H, D, O, and F stand for hot, diluted, oxygen, and fuel. The first two processes in Table 3.2 are characterized by

Table 3.2 Mild combustion configurations

Configuration Name	Description
HODO	Oxidant is hot and diluted
HFDO	Fuel is hot and oxidant is diluted
HFDF	Fuel is hot and diluted
HODF	Oxidant is hot and fuel is diluted

dilution of the oxidant. In the first case, the oxidant is also heated to high temperature, whereas in the second process, the fuel is heated. The last process is less interesting than the other three because the fuel is preheated to such temperatures that the fuel undergoes a classical pyrolysis, which may generate heavier and denser aromatic products. Consequently, the last process will not be discussed in this chapter because it would likely generate undesired pollutants.

The first process, on the other hand, has the potential to reduce several pollutants, such as nitrogen oxides and carbonaceous pollutants (Cavaliere and de Joannon, 2004). It is analyzed here in detail when the oxidant is diluted to an oxygen molar fraction of 0.05 and is heated up to a temperature of 1300 K. The results are reported in Figure 3.10 with the same line conventions as those reported before. The temperature over the entire range of frozen mixture fractions, plotted as a solid line, is always higher than in standard conditions, and decreases continuously from 1300 to 300 K, which is the fuel temperature. The maximum and equilibrium temperatures are similar in their overall trend to those in standard conditions, but the maxima of both are shifted toward lean conditions and are lowered to values of 1800 and 1700 K. These two correlated features are of great interest because even in the case of a diffusion flamelet stabilized around the stoichiometric value, as is similar to standard combustion, a significant change takes place. The maximum occurs in conditions where the mixture fraction gradients are lower than those in standard conditions. As a result, the dissipation rate of the scalars is also lower which inhibits flame quench. In contrast, the autoignition temperatures (drawn with a dotted line) are higher than the frozen temperature over quite a large mixture fraction range. This means that whenever extinction occurs, a mixture is formed

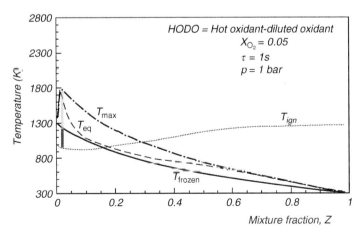

Figure 3.10 Frozen inlet temperature, maximum temperature, autoignition temperature, and equilibrium temperature as a function of the mixture fraction *Z* for the hot oxidant-diluted oxidant (HODO) process.

that can evolve spontaneously toward its oxidative state. In other words, whatever the local combustion regime may be, a reactive progression is always possible, and multiple combustion processes can take place by passing from the feedback mode, as occurs in diffusion flames to diffusion ignition oxidation. This is in contrast to events at standard conditions where multiple probable processes shift from the diffusion flame to the frozen condition with resulting noisy combustion. The local competition between these two processes should be evaluated case by case, taking into account the level of dilution and of oxidant preheating. In any case, the maximum temperatures in the MILD combustion examples are always lower than the maximum temperatures in the undiluted case, with the aforementioned positive effect in the reduction of both pollutant formation and of the temperature range in which the process advances.

Whether these characteristics are general for hot diluted oxygen conditions can be evaluated by analyzing Figure 3.11. This figure reports both the stoichiometric mixture fraction values, Z_s, and the stoichiometric temperature reached by the system in this condition, T_s, as a function of the oxygen mass fraction in the oxidant with solid and dashed lines, respectively. The two curves represent plots of Equations (3.6) and (3.7):

$$Z_s = (1 + v_s Y_{f_1}/Y_{ox_2})^{-1}, \tag{3.6}$$

$$T_s = \frac{Y_f \Delta H_c}{c_p^{ox} Y_{ox} + c_p^{dil} Y_{dil} + c_p^f Y_f} + T_{frozen/s}, \tag{3.7}$$

where $T_{frozen/s}$ is defined as

$$T_{frozen/s} = \frac{(Y_{ox_2} c_p^{ox} + Y_{dil_2} c_p^{dil})T_1 + Y_{ox_2}/v_s c_{pf} T_2}{1 + Y_{ox_2}/v_s}, \tag{3.8}$$

and v_s is the stoichiometric mass ratio, Y_{f_1} is the fuel mass fraction in the fuel stream, Y_{ox_2} is the oxidant mass fraction in the oxidant stream, Y_f, Y_{dil}, and Y_{ox} are the local mass fractions of fuel, diluent, and oxidant, respectively; ΔH_c is the heat of combustion of the fuel, and the c_p values are the specific heats of the different species.

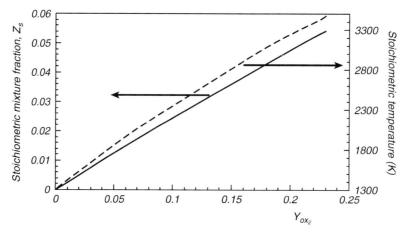

Figure 3.11 Stoichiometric mixture fraction (on the left axis) and the related temperature (on the right axis) as a function of oxygen mass fraction in the diluted stream.

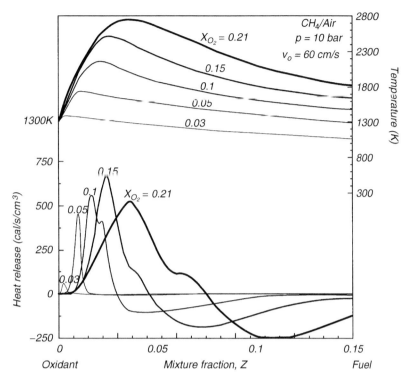

Figure 3.12 Temperature and heat release rate as a function of the mixture fraction Z for fixed strain rates in a HODO configuration.

The behavior described for such a configuration can be easily observed in the temperature and heat release profiles as a function of mixture fraction for a hot oxidant-diluted oxidant (HODO) case, as reported in Figure 3.12. The profiles refer to a methane flame at 10 bar with an initial velocity of 60 cm/s for a preheated, diluted oxidant stream of 1300 K and an undiluted fuel at 300 K. Clearly, the maxima of temperature (upper part of the figure) and heat release (lower part of the figure) occur at lower mixture fraction by increasing the dilution level, thus shifting toward the oxidant stream. At the same time, the flame structure also changes, as the shape of the heat release profiles show. In standard conditions, that is for an oxygen molar fraction of 0.21, two main regions can be recognized. The first flame zone lies between $Z = 0$ and $Z = 0.07$ where oxidation is the main reactive process, as the positive heat release rate testifies. In contrast, the second flame region for Z higher than 0.07 is found where, as the heat release rate demonstrates, an endothermic process overcomes the fuel oxidation. The extent of this second region decreases with decreasing oxygen content until it completely disappears, such as shown by the heat release profiles at $X_{O_2} = 0.05$ and 0.03. Moreover, the profiles at $X_{O_2} = 0.21, 0.15$, and 0.1 reach two maxima in the region where the heat release rate is positive, which is representative of the oxidation process occurring in two steps: fuel conversion principally to CO followed by oxidation to CO_2.

The characteristics of MILD combustion change when diffusion ignition takes place in a structure where the diluted fuel diffuses into the non-diluted hot oxidant.

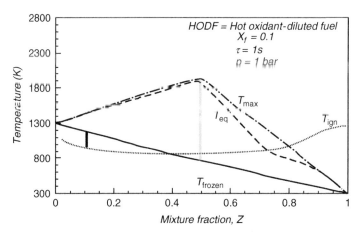

Figure 3.13 Frozen inlet temperature, maximum temperature, autoignition temperature, and equilibrium temperature as a function of the mixture fraction Z for the hot oxidant-diluted fuel (HODF) process.

The temperature-mixture fraction plot reported for this case in Figure 3.13 shows this change when the oxidant is heated up to the same temperature as before and the fuel molar fraction is 0.1 due to dilution in nitrogen. The frozen temperature profile is the same as that in Figure 3.9, whereas the maximum temperature profile is altered significantly with the shift of its maximum toward the richer region. In this specific case, the location of the maximum is at $Z = 0.5$, at which the maximum temperature is 1900 K, but both quantities depend on the dilution level, as is illustrated in Figure 3.14. Similar to Figure 3.11, and according to Equations (3.6) and (3.7), Figure 3.14 illustrates the dependence of the stoichiometric mixture fraction and the maximum temperature on the dilution level of the fuel. The shift of the stoichiometric value is toward the rich side, and is therefore in the direction of decreasing frozen temperature. This means that, jointly with the moderate temperature increase due to dilution, the variation in maximum temperature is not very steep. In fact, for this condition, there are two concurrent factors that both depend on large amounts of dilution. The first is the decrease in temperature increment, and the second is that the temperature increment has to be added to the initial frozen value, which has also decreased. This peculiar trend can be extended to a condition where the maximum temperature at the stoichiometric value is lower than the value on the oxidant side. In any case, there is a tendency toward a strong homogenization of the temperature over a wide mixture fraction range. This means that the oxidation process is controlled by the temperature since the kinetic characteristic times are the same for both the very lean side and the stoichiometric condition. In other words, the ignition diffusion process and feedback combustion have the same probability to occur. It is also of interest that the equilibrium temperature is nearly the same in this condition. This temperature closely follows the maximum temperature in Figure 3.13 for mixture fraction values up to $Z = 0.5$ and is only slightly lower for richer conditions. This means that, regardless of the kinetics involved in the process, maximum oxidation is favored over the formation of pyrolytic partially endothermic compounds.

Comparing profiles of the ignition and frozen temperatures is also informative. The crossover of the two curves occurs at a mixture fraction of 0.4 and is expected to occur at lower mixture fractions by increasing the dilution. This means that richer conditions

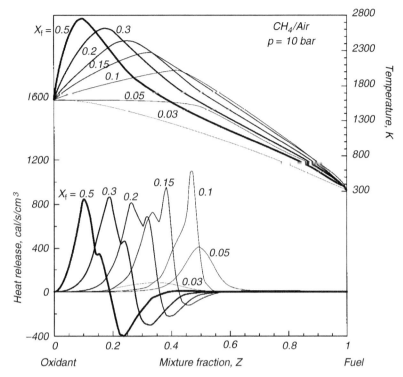

Figure 3.14 Temperature and heat release rate as a function of the mixture fraction Z for fixed strain rates in a HODF configuration.

than this value cannot spontaneously react and that in limiting conditions, autoignition is favored relative to the oxidation process at stoichiometric conditions. This is particularly true for processes in very turbulent regimes where quenching extinction of the diffusion flame may occur so that the only possible oxidation takes place at a mixture fraction lower than the stoichiometric condition; that is, only lean or super lean combustion takes place with enhanced inhibition of partial oxidation species.

The example reported in Figure 3.13 is relative to conditions in which the mass flow rate of the fuel is comparable to the mass flow rate of the oxidant. This is an advantage for the mixing of the two streams because they can be injected in the reactor with comparable momentum, but it can be a disadvantage because it is difficult to dilute the fuel to a high degree. For instance, these results were obtained with a mass diluent-fuel ratio on the order of ten, corresponding to an overall heating value on the order of 1000 kcal/kg. On the other hand, low dilution may be required anyway in the event a low heating fuel must be burned.

Examples of flame structures obtained in the hot oxidant-diluted fuel (HODF) configuration are reported in Figure 3.14. As with the HODO configuration, temperature (upper part of the figure) and heat release rate (lower part of the figure) profiles refer to a methane flame at a pressure of 10 bar and an initial velocity of 60 cm/s. As is also implied in Figure 3.15, the stoichiometric mixture fraction and heat release profiles shift toward the fuel side with increasing fuel dilution. Moreover, the flame structure changes with the fuel molar fraction. For $X_f = 0.5$, the heat release rate is positive up to

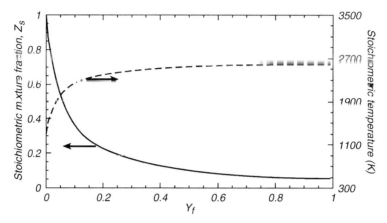

Figure 3.15   Stoichiometric mixture fraction (on the left axis) and the related temperature (on the right axis) as a function of fuel mass fraction in the diluted stream for HODF configuration.

$Z = 0.18$ and two maxima are present. These maxima are related to the successive steps of the oxidative process. For higher mixture fraction, the heat release rate becomes negative. By decreasing X_f, the maxima move toward the fuel side and become closer until they merge into a single peak. Simultaneously, the region corresponding to negative heat release rates reduces its size until it disappears.

The last HDDI process analyzed is obtained when the fuel is injected at high temperature and is diluted. This process is classified as a hot fuel-diluted fuel (HFDF) process. Figure 3.16 shows the behavior of the characterizing temperatures for an inlet fuel temperature of 1300 K and a fuel molar fraction of 0.1. The trend of the maximum temperature is different from previous cases: it increases steeply on the lean side, whereas it decreases gradually on the rich side. The inlet–ignition differential shows the opposite behavior where it is negative on the lean side and is positive on the rich side. In some respects these conditions are not favorable for diffusion ignition oxidation because the conversion should be restricted to the rich side. Nevertheless, the conditions

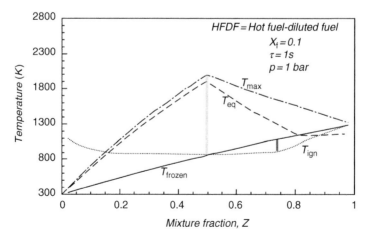

Figure 3.16   Frozen inlet temperature, maximum temperature, autoignition temperature, and equilibrium temperature as a function of the mixture fraction Z for a hot fuel-diluted fuel (HFDF) configuration.

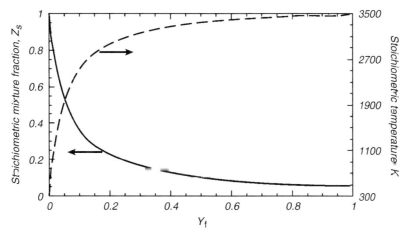

Figure 3.17 Stoichiometric mixture fraction (on the left axis) and the related temperature (on the right axis) as a function of fuel mass fraction in the diluted stream for HFDF configuration.

are better than those in traditional combustion. The temperatures do not attain high values of feedback combustion, and reforming of the fuel on the rich side can be achieved which helps in the complete oxidation of the fuel. In particular, this feature can be exploited for nitrogen oxide reduction in a reburning process, and in the suppression of recombination kinetics which are ter-molecular in nature because they reduce the concentration of any reactive species due to dilution.

This behavior is general for a wide range of preheat and dilution conditions, but is different from previous cases since the stoichiometric mixture fraction shifts with dilution toward the rich side where the frozen temperature also increases. These trends can be better analyzed with the aid of Figures 3.16 and 3.17, which shows the same quantities as Figures 3.13 and 3.15. The only difference is that the diluted fuel is preheated instead of the oxidant. In Figure 3.17, the curve related to the stoichiometric mixture fraction is the same as that in Figure 3.15. Figure 3.16 shows that in the HFDF case the maximum temperature increases for low values of dilution and, only for values higher than 0.5, starts to decrease.

3.3.5. HFFI IN PLUG FLOW REACTOR

The HFFI process is created when an HC flows in a plug flow reactor with inlet and outlet temperatures similar to those described for HCCI. In the plug flow reactor, the radial composition, temperature, and velocity are uniform and the only possible change is along one spatial direction. Under these conditions, the charge autoignites after the ignition delay and oxidation proceeds within the reactor controlled only by the chemical kinetics, since only convection plays the role of displacing the charge. This is analogous to HCCI because what develops as time progresses in the HCCI is the same as can be found along the axial direction of the HFFI. The only difference is that HCCI is a constant volume process, whereas HFFI is usually quasi-isobaric. This difference also allows for a variety of applications because the first is suitable for alternative reciprocating engines, whereas the second is suitable only for continuous flow reactors. Furthermore, HFFI is difficult to implement under practical conditions because mixing should be instantaneous, and in principle, the mixture should undergo ignition just as it is formed. In practice, the fuel can be mixed with the oxidant in characteristic times that are very short in comparison with those of autoignition, so that a quite reasonable approximation of the required properties is feasible. This makes the process interesting particularly for reactors used in laboratories because an accurate control of the mixing time is possible only in this framework. The appeal is also due to the fact that the spatial evolution of the process allows the sampling of reacting species in a more reliable way than for HCCI, where some form of time resolved sampling is needed. Details are available in the literature for the design of HFFI processes of gaseous fuels (Sabia *et al.*, 2005), in which very fast mixing process is obtained, as well as for practical use in pyrolytic (Lam, 1988) and oxidative conditions (Hermann *et al.*, 2005; Masson *et al.*, 2005).

3.4. PROCESSES AND APPLICATIONS OF MILD COMBUSTION IN GAS TURBINES

As described in Chapter 5, gas turbines are characterized by some constraints which are difficult to overcome. In particular, the turbine inlet temperature and the compressor/turbine efficiency are limited by the material robustness of the blades and by mechanical tolerances and lubrication problems. A practical high performance limit for the temperature is around 1573 K, whereas a more ambitious feasible limit can be around 1673 K. Compressor/turbine efficiencies can reach values around 0.7, and even higher values (up to 0.85) are possible but costly. The compressor outlet pressure, which is also the pressure in the isobaric combustion chamber, can vary between 3 and 30 bar, while the inlet combustion chamber temperatures, consistent with these values, can range between 200°C and 600°C.

Therefore, inlet and outlet temperatures of the gas turbine combustion chamber are suitable for MILD combustion applications as long as it is recognized that autoignition of mixtures is favored at high pressures. Specific temperature and pressure limits must be determined for each application, and these should be discussed with reference to a variety of possible configurations to ensure their feasibility. These configurations fall into two categories: MILD combustion with external control (independent source, external recirculation, and sequential combustion), and MILD combustion with internal recirculation. Each category is discussed in the following subsections, and directions for possible applications are also considered.

3.4.1. MILD COMBUSTION WITH EXTERNAL CONTROL

The dilution of the oxidant with inert gases can be achieved in open and closed gas turbine cycles by means of species that are not related to the gas turbine itself. In principle, they can be any kind of gas, but for economic reasons, a feasible choice for open cycles is flue gas from a combustion source. This makes the gas turbine outlet flow a possible candidate, as in sequential combustion and in systems with external recirculation. An analysis of a system with an independent source of inert flows demonstrates the theoretical limitations of such a system. To use flue gas as this source of diluent, it must be obtained from a relatively clean combustion process so that it is free from particulates and corrosive gases that could degrade the operation and life of the compressor and turbine. Under these constraints, the flue gas and the oxidant, say air, can be mixed, compressed, and distributed to the combustion chamber inlet with the oxygen content consistent with ignition and maximum allowable temperatures for MILD combustion.

The simplest reactor configuration pertains to the most advanced type of gas turbine. High temperature and pressure, which favor autoignition, are created at the combustion chamber inlet to achieve very high compression ratios. These elevated ratios are convenient not only for very high compressor/turbine efficiencies, but also for high turbine inlet temperatures. This is shown in Figure 3.18 (and Chapter 5) where the efficiencies of a Brayton cycle, which is based upon thermodynamic conditions sketched in the inset of the figure, are reported for two different final combustion temperatures and two values of the compressor/turbine efficiency. The inset shows on a temperature-entropy diagram, the isentropic compressor (1–2s) and turbine (3–4s) paths, which shift to the 1–2 and 3–4 paths in order to take into account the real non-isentropic transformation. Combustion takes place at nearly constant pressure along the path 2–3.

Efficiencies depend on the gas turbine compression ratio p_2/p_1, and are reported in Figure 3.18 for two temperatures T_3 and for constant compressor and turbine efficiency, η_c and η_t, which are assumed to be constant and equal to the isentropic efficiency η_{iso}. The efficiencies are given by Equation (3.9), obtained by simple relations (Hernandez *et al.*, 1995) where $\tau = T_1/T_3$ is the temperature ratio, and $\pi = p_3/p_4 = p_2/p_1$ is the compression ratio.

$$\eta = \frac{\left(\eta_t - \frac{\tau}{\eta_c}\pi^{(\gamma-1)/\gamma}\right)\left(1 - \frac{1}{\pi^{(\gamma-1)/\gamma}}\right)}{\left\{1 - \tau - \frac{\tau}{\eta_c}\left[\pi^{(\gamma-1)/\gamma} - 1\right]\right\}(1 - \eta_r) + \eta_r\eta_t\left(1 - \frac{1}{\pi^{(\gamma-1)/\gamma}}\right)}. \tag{3.9}$$

All of the curves in Figure 3.18 reveal that the efficiency maxima are situated at compression ratios ranging between 3 and 25. For higher values of compression ratios, the efficiencies decrease continuously. Higher final temperature values of the combustion process, T_3, shift the efficiency curve toward higher values and make the compression

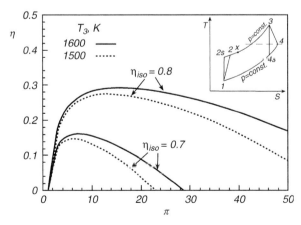

Figure 3.18 Efficiency of Brayton cycle as a function of compression ratio for two final combustion temperatures (T_3) and two compressor/turbine efficiencies (η_{iso}).

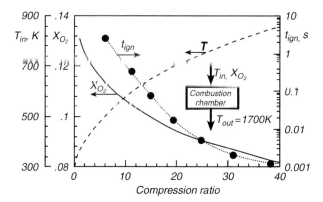

Figure 3.19 Temperature and oxygen concentrations at combustion chamber inlet versus compression ratio for a fixed inlet turbine temperature on the left axes and the related ignition delay time on the right axis.

ratio go up to 35–40 which is reasonable for practical applications. Limitations on the net power output suggest limiting compression ratios range between 30 and 35.

Figure 3.19 illustrates the relevance of high compression ratios and the consequent need for high values of T_3 and η_{iso} for suitable use in MILD combustion. The figure reveals the dependence of both the inlet temperature and the oxygen concentration on compression ratio at inlet combustion chamber pressure, which is the same as that of the compressor outlet. The first quantity, sketched with a short dashed line, is obtained by means of the isentropic compression relation $\pi^{(\gamma-1)/\gamma}$, and the oxygen concentration, represented by a continuous line, is obtained by means of an enthalpy balance between the inlet and the outlet of the combustion chamber, as is displayed schematically in the inset of the figure. The example refers to decane fuel mixing with air and flue gas. This mixture burns between the inlet temperature values and an outlet temperature value fixed at 1700 K. The autoignition delay time is calculated according to the procedures reported in the paper by de Joannon *et al.* (2002) and it is reported in the figure a dotted line by referring to the axis on the right side of the figure.

The comparison of these autoignition delays with those in non-diluted cases confirms that the beneficial effect of increases in pressure and temperature is partially counter-balanced by the dilution itself. This has the effect of decreasing the oxygen content in the oxidant stream. It has been shown (de Joannon *et al.*, 2002) that this parameter affects the autoignition delay following a power law with an exponent of 0.3 for temperatures lower than 700 K, of around unity for temperatures higher than 1000 K, and of around 2 in a narrow temperature range centered around 900 K.

Another possible application of MILD combustion is in regenerative gas turbines. This configuration allows for increasing inlet temperature in the combustion chamber for low values of compression ratio. It is well known that the Brayton cycle follows the path drawn in the inset of Figure 3.18. The isobaric heating of the compressed flow occurs between the T_2 and T_x temperatures by means of a counter-flow heater fed by the turbine flue gas with an efficiency given by Equation (3.10):

$$\eta_r = \frac{T_x - T_2}{T_4 - T_2},\qquad(3.10)$$

where η_r is 0 for the non-regenerative case analyzed before and is unity in the upper limit of regeneration when T_x is equal to T_4. This is the maximum temperature the

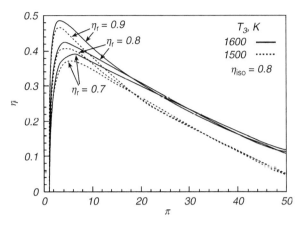

Figure 3.20 Efficiency of Brayton cycle as function of the compression ratio for different values of regeneration efficiency.

compressed flow can reach because it is heated by the flue gas, which is discharged at this temperature. The gas turbine efficiencies for this condition can be calculated by means of Equation (3.9) for regenerator efficiency $\eta_r = 0.7, 0.8$, or 0.9. Their values are plotted in Figure 3.20 for fixed turbine inlet temperatures $T_3 = 1500$ and 1600 K and for a fixed compressor-turbine efficiency of 0.8. The six curves in Figure 3.20 for the regenerative gas turbine are similar. They increase steeply at very low compression ratios and reach a maximum at a pressure range between 3 and 7 bar, then they decrease slightly, crossing a common value at a compression ratio around 20.

In an internal recirculation system, the diluent comes from within the same unit, and the system must be designed to minimize energy losses related to the use of recirculation. In particular, the high temperature of the flue gases means that higher compression work is needed compared to that for cold gases in traditional systems without recirculation. In order to overcome this problem, the two streams need to be at the same pressure and temperature. Therefore, the oxidant, for instance air, and the flue gases must be compressed in different units and heated in separate recuperator and heat recovery systems. Figure 3.21 provides a schematic of a plant layout that applies the concepts for a semi-closed combined cycle gas turbine proposed by Camporeale *et al.* (2004). Air is compressed in the low pressure compressor C1 and heated in a recuperator R2. Then it is mixed with recirculated flue gases which are compressed with the high pressure compressor C2 and heated in the recuperator R1. The mass flow, the compressor, and the heaters are adjusted in such a way that the oxygen concentration is 10% and the temperature and pressure are 1000 K and 20 bar, respectively. The combustion gases at 1700 K are expanded beforehand in the cooled-blade turbine T1, which first feeds the recirculation loop and the heat recovery steam generator HRSG1 with recirculation ratio 0.6. It then feeds the low pressure turbine T2 and the heat recovery steam generator HRSG2. The thermodynamic cycle of this plant has been assessed to be around 0.6 for a power corresponding to 100 kg/s under realistic efficiency assumptions of the components.

Another potential application of MILD combustion is in closed cycle turbine systems. Here, the fluid which operates the cycle is a gas which is usually externally heated by means of a heat exchanger fed by a fluid at high temperature. It is possible to

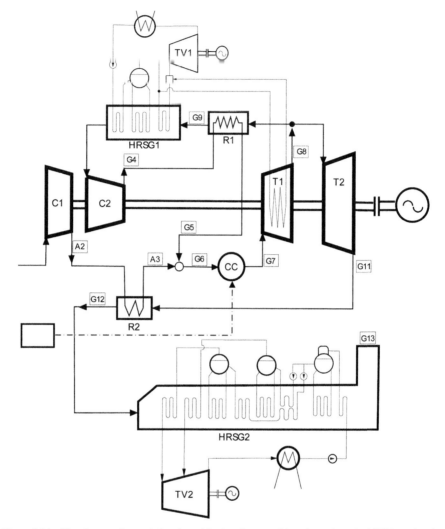

Figure 3.21 Plant layout of a semi-closed combined cycle gas turbine plant where the MILD combustion concept is used (adapted from Camporeale *et al.*, 2004).

substitute the heater with an internal heating system based on MILD combustion, when the pressure and temperature in the heater and the fluid are suitable. The temperature and pressure would have to be sufficiently high for autoignition to occur in a time consistent with the allowable residence time in the engine, so that the combustion products can be separated by the working fluid. This approach permits the reduction of pressure drop along the cycle gas flow and shifts the maximum temperature in the heater from the external to the internal side for an increase of the total efficiency of the system. A possible application is steam reheating in a Hirn cycle, which is illustrated by the temperature-entropy plot of Figure 3.22 (Milani and Saponaro, 2001).

According to the figure, the first and second heating of the fluid up to the same temperature (around 832 K or 560°C) is performed at high (point 6 at 210 bar) and low (point 8 at 20 bar) pressure by means of the same heat exchangers used in Rankine cycles and fed with flue gas generated by traditional combustion systems. In contrast,

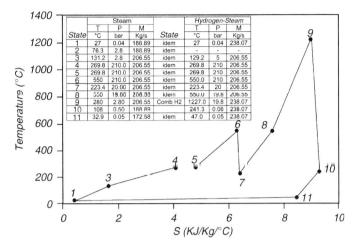

State	Steam T °C	P bar	M Kg/s	State	Hydrogen-Steam T °C	P bar	M Kg/s
1	27	0.04	188.89	idem	27	0.04	238.07
2	76.3	2.8	188.89	idem	-	-	-
3	131.2	2.8	206.55	idem	129.2	5	206.55
4	269.8	210.0	206.55	idem	269.8	210	206.55
5	269.8	210.0	206.55	idem	269.8	210	206.55
6	550	210.0	206.55	idem	550.0	210	206.55
7	223.4	20.00	206.55	idem	223.4	20	206.55
8	550	19.80	206.55	idem	550.0	19.8	206.55
9	280	2.80	206.55	Comb H2	1227.0	19.8	238.07
10	108	0.50	188.89	idem	241.3	0.06	238.07
11	32.9	0.05	172.58	idem	47.0	0.05	238.07

Figure 3.22 Hirn cycle in \raster="fig22"temperature-entropy plot (Milani and Saponaro, 2001).

the subsequent heating of the steam is performed by hydrogen oxidized with pure oxygen in the steam stream. The flow rate of the reactants is adjusted in the figure in order to reach a temperature of 1503 K (1230°C) (point 9). Finally, expansion pushes the pressure down to 0.06 bar and the temperature down to 513 K (240°C), and is followed by a partial cooling in order to fix the condenser conditions to a temperature and pressure of 320 K (47°C) and 0.05 bar, respectively. Excess steam produced by combustion is discharged after the condenser and the remainder of the steam is recirculated through the cycle.

This type of cycle is particularly appropriate for repowering applications, doubling the original power and simultaneously increasing efficiency. In the example presented in Figure 3.22 and analyzed by the authors (Milani and Saponaro, 2001), the efficiency is 0.6, which should be compared to that of a plant without repowering, which is 0.4. Another interesting application of this type of cycle is when hydrogen is produced in a coupled plant of coal hydro-gasification with separation and sequestration of carbon dioxide. Research activity is intense in this field particularly in Italy (Calabr *et al.*, 2005) in order to exploit hydrogen economically and to reduce greenhouse effects. Of course, critical steps must be faced in order to develop feasible plants based on this type of cycle. In some cases, the pressure and temperature can be adjusted to relatively high levels. This is the case for the supercritical steam turbine described in Figure 3.23 by a dashed line, where the pressure and temperature can reach values on the order of 10 bar and 800 K, respectively.

With these conditions, injecting fuel and oxidant into the steam flow should allow the mixture to reach its ignition temperature. The ignition delay time depends on the type of fuel, as is shown in Figure 3.23, where the temporal evolution of temperature is reported for two fuels, hydrogen and methane, and for two concentrations of the fuels at stoichiometric conditions. For a fixed fuel type, the autoignition delay is nearly the same for the two concentrations, whereas the maximum temperature increases with concentration. This shows how the power input can be adjusted according to the needed power output. Since the plots represent only one set of specific conditions, they need to be further validated because the kinetics of ter-molecular reactions with the chaperon effect based on water have not yet been completely assessed, and the numerical predictions are

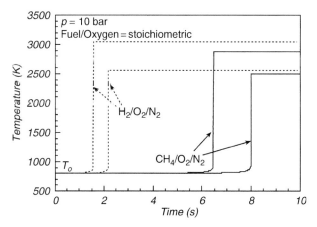

Figure 3.23 Temporal evolution of temperature for hydrogen and methane and for two oxygen concentrations at pressure $p = 10$ bar and initial temperature $T = 800$ K.

based on models developed in relatively low concentrations of water. Preliminary studies in shock tubes and plug flows show that the expected enhancement of recombination in the reaction, as shown in Equation (3.11), plays an effective role and tends to depress the reactivity of the system. Consequently, further studies are needed and are under consideration (Sabia *et al.*, 2006) in this field.

$$H + O_2 + H_2O \rightarrow HO_2 + H_2O, \tag{3.11}$$

For the situation where an internal heating system is based upon the use of hydrogen, the final combustion product is water, so part of the water must be removed and may eventually be used for other purposes. In contrast, the methane system produces water and carbon dioxide, the latter of which must also be removed by degasification after steam condensation. The employment of other fuels is also feasible, but the fuel has to be particularly clean. Otherwise, a cleaning system must be added in order to reuse the steam in the closed cycle. Nonetheless, it is of interest to explore this possibility because the separation of solid as well as solute species in the final condensed polluted water is generally easier than in gaseous flows. As a result, the use of this more complex system becomes more attractive from economic and environmental standpoints.

Applications of water-diluted combustion have not yet appeared for gas turbines even though a limited amount of steam injection has been utilized, and some furnaces at atmospheric pressure have exploited "synthetic air" formed by the addition of oxygen in steam in the same percentage as the air. The combustion process is of some interest because it occurs without the presence of nitrogen, and it is therefore impossible to form any kind of nitrogen oxide or other species in which nitrogen may also be present. Furthermore, water may be beneficial in incineration processes because it may be a source of OH radicals, which are very reactive in the oxidation processes of any gaseous or condensed organic species.

Although there is a lack of facilities, and consequently combustors, that are actually built, there are some interesting studies that have shown the practical feasibility of such cycles. One of these studies describes the analysis of a 400 MW (electric) zero-atmospheric emission power plant, in which the working fluid is water and the carbon dioxide is removed and injected into a well for sequestration (Martinez-Frias *et al.*, 2003). A schematic of the core of this plant is shown in Figure 3.24, in which the numbers

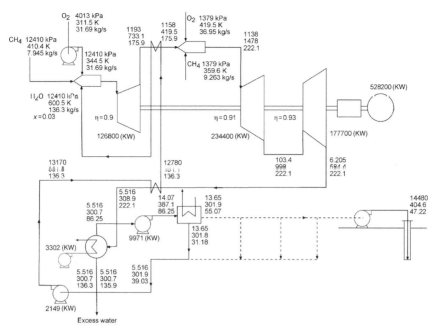

Figure 3.24 Zero-atmospheric emission power plant with water as working fluid (adapted from Martinez-Frias *et al.*, 2003).

reported along the connecting branches are respectively the pressure (bar) and the temperature (K) and the mass flow (Kg/s). The first unit is kept at 124 bar and is fed by O_2/CH_4 at a nearly stoichiometric ratio as well as H_2O at 600 K with mass flow rates so high that the outlet temperature is less than 1100 K (not reported in figure). Maximum temperature is quite mild due to the large mass flow ratio between the diluent and the fuel, and autoignition can take place because the steam is partially heated by the reheater placed just downstream of the first high-pressure turbine. A second combustion stage, fed by oxygen and methane, brings the working fluid from 400 K up to 1500 K at a chamber pressure of 11 bar. The inlet temperature and pressure are slightly lower than for autoignition, but neither slight adjustments in operating conditions nor exchanging in place the reheater with the combustion chamber seem to significantly change the overall plant efficiency. At the end, the working fluid consists of a 0.8 mass fraction of steam and 0.2 mass fraction of carbon dioxide, such as results from mass balance on combustion chamber. It enters intermediate and low-pressure stages of the turbine system and converts its thermal power into mechanical power. The exhaust flows through a preheater and reaches the condenser where a great part of the water and carbon dioxide is separated, but still contains some moisture. The subsequent seven-stage compression and intercooling systems allow a final separation and compression of the carbon dioxide up to 150 bar for enhanced recovery of oil or coal-bed methane, or sequestration.

A similar concept is applied in the so-called Graz cycle (Heitmeir and Jerticha, 2003), developed via cooperation with Japanese public and private research centers. This cycle is illustrated in Figure 3.25 by means of a temperature-entropy plot. Oxy-fuel combustion takes place along the isobaric branch at 40 bar inside a mixture of steam and carbon dioxide, with inlet and outlet temperatures of 600 and 1700 K, respectively. Expansion, cooling, and re-expansion of the working fluid takes place in the 2–3, 3–4, and 4–5 branches. The two

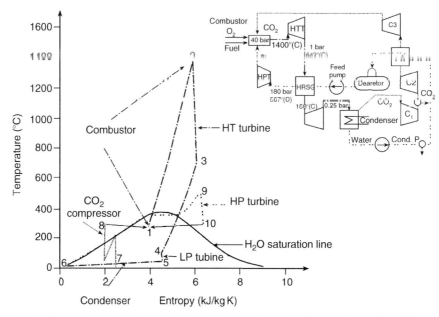

Figure 3.25   Graz cycle on temperature-entropy plot (adapted from Heitmeir and Jerticha, 2003).

species then separate in the condenser where the water undergoes compression of the working fluid in the liquid phase as well as heating and expansion similar to other H_2O cycles, whereas the CO_2 is compressed for partial recycling and partial sequestration.

Inert gases other than water can be used in closed gas turbine cycles. For instance, many working fluids have been studied for heat exploitation applications pertaining to nuclear reactors due to security restrictions or to relatively low temperature ratios (Dostal *et al.*, 2001), and to solar-assisted cycles (Popel *et al.*, 1998). Specifically, a supercritical carbon dioxide Brayton cycle has been proposed (Popel *et al.*, 1998; Mathieu and Nihart, 1999) in order to make use of the large heat capacity and intense heat transfer capabilities of this fluid with respect to other fluids at supercritical conditions, which are reached for carbon dioxide at much lower temperatures than for water. Other examples, that include the MATIANT cycles (Mathieu and Nihart, 1999), are used not only for nuclear applications, but also for removing carbon dioxide from combustion processes in the liquid state for reuse or final storage. They are based on several cycles like Rankine, Ericsson, and Brayton with or without regeneration, but all of them employ CO_2 as the main working fluid (Mathieu *et al.*, 2000). One of these cycles, as can be seen in Figure 3.26, is a regenerative Ericsson-type CO_2 cycle (Sundkvist *et al.*, 2001). The fluid undergoes multi-step inter-cooled compression in the 1–2 branch, and is then heated in a regenerator up to 900 K – the same temperature as combustion chamber inlet conditions. This elevated temperature, coupled with a very high pressure of 110 bar, makes the mixture autoignitable, while oxygen and fuel are injected in very dilute conditions so that the final temperature is only 1600 K. After the expansion in the turbine (4–5), there is a second combustion process in the reheater branch (5–6), which lies in the MILD combustion regime since the temperature is between 1400 and 1600 K at 40 bar. Finally, the expansion in the 6–7 branch is followed by cooling of the working fluid in an internal regenerative exchanger (7–8) and water-cooled system that is also used to extract water produced in the combustion process.

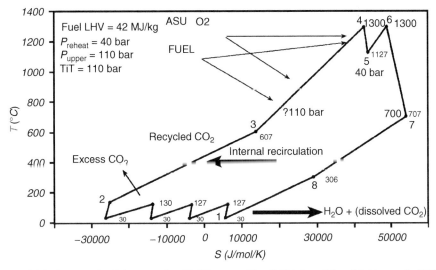

Figure 3.26 Regenerative Ericsson cycle with CO_2 as working fluid (adapted from Mathieu *et al.*, 2000).

The last three water and carbon dioxide cycles are of interest not only for their capability for separating CO_2 in efficient ways, but also because they are ultra low pollution generators. Besides maintaining the low levels of NO_x and soot production common to all other MILD combustion systems (Cavaliere and de Joannon, 2004), it is also possible to remove a small or large part of the produced water. This, in turn, permits many water soluble pollutant species to condense and a great part of the condensed species to be collected as either an emulsion of insoluble condensable pollutants or as a dispersion of solid particulates.

The last example of external control of MILD combustion conditions is the sequential combustion used in some recently designed advanced turbines (www.power.alstom. com), illustrated in Figure 3.27. The first combustion chamber on the left of the figure confines an air-diluted system designed for low NO_x emission. The central part of the system consists of a turbine that extracts part of the enthalpy content so that the stream outlet is formed by partially exhausted gas at temperatures comparable to and for compositions with lower oxygen content than found in standard turbines. This stream feeds the last section of the plant depicted on the right side of Figure 3.27.

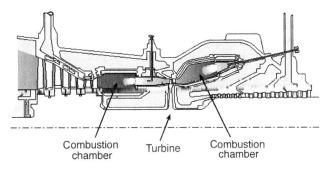

Figure 3.27 Layout of sequential combustion gas turbine (www.power.alstom.com).

3.4.2. MILD COMBUSTION WITH INTERNAL RECIRCULATION

The gas turbine configurations described in the previous subsection refer to well-defined conditions that are required in the design of the system and are controllable during plant operation. These plants are constructed because they can be accurately repaired and tested. In contrast, alternative combustion configurations produce local MILD combustion conditions, such as high initial temperatures and high levels of dilution, by means of fluid dynamic constraints. The fluid motion creates such intense internal recirculation that the mass flow rate of recirculated flow is comparable to or higher than that of the inlet fresh mixture. This type of process coupled with high inlet temperatures has been named flameless (Wünning and Wünning, 1997; Milani and Wünning, 2002) because it often yields colorless (Wünning and Wünning, 1997; Gupta, 2002) and noiseless (Wünning and Wünning, 1997; Gupta, 2002; Hasegawa *et al.*, 2002) combustion processes. A more rigorous definition of this type of process is difficult to generate because the way to create internal recirculation is different, and determining the location of MILD combustion conditions *a priori* is challenging.

In applications involving only gas turbines, an additional constraint of the final outlet temperature exists, which cannot be higher than the cooled metal can resist, that is a temperature around 1673 K (1400°C) under present technological limitations. This can be obtained only by means of external dilution with an inert species as was analyzed in the previous subsection, or with a lean mixture (since employing rich mixtures is not compatible with constraints linked to pollution and power maximization). Thus, the lean condition is the only plausible choice to be coupled with internal recirculation, and requires that only specific processes be considered locally in MILD combustion regimes. These are analyzed in this subsection for two possible cases: lean premixed and rich diluted quench lean (RDQL) processes.

Lean premixed processes can take advantage of the fact that the amount of excess air is high enough for the mixture to be outside the flammability limit. Figures 3.28 and 3.29 illustrate this point. The first figure is a plot of the adiabatic flame temperature versus the equivalence ratio of a methane–air mixture. The maximum corresponds to the stoichiometric mixture, and the values decrease more steeply on the lean side down to

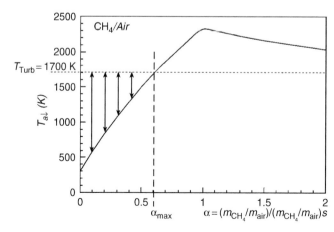

Figure 3.28   Adiabatic temperature versus equivalence ratio for air/methane system.

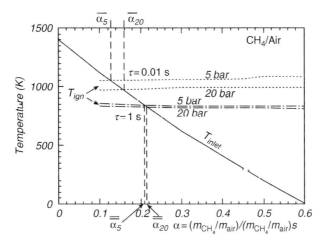

Figure 3.29 Difference between the maximum allowable temperature and the adiabatic flame temperature (reported with arrows in Figure 3.28) as a function of α.

zero than on the rich side. In this specific example pertaining to methane, the maximum allowable equivalence ratio is $\alpha_{max} = 0.6$, since it implies an adiabatic flame temperature of 1700 K, that is the maximum allowable temperature in the turbine. The difference between this temperature and the adiabatic flame temperature, denoted on the figure by arrows, is the increase in temperature that can be achieved by air compression or by heating through a regenerative heat exchanger. This temperature difference is shown in greater detail in Figure 3.29, where only the region in Figure 3.28 limited by equivalence ratios lower than α_{max} is displayed. This temperature is named T_{inlet} in figure since it is the only possible inlet combustion chamber temperature consistent with the aforementioned constraints.

Figure 3.29 also plots autoignition temperatures versus equivalence ratio in the combustion chamber at 5 and 20 bar for residence times of 0.1 and 1 s. The two curves intersect the T_{inlet} curve at $\bar{\alpha}_5$ and $\bar{\alpha}_{20}$ for a residence time of 0.01 s, and at α_5 and α_{20} for a residence time of 1 s. Considering only the first pair of values as examples, $\bar{\alpha}_5$ and $\bar{\alpha}_{20}$ divide the equivalence ratio domain into two parts. Mixtures with α lower than $\bar{\alpha}_5$ and $\bar{\alpha}_{20}$ (at pressures of 5 and 20 bar, respectively) can undergo autoignition, whereas mixtures with α higher than $\bar{\alpha}_5$ and $\bar{\alpha}_{20}$ cannot autoignite within 0.01 s time (dashed line). The comparison of these two ignition limits with the flammability limits α_5 and α_{20} show that, in the case of 5 bar, there is a range of equivalence ratios between $\bar{\alpha}_5$ and α_5 (that is about 0.5) in which neither autoignition nor deflagration can take place, since the inlet temperature is lower than the autoignition temperature and the equivalence ratio is lower than that at the flammability limit. Under these conditions, it is possible in principle to ignite the mixture with inert hot gases recirculated from the part of the combustion chamber where the combustion products flow. This is the same condition which was analyzed in Section 3.3.3 for HCDI systems. These types of processes take place in gas turbines where internal recirculation is significantly enhanced to yield a high level of mixing and subsequent heating of the fresh air–fuel mixture by means of the recirculated hot product.

The second process, which has been named RDQL, consists of two subprocesses: partial oxidation in rich diluted conditions and further oxidation of this partially oxidized mixture. The first process can be considered as a form of fuel reformation, in which the fuel is partially oxidized so that not only CO_2 or H_2O, but also CO, oxygenated species like aldehydes and chetons, and other pyrolytic species like acetylene aromatic and carbonaceous particles can be formed. The novelty of this concept with respect to other types of partial oxidation is the fact that the process develops in hot diluted conditions. In this manner, the maximum oxidation level of the fuel can be obtained since the low reaction rates are accelerated by the high temperatures. Furthermore, the products of the more condensed species are suppressed. In fact, these species are typically formed in recombination reactions which are ter-molecular in nature and are very efficient when reactant concentrations are relatively high. Therefore, the dilution of intermediate organic species significantly suppresses these types of reactions and favors lower order oxidation reactions. Indirect evidence of this tendency is given by the soot yield dependence on temperature, which attains a maximum for values around 2000 K, decreases for higher and lower temperatures, and is nearly negligible for temperatures lower than 1500 K. A more detailed analysis has been given by de Joannon *et al.* (2000), by whom the yield for several oxidation and pyrolysis products has been calculated in a well-stirred reactor fed with methane, oxygen, and nitrogen in rich diluted conditions. The second step of the RDQL process is the oxidation of the reformed fuel, which is mainly composed of CO, H_2, and H_2O for methane–air systems. These conditions are similar to those discussed for the HFDF system and therefore do not require further discussion.

3.5. CONCLUSION

The analyses reported so far have demonstrated that MILD combustion differs from deflagration, detonation, and explosion in their classical forms, as well as processes which generate diffusion flames. Therefore, MILD combustion deserves both to be analyzed independently and to be given a specific name. This name was proposed and used in recent literature (Cavaliere and de Joannon, 2004), and occurs when the process refers to maximum, frozen, and autoignition temperatures, which combine in such a way that the mixture can be autoignited and can yield a relatively small temperature increase comparable to the difference between the ignition and the standard temperatures. Such a broad definition can be made more specific in the case of certain processes in homogeneous mixtures (HBBI, HCCI, and HFFI) and in separated reactants (HDDI). This is a field that must be further studied in order to provide a clearer and narrower definition of MILD combustion, but the main characteristics are clear. It is a process that does not have a double state (frozen and burnt) and develops for a wide range of mixture fractions when it is restricted to the HDDI case. These characteristics are of interest in comparison to the diffusion flame structure where the mixture fraction range in which the reaction takes place is quite narrow and changes slightly with dissipation rate. Here, the quenching dissipation rate characterizes the whole reacting system when the fuel and oxidant are fixed. In contrast, the oxidation structure of MILD combustion shifts along the mixture fraction phase according to different boundary conditions (temperature and dilution). Therefore, it cannot be easily parameterized, at least under the present state of technology.

The processes described in the previous section have been used in some configurations of interest in gas turbine applications because of potential benefits with regard to low levels of pulsations in the combustion chamber as well as very low NO_x emission.

It is also possible to obtain the same high temperature, high diluted conditions with external recirculation of either the flue gas itself or some of its parts with the use of combustion products from a companion unit (sequential combustion), from a reservoir fed by an external combustion unit, or from inert species like water vapor or carbon dioxide. In this field, such technologies seem to be very promising in terms of efficiency increase and pollutant reduction, although the complexity of power plants is increased. From this point of view, Rankine, Brayton, and Hirn cycles can be combined in different ways so that MILD combustion conditions are created in the combustion process.

The final type of MILD combustion application mentioned here is the most attractive in terms of potential, but has not been pursued in a systematic way even though there are exceptional examples. The application is based on the fact that MILD combustion has been recognized as a kind of clean combustion, particularly in terms of soot and nitrogen oxide suppression (Cavaliere and de Joannon, 2004). It is still not clear whether this is true for nanoparticle particulates or for oxygenated compounds, but it is likely that selection of appropriate working conditions could ensure a minimization of these products. In particular, the choice of diluent flow may be decisive in the inhibition of these pollutant categories, since it is possible that there is a reactive species that favors complete oxidation.

This last property further pushes the process toward clean combustion in the sense that the pollutants are not just suppressed in their formation, but are also destroyed if they are present individually (Sabia *et al.*, 2007). This can hold true for species containing carbon, hydrogen, and nitrogen atoms because they can be oxidized or reduced to carbon dioxide, water, or molecular nitrogen. Additionally, inorganic atoms different from nitrogen cannot be transformed in the aforementioned species, but can be released in the form of species that are less noxious or easier to split as a result of the chemical reaction they undergo, or when they interact positively with other diluent species or combustion products. Thus, it is suitable to name the process that is capable of such a transformation as clearing combustion. In this case, "to clear" is used in the same sense as that used in the separation of particulate in liquid streams. The fact that MILD combustion can lead to clearing combustion is based on properties that develop over a narrow temperature range in inert diluted flow.

REFERENCES

Aris R (1969) *Elementary Chemical Reactor Analysis*. Prentice Hall, Englewood Cliffs, NJ.

Beér, J.M. (2000). Combustion technology developments in power generation in response to environmental challenges. *Prog. Energy Combust. Sci.* **26**, 301–327.

Blouch, J.D., Sung, C.J., Fotache, C.G., and Law, C.K. (1998). Turbulent ignition of nonpremixed hydrogen by heated counterflowing atmospheric air. *Proc. Combust. Inst.* **27**, 1221–1228.

Böhm, H., Hesse, D., Jander, H., Lüers, B., Pietscher, J., Wagner, H.G.G., and Weiss, M. (1988). The influence of pressure and temperature on soot formation in premixed flames. *Proc. Combust. Inst.* **22**, 403–411.

Calabr, A., Deiana, P., Fiorini, P., Stendardo, S., and Girardi, G. (2005). Analysis and optimization of the ZECOMIX high efficiency zero emission hydrogen and power plant. *Second International Conference on Clean Coal Technologies for Our Future*, Castiadas (Cagliari), Sardinia.

Camporeale, S.M., Casalini, F., and Saponaro, A. (2004). *Proceedings of the 7th Biennial Conference on Engineering Systems Design and Analysis*, ESDA2004-58472.

Cavaliere, A. and de Joannon, M. (2004). Mild combustion. *Prog. Energy Combust. Sci.* **30**(4), 329–366.

Cavaliere, A. and Ragucci, R. (2001). Gaseous diffusion flames: Simple structure and their interaction. *Prog. Energy Combust. Sci.* **27**, 547–585.

Chen, R. and Milovanovic, N. (2002). A computational study into the effect of exhaust gas recycling on homogeneous charge compression ignition combustion in internal combustion engines fuelled with methane. *Int. J. Therm. Sci.* **41**, 805–813.

Crest 3rd International Symposium on High Temperature Air Combustion and Gasification (2000), Yokohama.

Daribha, N., Candel, S., Giovangigli, V., and Smooke, M.D. (1980). Extinction of strained premixed propane–air flames with complex chemistry. *Combust. Sci. Technol.* **60**, 267.

de Joannon, M., Langella, G., Beretta, F., Cavaliere, A., and Noviello, C. (1999). Mild combustion: Process features and technological constrains. *Proceedings of the Mediterranean Combustion Symposium*, pp. 347–360.

de Joannon, M., Saponaro, A., and Cavaliere, A. (2000). Zero-dimensional analysis of methane diluted oxidation in rich conditions. *Proc. Combust. Inst.* **28**, 1639–1646.

de Joannon, M., Cavaliere, A., Donnarumma, R., and Ragucci, R. (2002). Dependence of autoignition delay on oxygen concentration in mild combustion of heavy molecular paraffin. *Proc. Combust. Inst.* **29**, 1139–1146.

de Joannon, M., Cavaliere, A., Faravelli, T., Ranzi, E., Sabia, P., and Tregrossi, A. (2007). Analysis of process parameters for steady operations in methane Mild combustion technology. *Proc. Combust. Inst.* **31**.

Dostal, V., Hejzlar, P., Driscoll, M.J., and Todreas, N.E. (2001). A supercritical CO_2 Brayton cycle for advanced reactor applications. *Trans. Am. Nucl. Soc.* **85**, 110.

Douce, F., Djebaïli-Chaumeix, N., Paillard, C.E., Clinard, C., and Rouzaud, J.N. (2000). Soot formation from heavy hydrocarbons behind reflected shock waves. *Proc. Combust. Inst.* **28**, 2523–2529.

Flowers, D.L., Aceves, S., Westbrook, C.K., Smith, J.R., and Dibble, R. (2001). Detailed chemical kinetic simulation of natural gas HCCI combustion: Gas composition effects and investigation of control strategies. *J. Eng. Gas Turb. Power* **123**, 433.

Flowers, D.L., Aceves, S.M., Martinez-Frias, J., and Dibble, R.W. (2002). Prediction of carbon monoxide and hydrocarbon emissions in iso-octane HCCI engine combustion using multi-zone simulations. *Proc. Combust. Inst.* **29**, 687–694.

Fogler, H.S. (1986). *Elements of Chemical Reaction Engineering*. Prentice Hall, Englewood Cliffs.

Griffiths, J.F. and Barnard, J.A. (1995). *Flame and Combustion*. Blackie Academic and Professional, Glasgow.

Gupta, A.K. (2002). *High Temperature Air Combustion: From Energy Conservation to Pollution Reduction*. CRC Press, Boca Raton, Florida.

Hasegawa, T., Mochida, S., and Gupta, A.K. (2002). Development of advanced industrial furnace using highly preheated combustion air. *J. Prop. Power* **18**(2), 233.

Heitmeir, F. and Jerticha, H. (2003). Graz cycle – An optimized power plant concept for CO_2 retention. *First International Conference on Industrial Gas Turbine Technologies*, Brussels.

Hermann, F., Orbay, R.C., Herrero, A., and Klingmann, J. (2005). Emission measurements in an atmospheric preheated premixed combustor with CO_2 dilution. *Proceedings of the European Combustion Meeting*, Louvain-la-Neuve.

Hernandez, A.C., Medina, A., and Roco, J.M. (1995). Power and efficiency in a regenerative gas turbine. *J. Phys.* **28**, 2020–2023.

HiTAC (2000). *Symposium of High Temperature Air Combustion and Applications*, Hsinchu.

HiTAC (2002). *Proceedings of the 5th International Symposium on High Temperature Air Combustion and Gasification*, Yokohama.

Howell, J.R., Hall, M.J., and Ellzey, J.L. (1996). Combustion of hydrocarbon fuels within porous inert media. *Prog. Energy Combust. Sci.* **22**, 121–145.

Katsuki, M. and Hasegawa, T. (1998). The science and technology of combustion in highly preheated air. *Proc. Combust. Inst.* **27**, 3135–3146.

Kraft, M., Maigaard, P., Mauss, F., Christensen, M., and Johansson, B. (2000). Investigation of combustion emissions in a homogeneous charge compression injection engine: Measurements and a new computational model. *Proc. Combust. Inst.* **28**, 1195–1201.

Kuo, K.K. (1986). *Principles of Combustion*. John Wiley & Sons, New York.

Lam, F.W. (1988). The formation of polycyclic aromatic hydrocarbons and soot in a jet-stirred reactor. PhD thesis, Massachusetts Institute of Technology, Cambridge.

Libby, P.A. and Williams, F.A. (1982). Structure of laminar flamelets in premixed turbulent flames. *Combust. Flame* **44**, 287–303.

Libby, P.A. and Williams F.A. (1983). Strained premixed laminar flames under non-adiabatic conditions. *Combust. Sci. Technol.* **31**(1–2), 1–42.

Martinez-Frias, J., Aceves, S.M., and Smith, J.R. (2003). Thermodynamic analysis of zero atmospheric emission power plant. *Proceedings of the AES-IMECE 2003, ASME, International Mechanical Engineering Congress Exposition*, Washington, DC.

Masson, E., Taupin, B., Aguilé, F., Carpentier, S., Meunier, P., Quinqueneau, A., Honoré, D., and Boukhalfa, A.M. (2005). An experimental facility at laboratory scale to assess the effect of confinement on flameless combustion regime. *Proceedings of the European Combustion Meeting*, Louvain-la-Neuve.

Mastorakos, E., Taylor, A.M.K.P., and Whitelaw, J.H. (1995). Extinction of turbulent counterflow flames with reactants diluted by hot products. *Combust. Flame* **102**, 101–114.

Masunaga, A. (2001). Application of regenerative burner for forging furnace. *Proceedings of the Forum on High Temperature Air Combustion Technology*, Japan.

Mathieu, P. and Nihart, R. (1999). Zero emission MATIANT cycle. *J. Eng. Gas Turb. Power* **121**, 116–120.

Mathieu, P., Dubuisson, R., Houyou, S., and Nihart, R. (2000). New concept of CO_2 removal technologies in power generation, combined with fossil fuel recovery and long term CO_2 sequestration. *ASME Turbo Expo Conference*, Munich.

Milani, A. and Saponaro, A. (2001). Diluted combustion technologies. *IFRF Combust. J.*, Article 200101, February, ISSN 1562–479X.

Milani, A. and Wünning, J. (2002). What are the stability limits of flameless combustion? *IFRF Online Combustion Handbook*, ISSN 1607–9116, Combustion File No: 173, http://www.exchange.ifrf.net/coop/imp.htm.

Ng, C.K.W. and Thomson, M.J. (2004). A computational study of the effect of fuel reforming, EGR and initial temperature on lean ethanol HCCI combustion. *SAE 2004 World Congress and Exhibition*, Detroit, SAE Paper 2004-01-0556.

Peters, N. (2000). *Turbulent Combustion*. Cambridge University Press, Cambridge.

Peters, N. (2001). Principles and potential of HiCOT combustion. *Proceedings of the Forum on High Temperature Air Combustion Technology*.

Popel, O.S., Frid, S.E., and Shpilrain, E.E. (1998). Solar thermal power plants simulation using the TRNSYS software Pr3-599. *Proceedings of the 9th SolarPACES International Symposium on Solar Thermal Concentrating Technologies*, Font-Romeu.

Puri, I.K. and Seshadri K. (1987). The extinction of counterflow premixed flames burning diluted methane–air, and diluted propane–air mixtures. *Combust. Sci. Technol.* **53**(1), 55–65.

Robertson, T. and Newby, J.N. (2001). A review of the development and commercial application of the LNI technique. *Proceedings of 4th High Temperature Air Combustion and Gasification*, Rome, paper no. 11.

Sabia, P., de Joannon, M., Ragucci, R., and Cavaliere, A. (2005). Mixing efficiency of jet in cross-flow configuration in a tubular reactor for Mild combustion studies. *Proceedings of the European Combustion Meeting*, Louvain-la-Neuve.

Sabia, P., Schießwhol, E., de Joannon, M., and Cavaliere, A. (2006). Numerical analysis of hydrogen Mild combustion. *Turk. J. Eng. Env. Sci.* **30**, 127–134.

Sabia, P., Romeo, F., de Joannon, M., and Cavaliere, A. (2007). VOC destruction by water diluted hydrogen Mild combustion. *Chemosphere*, **68**(2), 330–337.

Seshadri, K. (1983). The asymptotic structure of a counterflow premixed flame for large activation energies. *Int. J. Eng. Sci.* **21**, 103–111.

Smith, J.R., Aceves, S.M., Westbrook, C., and Pitz, W. (1997). Modeling of homogeneous charge compression ignition (HCCI) of methane. *ASME Internal Combustion Engine 1997 Fall Conference*, Madison.

Smooke, M.D., Crump, J., Seshadri, K., and Giovangigli, V. (1990). Comparison between experimental measurements and numerical calculations of the structure of counterflow, diluted, methane–air, premixed flames. *Proc. Combust. Inst.* **23**, 463–470.

Sundkvist, S.G., Griffin, T., and Thorshaug, N.P. (2001). AZEP – Development of an integrated air separation membrane – Gas turbine, *Second Nordic Minisymposium on Carbon Dioxide Capture and Storage*, Göteborg, http://www.entek.chalmers.se/~anly/symp/symp2001.html

Wang, R. and Cadman, P. (1998). Soot and PAH production from spray combustion of different hydrocarbons behind reflected shock waves. *Combust. Flame* **112**, 359–370.

Williams, F. (1985). *Combustion Theory*. Benjamin Cummings Menlo Park, CA.

Wünning, J. and Wünning, J. (1997). Flameless oxidation to reduce thermal NO formation. *Prog. Energy Combust. Sci.* **23**, 81.

www.floxcom.ippt.gov.pl.

www.power.alstom.com.

Xing-Cai, L., Wei, C., and Zhen H. (2005). A fundamental study on the control of the HCCI combustion and emissions by fuel design concept combined with controllable EGR. Part 2: Effect of operating conditions and EGR on HCCI combustion, *Fuel* **84**, 1084–1092.

Yamada, M., Onoda, A., Maeda, F., and Furukawa, T. (1999). Gas turbine combustion for gasified L BTU fuel *Proceedings of the 2nd International High Temperature Air Combustion Symposium*, Taiwan.

Yasuda, T, (1999). Dissemination project of high performance industrial furnace with use of high temperature air combustion technology. *Proceedings of the 2nd International High Temperature Air Combustion Symposium*, Taiwan.

Zheng, X.L., Bloush, J.D., Zhu, D.L., Kreutz, T.G., and Law, C.K. (2002). *Proc. Combust. Inst.* **29**, 1637–1644.

Chapter 4

Lean-Burn Spark-Ignited Internal Combustion Engines

Robert L. Evans

Nomenclature

BMEP	Brake mean effective pressure
BOI	Beginning of pilot injection
BSFC	Brake specific fuel consumption
BSNOx	Brake specific nitrogen oxides
BSTHC	Brake specific total hydrocarbons
COV	Coefficient of variation
C_p	Specific heat at constant pressure
C_v	Specific heat at constant volume
d	Bowl diameter
D	Cylinder bore
DD	Detroit Diesel
EOI	End of injection
HC	Hydrocarbon
HCCI	Homogeneous charge compression ignition
IMEP	Indicated mean effective pressure
MBT	Minimum advance for best torque
PSC	Partially stratified-charge
r	Compression ratio
r_c	Cutoff ratio
r_v	Volumetric compression ratio
TDC	Top-dead-center
tHC	Total hydrocarbon
WOT	Wide-open throttle
γ	Ratio of specific heats, C_p/C_v
$\Delta\theta_c$	Combustion duration
η	Efficiency
λ	Relative air–fuel ratio
ϕ	Equivalence ratio

4.1. INTRODUCTION

Transportation accounts for approximately 25% of total energy consumption in most industrialized countries, and since the fuel comes entirely from fossil fuels this represents an even greater proportion of total greenhouse gas emissions. In addition to greenhouse gas emissions, the emission of unburned hydrocarbons (HCs), particulate matter, and nitrogen oxides from transportation sources are of particular concern because they are released primarily in heavily populated areas, resulting in poor urban air quality. Improving the efficiency of internal combustion engines, therefore, as well as reducing the level of pollutant emissions, is of great interest for both light- and heavy-duty vehicles. Modern light- and medium-duty vehicles with spark-ignited engines are equipped with three-way catalytic converters which are very effective in reducing the output of unburned HCs, carbon monoxide, and nitrogen oxides. Medium- and heavy-duty vehicles usually employ diesel engines, and although these may be fitted with an oxidation catalyst to reduce carbon monoxide and unburned HCs, they do not affect the high level of nitrogen oxide emissions produced by these engines. There is a need, therefore, for improved engine technology which addresses the need both to increase thermal efficiency, and reduce engine-out emissions from these vehicles. As seen in Chapter 2, one way to simultaneously increase efficiency and reduce emissions is to use spark-ignited engines which operate with an air–fuel mixture that is much leaner than the normal stoichiometric mixture used in most such engines. This "lean-burn" strategy is particularly effective for reducing nitrogen oxide emissions since these are very temperature dependent, and lean mixtures burn with a lower flame temperature due to the heat sink provided by the excess air. Thermal efficiency is also increased with lean combustion, since the ratio of specific heats increases monotonically with excess air ratio. The efficiency of lean-burn spark-ignition engines during part-load operation is further increased compared to stoichiometric-charge engines, since a greater throttle opening is required to achieve a given power output. This larger throttle opening reduces the "pumping losses" inherent in any engine operating with a partly closed throttle.

The "leanness" of a lean-burn engine may be measured by the ratio of the actual air-to-fuel mass ratio to the stoichiometric air–fuel ratio in which there is just enough air to completely oxidize all of the fuel. In Europe, this ratio, the "relative air–fuel ratio," is usually denominated by λ (or RAFR), so that a value of $\lambda > 1$ denotes a lean mixture, and a value of $\lambda < 1$ is a rich mixture. In the United States, the fuel–air "equivalence ratio," ϕ, defined as the ratio of the actual fuel–air ratio to the stoichiometric fuel–air ratio is more often used, so that $\phi < 1$ indicates a lean mixture, and $\phi > 1$ a rich mixture. In this chapter, we will use both ϕ and λ to describe lean mixture composition, but of course one can be easily converted to the other since $\phi = 1/\lambda$. The efficiency of lean-burn engines continues to increase, and the emissions continue to decrease as long as the air–fuel ratio, and therefore λ, continues to increase. The burning rate of the mixture also decreases with increasing λ, however, and at some point the mixture becomes too lean to sustain complete combustion. Although it is not a precise number, this value of λ at which it is no longer possible to achieve reliable ignition and sustain complete combustion is referred to as the "lean limit." Before the lean limit is reached, however, there is usually a fairly small window of λ values in which intermittent misfiring occurs, together with an increase in the cycle-to-cycle variation in indicated (i.e., in-cylinder) mean effective pressure (IMEP) and in exhaust emission levels as a result of incomplete combustion during the misfiring cycles.

In this chapter, we will first examine the theoretical performance and efficiency of the ideal air-standard internal combustion engine. This will provide the fundamental background necessary to understand how lean air–fuel ratios can be used to improve engine performance. The following section will then examine the effects of lean mixtures on the combustion and emissions in actual spark-ignition engines, and will discuss the natural "lean limit" of combustion, beyond which point combustion becomes unstable, and emission levels begin to increase. Finally, two methods aimed at extending the lean limit of operation will be described, and the results of experimental verification of these techniques will be presented. The first method utilizes combustion chamber design to enhance turbulence generation, and therefore the burning rate, while the second technique provides a more stable and reliable ignition of very lean mixtures by introducing a partially stratified-charge (PSC) with a richer air–fuel mixture adjacent to the spark-plug electrodes. Further leaning of the mixture can be accomplished with homogeneous charge compression ignition (HCCI) strategies and with hydrogen (as described in Chapter 8), but this chapter concentrates on the lean combustion performance possible using conventional gasoline fuel and spark-ignition.

4.2. PERFORMANCE OF THE IDEAL INTERNAL COMBUSTION ENGINE

The overall thermal efficiency of any reciprocating internal combustion engine is primarily a function of three parameters:

1. Compression ratio
2. Air–fuel ratio
3. Combustion duration

The ideal efficiency of an internal combustion engine is usually determined using an "air-standard" analysis, in which pure air is used as the working fluid. A simple thermodynamic analysis of the ideal air-standard Otto cycle, which is a theoretical model for a spark-ignition engine, shows the efficiency to be

$$\eta = 1 - \frac{1}{r_v^{\gamma-1}} \tag{4.1}$$

where r_v = compression ratio and $\gamma = C_p/C_v$ (ratio of specific heats).

The "compression ratio," r_v, more accurately described as the "volumetric compression ratio," is the ratio between the maximum cylinder volume at bottom-dead-center and the minimum volume at top-dead-center (TDC). All of the heat is assumed to be added to the cycle at constant volume at TDC. This simple analysis shows that for high efficiency, the compression ratio (or, as shown in Chapter 2, the expansion ratio) should be as high as possible.

A similar analysis of the ideal air-standard Diesel cycle results in

$$\eta = 1 - \frac{1}{r_v^{\gamma-1}} \left\{ \frac{r_c^{\gamma} - 1}{\gamma(r_c - 1)} \right\} \tag{4.2}$$

where r_c = cutoff ratio.

In an ideal diesel engine, heat addition takes place continuously at constant pressure over a period between TDC and the "cutoff" point, which corresponds approximately to the time when fuel injection would end in an actual engine. The "cutoff ratio," r_c, for the ideal diesel cycle represents the ratio between the cylinder volume when the heat addition ends and the volume at TDC. The ideal thermal efficiency of the air-standard Otto cycle as a function of compression ratio is shown in Figure 4.1, compared to that for the air-standard Diesel cycle with various values of the cutoff ratio. Although, for a given compression ratio, the efficiency of the air-standard Diesel cycle is less than that of the Otto cycle, in practice, the diesel engine has a higher efficiency because it operates at a much higher compression ratio (typically 20:1 compared to 10:1 for the Otto cycle, which is limited by the peak pressure and engine knock).

The analysis also indicates that for high efficiency, the ratio of specific heats of the working fluid should be as high as possible. In practice, it turns out that γ for air (1.4) is greater than γ for the air–fuel mixture for typical HC fuels. This means that the value of γ will be higher for mixtures with more air (i.e., lean mixtures) than for rich mixtures. The analysis predicts then that thermal efficiency is higher for lean mixtures (mixtures with excess air) than for rich mixtures. Figure 4.2, taken from Heywood (1988), shows the theoretical efficiency for an ideal Otto cycle engine using fuel and air as the working fluid, rather than just air, as a function of the fuel–air equivalence ratio, ϕ, for a range of compression ratios, r, from 6:1 to 24:1. An equivalence ratio of 1.0 provides a stoichiometric mixture, while values greater than 1.0 are rich mixtures and values less than 1.0 are lean mixtures.

Figure 4.2 clearly shows the trend of higher thermal efficiency as the mixture becomes leaner. This much steeper drop in efficiency for ϕ greater than 1.0 is a result of the presence of unburned fuel in the mixture. In other words, for rich fuel–air mixtures, there is not enough oxygen present to support complete combustion of all of the fuel. This figure indicates another reason for diesel engine efficiency being

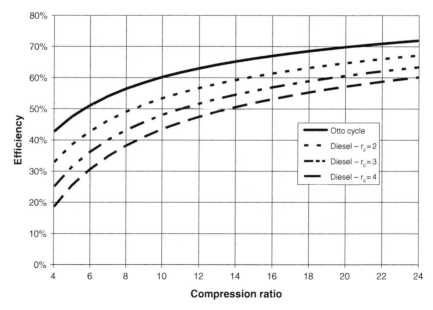

Figure 4.1 Variation of efficiency with compression ratio for a constant volume air-standard cycle.

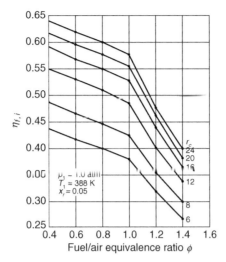

Figure 4.2 Variation of thermal efficiency with equivalence ratio for a constant volume fuel–air cycle with 1-octene fuel (Heywood, 1988).

greater than Otto cycle efficiency, since the un-throttled diesel engine always operates at a very lean overall fuel–air ratio, particularly at part load. It also shows one reason why some spark-ignition engine manufacturers have moved towards production of "lean-burn" engines in recent years.

The length of the burning time, or combustion duration, also has an effect on thermal efficiency. The ideal situation would be to release all the energy into the cylinder instantaneously at TDC of the compression stroke. Since all fuels have a finite burning rate, this is not possible, and the power output obtained for a given amount of fuel burned is reduced compared to the ideal cycle, resulting in a reduction in thermal efficiency. The results of these effects are shown schematically in Figure 4.3, taken from Campbell (1979), which shows the power output as a function of spark advance, for three different values of combustion duration, $\Delta\theta_C$. The combustion duration and spark advance are given in terms of crank angle degrees, and degrees before TDC, respectively. As the combustion duration is increased, the optimum value of the spark timing also increases, as shown in the diagram. The figure clearly shows that reduced

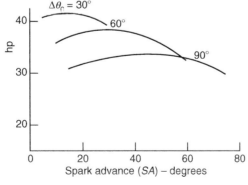

Figure 4.3 Influences of spark advance and combustion duration on power output (Campbell, 1979).

combustion duration results in higher power output, and therefore increased thermal efficiency.

Since burning rates are generally highest close to the stoichiometric air–fuel ratio, operating an spark ignition (SI) engine lean, with an equivalence ratio of less than one, results in increased combustion duration. As can be seen from Figure 4.3, this then reduces power output and thermal efficiency, thereby tending to counteract the increased efficiency of lean operation due to an increased ratio of specific heats, as seen in Figure 4.2. It is important, therefore, when choosing to operate an SI engine under lean-burn conditions to design the combustion system to provide a high burning rate.

4.3. ENGINE COMBUSTION AND EMISSIONS

The previous section described factors which affect the performance of idealized internal combustion engines. In this section, the combustion and emissions characteristics of the spark-ignition engine are discussed in more detail. The homogeneous charge spark-ignition engine, or Otto cycle engine, is in widespread use today, while the stratified-charge engine is used in more specialized applications, such as in the gasoline direct-injection (GDI) engine.

4.3.1. HOMOGENEOUS CHARGE SPARK-IGNITION ENGINES

A characteristic feature of the spark-ignition engine is that combustion occurs as a premixed flame, that is, a flame front moves through a mixture of fuel and air which has been premixed to be at, or very near, stoichiometric conditions. Ignition of the charge is by an electric spark, timed to occur so that maximum mean effective pressure (MEP) – pressure work done in one cycle divided by the cylinder displacement volume – occurs in the cylinder. The premixed air–fuel charge for a conventional spark-ignited engine is homogeneous in composition, providing a uniform equivalence ratio everywhere in the cylinder. This air–fuel ratio must be kept within the combustible limits of the mixture, somewhere between the rich limit and the lean limit of the particular fuel–air mixture being used. From a practical engine operation point of view, the rich limit is usually of little importance, since rich operation (more fuel than there is air available for complete combustion) results in unburned fuel, and is avoided. The lean limit is of practical importance, however, since as we have seen, lean operation can result in higher efficiency and can also result in reduced emissions. The lean limit is where misfire becomes noticeable, and is usually described in terms of the limiting equivalence ratio, Φ, which will support complete combustion of the mixture. In most engines, a value of $\Phi = 0.7$ is usually the leanest practical mixture strength. This value is consistent with the values reported in Table 2.2.

The requirement for a near constant air–fuel ratio at all operating conditions results in one of the main weaknesses of the spark-ignition engine. For part-load operation, as the supply of fuel is reduced, the supply of air must also be reduced to maintain the correct air–fuel ratio. In order to achieve this, the air supply must be throttled using the throttle valve. This throttling of the mixture results in additional pumping losses (work required to pump the mixture past a partially closed throttle). This throttling operation results in poor part-load efficiency of the spark-ignition engine compared to the un-throttled diesel engine, for example. The homogeneous fuel–air mixture always present in the cylinder results in another characteristic of the spark-ignition engine – knock. Knock occurs when unburned mixture self-ignites due to the increasing cylinder pressure as a result of combustion of the bulk of the mixture. Persistent knock causes very rough engine operation, and can cause engine failure if it is not controlled. This problem is exacerbated by high compression ratios and fuels which readily self-ignite at the temperature achieved following compression (low octane fuels). Knock is the principal reason why spark-ignition engines are usually limited to a compression ratio of less than approximately 10:1 with currently available fuels. This relatively low compression ratio results in lower thermal efficiency compared to diesel engines operating at approximately twice the compression ratio, as we have seen in the previous section.

4.3.2. STRATIFIED-CHARGE ENGINES

The stratified-charge engine is something of a hybrid between the homogeneous charge spark-ignition engine and the diesel engine. The concept is aimed at incorporating some of the design features of each engine in order to achieve some advantages of both. The result has been an engine more nearly like the spark-ignition engine, but one in which much leaner operation can be achieved and which is able to burn a wide variety of fuels. One example of a stratified-charge engine combustion chamber is shown schematically in plan view in Figure 4.4, taken from Benson and Whitehouse (1979). Air is introduced into the cylinder without fuel as in the diesel engine, although a throttle is still used to regulate the quantity of air, as in the spark-ignition engine. The inlet port is shaped so that the air has a strong swirling motion as it enters the combustion chamber, as shown in the figure. Fuel is introduced directly into the chamber through the fuel injection nozzle, mixed with the swirling air and then swept past the spark plug. The injection of fuel and firing of the spark plug are timed so that the local air–fuel ratio near the spark plug is nearly stoichiometric at the instant the spark is discharged. The mixture then burns near the spark plug and downstream of the spark plug as a premixed flame, while in the rest of the cylinder the mixture is very lean. The fact that fuel is introduced at one point in the chamber and burnt immediately means that the overall air–fuel ratio can be very lean, while the local ratio is near stoichiometric, and is a direct result of the stratified nature of the charge. The stratified-charge engine is thus able to operate over a much wider range of air–fuel ratios than a

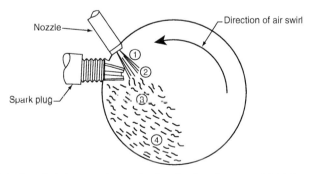

Figure 4.4 Stratified-charge engine combustion chamber (Benson and Whitehouse, 1979).

conventional spark-ignition engine. This results in higher efficiency due to leaner overall air–fuel ratios and reduced pumping losses, since less throttling is required.

Mention should be made of some pre-chamber engines, such as the Honda CVCC which might be called PSC engines. These engines usually employ a second intake valve and fuel system to introduce a rich mixture into a small pre-chamber. This mixture is ignited with a spark plug, and the hot combustion products pass as a jet into the main chamber which has been filled with a very weak mixture through the main inlet valve and fuel system. The expanding flame front from the pre-chamber is able to ignite a much leaner mixture in the main chamber than can a conventional spark plug. These engines are then able to operate at leaner overall air–fuel ratios and higher efficiency than conventional spark-ignition engines. More recently, the PSC approach has been developed without the use of a separate pre-chamber, and these will be discussed in more detail in a later section.

4.3.3. Spark-Ignition Engine Emissions

The emission levels of a spark-ignition engine are particularly sensitive to air–fuel ratio. This can be seen in Figure 4.5, taken from Heywood (1988), which shows schematically the level of emissions from a spark-ignition or Otto cycle engine as a function of relative air–fuel ratio. At rich air–fuel ratios, with ϕ greater than 1.0, unburned HC levels are high since there is not enough air to completely burn all the fuel. Similarly, CO levels are high, because there is not enough oxygen present to oxidize the CO to CO_2. For lean mixtures, with ϕ less than 1.0, there is always excess air available, so that CO almost completely disappears, while HC emissions reach a minimum near $\phi = 0.9$. For ϕ less than about 0.9, some increased misfiring occurs because of proximity to the lean misfire limit, and HC emissions begin to rise again. The main factor in production of NO is combustion temperature: the higher the temperature,

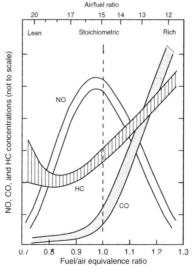

Figure 4.5   Emissions as a function of fuel–air equivalence ratio ϕ (Heywood, 1988).

the greater the tendency to oxidize nitrogen compounds into NO. Since the combustion temperature is at a maximum near stoichiometric conditions where $\phi = 1.0$, and falls off for both rich and lean mixtures, the NO curve takes the bell shape shown in Figure 4.5.

In addition to the higher efficiency discussed previously, an examination of Figure 4.5 clearly shows another benefit of lean operation. At values of ϕ less than about 0.9, CO production is negligible, HC emissions are near the minimum level, and NO emissions are greatly reduced. As ϕ decreases even further, there is a trade-off between further reduction in NO emissions and an increase in HC emissions. Lean operation is, in fact, an excellent control strategy for reducing emissions, and is the reason why some engines operating with very lean overall mixtures could meet early emission standards without the need for exhaust gas clean-up devices.

4.4. EXTENDING THE LEAN LIMIT OF OPERATION

In order to further increase thermal efficiency and reduce emissions from lean-burn engines, it would be useful to develop techniques aimed at extending the practical lean limit of combustion. Since lean mixtures are normally difficult to ignite, and have a much slower burning rate than do stoichiometric mixtures, any technique which may provide a stronger initial flame kernel, or increase the air–fuel mixture burning rate, should be beneficial. As discussed in Chapter 2, turbulence is one of the most important parameters in determining burning rate in homogeneous charge mixtures, and so anything which might enhance the turbulence levels in the mixture during the combustion event should be helpful. A stronger initial flame kernel produced following the spark-ignition event should also be effective in igniting a very lean mixture which may not otherwise ignite, or which may result in incomplete combustion. In the following sections, we will describe two techniques which have been developed over a number of years at the University of British Columbia in order to extend the lean limit of operation of a spark-ignited, homogeneous-charge engine. These two methods are meant to represent the kinds of approaches that can be used to create practical lean-burn spark-ignited engines. The first technique, aimed at increasing turbulence generation in the combustion mixture just before ignition and during the combustion process, involved a new combustion chamber designed specifically for lean-burn engines. The second technique used a "PSC" concept to produce a small pocket of relatively rich mixture near the spark plug so that it would ignite more readily than the main, very lean, combustion charge. Both of these techniques have been used to extend the lean limit of operation of natural gas-fueled spark-ignition engines, resulting in higher thermal efficiency and lower levels of exhaust emissions.

4.4.1. EXTENDING THE LEAN LIMIT THROUGH INCREASED TURBULENCE GENERATION

Combustion chamber design is one of the key factors affecting the performance of a lean-burn spark-ignition engine. In particular, it is well known that the level of turbulence in the chamber just prior to ignition and during the combustion process has an important impact on the burning rate of the air–fuel mixture. The level of turbulence in the chamber can be influenced by the chamber design through the degree of swirl imparted to the mixture during the intake process, by tumble motion generated in a four-valve engine design, and by the squish motion generated as the piston nears TDC. In this chapter, the results of a study of squish-generated charge motion in the combustion chamber are reported for several different chamber designs. High levels of turbulence generation lead to faster burning rates which can result in improved thermal efficiency and reduced levels of exhaust emissions. It is important, however, to realize that the scale of turbulence generated in the chamber, not just the magnitude of the velocity fluctuations, is a key factor in optimizing the combustion rate.

As mentioned earlier, when the air–fuel ratio is increased towards the lean limit of combustion, misfire cycles occur and incomplete combustion may result in an increase in unburned HC emissions. A further disadvantage of lean operation is that the burning rate is reduced compared to combustion under stoichiometric conditions. The situation is made worse if a slow-burning fuel, such as natural gas, is used in an attempt to further reduce exhaust emissions. The reduction in burning rate results in an increase in the overall combustion duration, which in turn leads to increased heat transfer losses to the cylinder walls and a decrease in the overall engine thermal efficiency. The increased combustion duration under lean conditions requires the spark timing to be advanced compared to stoichiometric operation, which may in turn lead to knock and an increase in unburned HC formation. To successfully implement a lean-burn strategy to minimize exhaust emissions and maximize thermal efficiency, therefore, the combustion chamber should be designed to enable the fastest possible combustion rate to be achieved under all operating conditions. In practice, this means utilizing the chamber geometry to generate high levels of turbulence just prior to ignition and during the early combustion phase.

Turbulence in spark-ignition engines is generated by the shear flow which occurs during the intake process, and by the squish motion as the piston approaches the cylinder head just before TDC. A review of the literature shows the progression of engine investigations of turbulence from the early 1920s to present day. Semenov's (1963) studies of turbulent gas flow in piston engines are regarded as fundamental. Hot wire anemometry (HWA) studies in a motored engine with a flat top piston showed that shear flow through the intake valve was the primary source of turbulence generation. In most cases, however, much of the turbulence generated during the intake stroke decays before the combustion process is started. Squish-generated turbulence, on the other hand, is generated during the last phase of the compression stroke, just before it is most needed to influence the combustion process. Andrews *et al.* (1976) have demonstrated in combustion bomb studies that burning velocity is nearly a linear function of turbulence intensity. Nakamura *et al.* (1979) discovered that the introduction of a small diameter jet of air into the combustion chamber through a third valve, just before ignition, resulted in a significant increase in burning rate and a substantial improvement in engine performance. The air jet evidently resulted in an enhancement of the turbulence field near the spark plug, leading to a significant increase in the combustion rate.

Heywood (1988) has shown that turbulence generation by combustion chamber design, with an emphasis on squish, should provide a fast-burn rate for high efficiency and good emission control. He also noted that the chambers should be geometrically simple, and practical to manufacture. Gruden's (1981) investigation of combustion chamber layout in modern passenger cars showed that a combustion chamber located in the piston crown was the simplest way to meet these requirements. In line with

Young's (1980) optimization criterion of an open and compact chamber, Gruden (1981) concluded that the dimensions and the positions of the quench areas and the quench distance, that is, maximum flame travel, were important. Overington and Thring (1981), using a Ricardo Hydra engine with variable compression ratio and combustion chambers in the head and piston, showed a 2–5% improvement in fuel economy for the chamber in the crown over that of a chamber contained in the head. This improvement resulted from the higher turbulence generation induced by squish motion.

Evans (1986) proposed a variation of the standard bowl-in-piston chamber, described as a "squish-jet" chamber, which enhances the squish effect by using channels in the top of the piston to trap fluid and generate a series of jets. Evans and Cameron (1986) evaluated this concept experimentally with HWA velocity measurements and combustion pressure measurements in a Cooperative Fuel Research (CFR) engine. Initial results showed increased peak pressure and reduced combustion duration compared to a standard bowl-in-piston geometry. Subsequent investigations carried out by Dymala-Dolesky (1986), Evans and Tippett (1990), and Mawle (1989) further illustrated the influence of combustion chamber design on turbulence enhancement in a fast, lean-burning engine. Potential was found for the squish-jet action to improve engine efficiency and increase the knock limit. Reduced coefficient of variance (COV) of indicated mean effective pressure (IMEP) and reduced ignition advance requirements were also exhibited.

The goal of the work reported in the following subsections is to optimize a "fast, lean-burn" combustion system for spark-ignited natural gas engines which would enable emission regulations to be met without the use of exhaust gas cleanup equipment. A family of "squish-jet" pistons was developed and tested over a wide range of air–fuel ratios under full-load operating conditions. The level of turbulence was controlled by the geometrical features of the combustion chamber: that is by varying the piston clearance height, piston crown recess depth, and number and width of "jet" slots. The fuel used was natural gas, a slow-burning fuel compared to gasoline, which makes the use of a fast-burn type of chamber even more important. Several versions of the squish-jet chamber were then compared to conventional bowl-in-piston chambers of the type normally found in diesel engines which have been converted to spark-ignited operation on natural gas.

All experimental work was conducted in a Ricardo Hydra single-cylinder research engine using natural gas fuel. In addition to the standard engine instrumentation, a complete exhaust emissions measurement facility, as described by Blaszczyk (1990), was used for all of the tests. The natural gas flow rate was obtained with an MKS mass flowmeter, model number 1559A-100L-SV, which provided a readout in standard litres per minute (flow corrected to standard conditions). A lambda sensor mounted in the exhaust pipe was used to check the air–fuel ratio of the mixture being burned in the engine. The sensor was connected to an AF Recorder, model number 2400E, and to the data acquisition system. The lambda sensor and CO_2 emissions provided a check on the air–fuel ratio determined from the air and fuel flow meters. All of the data reported were obtained at steady state operating conditions over a wide range of air–fuel ratios, from stoichiometric to the lean limit, at 2000 rpm. The main engine operational parameters, such as performance and emissions, were recorded at all operating points, while for selected cases cylinder pressure data were also recorded and analyzed. All test procedures conformed to SAE standard J1829.

The Ricardo Hydra engine configuration has a flat cylinder head and uses piston inserts with combustion chambers machined into them. This configuration was chosen to be representative of most medium-duty direct-injection diesel engines, which are commonly used for conversion of trucks and buses to natural gas operation. The piston was cast from aluminum alloy A356 heat-treated to T6, which is the same material as that used in the cylinder head, and has two compression rings and one oil control ring. Since many different piston-top geometries were required for testing, a single-piston "body" was used, with separate removable piston crown inserts which were threaded for attachment to the main piston body. This permitted rapid changes in piston crown geometry to be made by simply unscrewing one insert and replacing it with another, leaving the same piston body in place in the cylinder. This procedure also ensured exactly the same piston skirt frictional characteristics for each different piston geometry tested.

The combustion chamber design investigated was based on the principle of using squish motion to generate a series of jets directed towards the center of the chamber just prior to ignition. An improved family of combustion chambers utilizing this concept, and referred to as "squish-jet" combustion chamber designs, was described by Evans (1991). The squish-jet chamber reported here (UBC#1C) is similar, but not identical to that described by Evans and Blaszczyk (1993), and is shown in the piston insert drawing of Figure 4.6. The purpose of the cut-out areas, or "pockets," machined into the piston crown to a depth of 3.0 mm is to trap mixture in these areas which is then ejected through the narrow exit passages as the piston nears TDC. In this way, a series of jets is generated, and these jets, colliding near the center of the chamber, break down into small-scale turbulence. By carefully choosing the dimensions of both the pocket in the piston crown, and the outlet passages, the scale of turbulence, as well as the intensity, can be controlled. Figure 4.7 shows a conventional bowl-in-piston chamber, typical of that used in a Detroit Diesel (DD) two-stroke diesel engine, labeled as "DD" in the figures, which was used for comparison with the squish-jet case. The squish area for

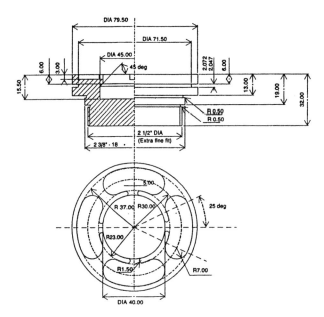

Figure 4.6 UBC#1C "squish-jet" combustion chamber.

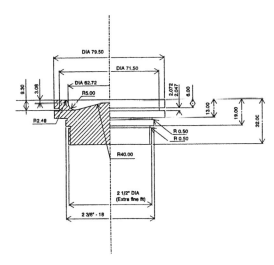

Figure 4.7 DD combustion chamber.

each piston was determined as a percentage of cylinder cross-sectional area. For the simple bowl-in-piston, this is determined from

$$\%S = 100\%(D^2 - d^2)/D^2 \tag{4.3}$$

where D is the cylinder bore and d is the bowl diameter. The squish area of the enhanced squish UBC#1C combustion chamber, including the cut-out areas, was 75%, and the clearance height (the distance between the top of the piston crown and the cylinder head at closest approach) was 3.1 mm. The DD piston configuration had a squish area of 30% and a clearance height of 6.8 mm, resulting in a fairly quiescent chamber design. The large clearance height for the DD piston was required to achieve the fixed compression ratio of 10.2:1, which was used for both combustion chamber designs.

4.4.1.1. Combustion Chamber Comparison

Figures 4.8–4.11 compare the performance of the Ricardo Hydra engine with the UBC#1C combustion chamber, and the DD chamber. All results are shown for the standard test conditions of wide-open throttle (WOT) at 2000 rpm as a function of the relative air–fuel ratio from stoichiometric ($\lambda = 1.0$) to near the lean limit of combustion. The relationship between Minimum advance for Best Torque (MBT) ignition timing and λ for both combustion chambers is shown in Figure 4.8. The MBT ignition timing increases with increasing λ, due to the decreasing burning velocity of lean mixtures. Ignition timing for the UBC#1C chamber was 5–10° less advanced than for the DD chamber over the full range of air–fuel ratios, with the greatest difference occurring at the leanest operating point. This reduction in MBT spark advance provides clear evidence of the faster burning rate achieved by the enhanced squish combustion chamber design.

The relation between brake specific fuel consumption (BSFC) and λ, shown in Figure 4.9, indicates considerably better fuel utilization for the UBC#1C than for the DD chamber. On a mass basis, there is an indicated 4.8% reduction in natural gas consumption per kilowatt hour with the UBC chamber compared to the DD chamber at

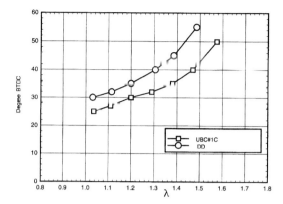

Figure 4.8 MBT ignition advance as a function of λ.

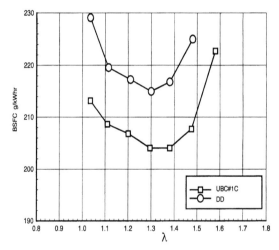

Figure 4.9 Brake specific fuel consumption as a function of λ.

the minimum specific fuel consumption point of $\lambda = 1.3$. These results demonstrate a significant improvement in overall thermal efficiency over the complete air–fuel range. The variation of brake specific total hydrocarbons (BSTHC) with λ is shown for both chambers in Figure 4.10. The values shown are total HC emissions, a large part of which are known to be methane emissions when natural gas is the fuel. As methane is not a very reactive gas, it does not contribute significantly to smog formation, and so is not considered in the emission regulations by most authorities. It is a greenhouse gas, however, and its emission will likely come under increased scrutiny if it becomes a significant transportation fuel. HC emission levels are quite sensitive to quench area, and are therefore normally expected to increase with an increase in squish area. For the two chambers considered here, however, the relatively large clearance height, particularly for the DD chamber, can be expected to greatly reduce any quench effects. The results shown in Figure 4.10 do show a slight increase in total HC emissions for the UBC#1C chamber over most of the λ range, although the general level is quite low at most operating points. At the highest values of λ, the emissions for the DD chamber

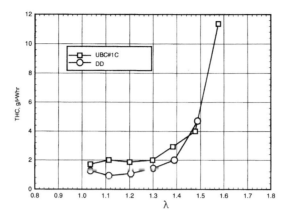

Figure 4.10   Brake specific HC emissions as a function of λ.

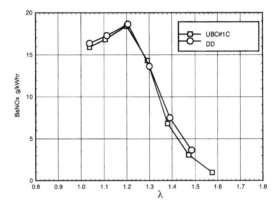

Figure 4.11   Brake specific NO_x emissions as a function of λ.

increase rapidly as the lean misfire limit is approached before that of the UBC#1C chamber.

The relationship between brake specific nitrogen oxides ($BSNO_x$) and λ for both combustion chamber designs is shown in Figure 4.11. The NO_x emission at all values of λ are very similar for both cases, showing the highest levels near the maximum efficiency point, as expected. The reduction in the required MDT spark advance evidently results in a moderation of the peak combustion temperature, thereby reducing NO_x emissions compared to what might otherwise be expected with a faster burning rate.

4.4.1.2.   Comparison of Squish and Swirl Effects

The results of experiments conducted at Ortech International in Toronto, Canada, as described by Goetz *et al.* (1993), using a single-cylinder version of the Cummins L-10 engine are shown in Figures 4.12 and 4.13. The experiments were designed to compare the effectiveness of the UBC squish-jet design in comparison to a high swirl design during lean operation. In total, there were five different experimental configurations tested. The base case consisted of the standard L-10 engine configuration, with a quiescent combustion chamber design. This was used for comparison with two variations

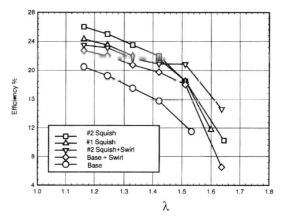

Figure 4.12   Efficiency as a function of λ.

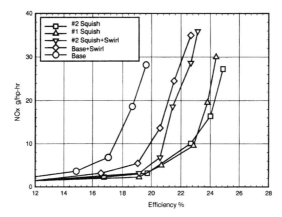

Figure 4.13 NO_x versus efficiency trade-off.

of the UBC squish-jet design, shown as "#1 Squish" and "#2 Squish." The "#1 Squish" combustion chamber was similar to that shown in Figure 4.6, while the "#2 Squish" case utilized a tangential squish-jet outlet to provide a swirl component to the squish motion. Both of these chamber designs are described in more detail by Goetz *et al.* (1993). In addition, a swirl deflection plate was used in the intake port to increase the swirl ratio to 2.5 for both the "Base + Swirl" configuration and the "#2 Squish + Swirl" configuration.

Thermal efficiency as a function of λ is shown in Figure 4.12 for the five cases tested. It can be seen that the base case has the lowest efficiency over the whole range of λ, while all of the cases in which the in-cylinder flow field has been enhanced exhibit significantly higher levels of efficiency. Up to λ values of approximately 1.4, the two squish-jet cases resulted in higher efficiency than the two swirl-enhanced cases, with the UBC #2 Squish case showing the best thermal efficiency. It is also interesting to note that the addition of swirl to the best squish-jet case actually results in a reduction of thermal efficiency up to $\lambda = 1.4$, while at values of λ greater than 1.4 the addition of swirl to the #2 Squish case yields the highest efficiency. It would appear that at the lower air–fuel ratios the addition of a strong swirl motion causes a loss in efficiency due to an increased heat transfer rate at the cylinder walls, while at the higher values of λ the effect of the increased mixture motion in increasing the burning rate more than makes up for the losses due to increased heat transfer.

The important trade-off between $BSNO_x$ emissions is illustrated for the five combustion chamber configurations in Figure 4.13. Once again the base case L-10 engine configuration provides the poorest performance, while the two squish-jet cases exhibit the best performance over the complete range of air–fuel ratios. It seems that the mixture motion near the center of the chamber provided by the squish-jet cases is much more effective than that generated near the cylinder walls by the swirl cases and which tend to increase heat transfer losses.

4.4.2. EXTENDING THE LEAN LIMIT THROUGH PARTIAL STRATIFICATION

In the quest for increased efficiency and reduced exhaust emissions, Internal Combustion (IC) engine combustion technology appears to be converging. Engine development is moving ahead in some cases by deliberately selecting the best features found in HCCI engines, direct-injection engines, and lean-burn homogeneous charge engines. In fact, automotive engines of the future may spend part of their time operating in each of these modes, depending on the particular speed and load conditions, and the emission requirements. This section describes the development of a PSC engine combustion concept, as described by Evans (2000), aimed at extending the lean limit of spark-ignition engines in order to reduce NO_x emissions and increase part-load thermal efficiency.

Lean-burn natural gas engines offer higher efficiency and lower NO_x than stoichiometric natural gas engines, and much lower particulate emissions than diesel engines. To a certain extent, a "lean strategy" allows control of the engine load without throttling by changing the air/fuel ratio. The prior section described the concept of increasing turbulence locally to extend lean operation, and this section describes the approach of locally enriching the mixture to establish a robust flame initiation. In particular, a new PSC concept is introduced to accomplish this task. In this approach, a very small quantity of fuel (typically less than 5% of the total fuel introduced into the engine) is injected adjacent to the spark-plug electrodes just before ignition. This results in the formation of a relatively rich mixture at the spark plug, providing rapid and reliable ignition, and the formation of a strong initial flame kernel. PSC requires no modification to base engine design, but only to the method of introducing fuel to the combustion chamber. Natural gas has been used for injection through the spark plug and for the main homogeneous fueling, both because it is a relatively "clean burning" fuel and because PSC is particularly suited to gaseous fueling. Natural gas is a promising fuel, particularly for medium- and heavy-duty engines.

As in the previous section, the experimental work reported here was conducted using a Ricardo Hydra single-cylinder research engine. The spark-ignition engine is fueled by natural gas, and has a flat fire-deck cylinder head and a bowl-in-piston type combustion chamber. It is fully instrumented and high-speed measurement of in-cylinder pressure provides data for accurate combustion analysis. A comprehensive emissions bench enables measurement of the following exhaust gas levels: NO_x, CO, total HCs, CH_4, O_2, and CO_2. For all test conditions, the spark timing was set to minimum spark advance for best torque. A test speed of 2500 rpm was used for initial measurements of engine performance and emissions, while 2000 rpm was used for further experiments

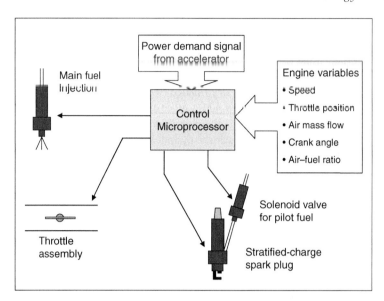

Figure 4.14   Schematic of the PSC control system.

aimed at optimization of the PSC system. For the purposes of this research, the lean limit has been defined as the point at which the coefficient of variation of gross COV IMEP becomes greater than 5%. The relative air–fuel ratio was calculated using the overall fuel and air mass flow rates.

A schematic of the PSC ignition concept is shown in Figure 4.14. During operation in a motor vehicle, control of the main fuel injection, the pilot fuel injection through (or near) the spark plug, and the spark timing itself would all be controlled by the fuel management microprocessor. Following a change in engine speed or load, signaled by a change in the "accelerator" pedal position, the fuel management system would adjust the quantity of main fuel to be injected, the quantity of pilot fuel and the timing of the injection of pilot fuel relative to the spark timing, and the spark timing itself. Finally, the electronically operated throttle position would be adjusted to obtain the air–fuel ratio of the homogeneous lean charge required for that particular operating condition.

Figure 4.15 shows a drawing of the modified spark plug used to implement the PSC system, as described by Evans *et al.* (1996). A passage was drilled in the metal cladding of the spark plug to accommodate a small section of capillary tubing used to introduce the pilot fuel. The capillary tube connects to a fine slot in the threads, and ultimately the PSC fuel is delivered through a very small hole into the space around the ground electrode of the plug. It was recognized that any production system might look quite different, but the approach shown worked well for the initial experiments described here.

The PSC system uses a high-pressure natural gas line (200 bar), which is regulated to lower pressures (around 20–50 bar) to supply the spark-plug injector mounted in the research engine. The PSC injection flow rate is measured using a low range thermal mass flow meter. A fast response solenoid valve is used to control the injection timing and quantity. After the solenoid, a capillary tube with 0.53 mm internal diameter channels the PSC fuel to the plug injector, and a check valve installed close to the plug prevents back-flow of combustion gases. Software control of both the spark timing and

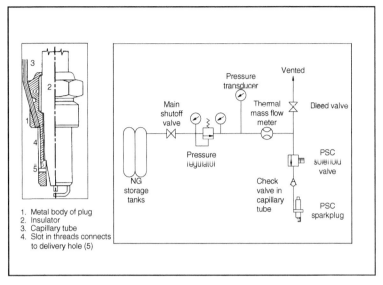

Figure 4.15 The PSC spark-plug-injector and system schematic.

the PSC injector settings is implemented through a Labview interface. The spark-plug/injector design is critical to the operation of the PSC engine. It must provide a means of precisely directing fuel into the region of the spark-plug electrodes, without compromising the quality or reliability of the spark.

The PSC concept was investigated by Reynolds and Evans (2004), who conducted two sets of experiments with a Ricardo Hydra single-cylinder engine, one at 2500 rpm, and one at 2000 rpm. The initial experiments at 2500 rpm were designed to optimize the PSC operating conditions, and the final set of data at 2000 rpm was used to examine the ignition and combustion process in more detail. All of the measurements were taken with a WOT, except those presented in Figure 4.20, where throttling was used to make an evaluation of the NO_x-reducing potential of the PSC approach. The engine load was varied by changing the total fuel flow (the main fuel aspirated into the intake manifold plus the pilot fuel injection quantity), hence altering the air–fuel ratio.

4.4.2.1. Performance and Emissions Results at 2500 rpm

The initial test data at 2500 rpm was primarily used to find optimum values for the timing of the pilot fuel injection with respect to the spark timing. A set of baseline data were obtained by running the engine as a purely homogeneous charge engine, with the PSC disabled. For the test with the PSC system operational, the pilot fuel injection flow rate was limited to 14 g/h by the original pilot injector solenoid used for these studies. This value, which corresponds to approximately 0.2 mg/cycle at 2500 rpm, or only 1% of total fuel flow, was therefore used for all tests. Tests were conducted at three different values of the "End of Injection" (EOI) measured relative to the MBT spark timing: 5° before the spark timing (EOI at MBT-5), 10° before the spark (EOI at MBT-10), and 15° before the spark (EOI at MBT-15). As there was no benefit from using the PSC strategy for air–fuel ratios richer than $\lambda = 1.4$, data is only shown from this value out to the lean limit of operation.

Figure 4.16 shows a plot of BSFC versus λ. It can be seen that there was little change in BSFC for λ values less than about 1.6 for all pilot fuel injection timing values. An improvement is only evident when the mixture is so lean ($\lambda > 1.5$) that the spark is insufficient to initiate reliable combustion without PSC fuel addition. Nonetheless, a significant BSFC improvement from the baseline case, a reduction on the order of 7%, was observed at $\lambda = 1.65$ with PSC. Although, the most advanced pilot injection timing provides the greatest reduction in BSFC, all timing values provided a significant extension of the lean limit of operation.

The NO_x emissions as a function of λ are shown in Figure 4.17 for both homogeneous charge operation and for operation with the PSC system. The use of PSC at mixtures richer than about $\lambda = 1.7$ is higher for all of the pilot injection timing cases, but appear to be comparable to the homogeneous charge value at the leanest conditions.

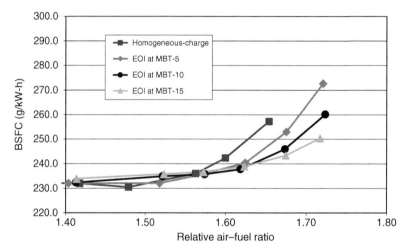

Figure 4.16   BSFC versus relative air–fuel ratio, λ, at 2500 rpm.

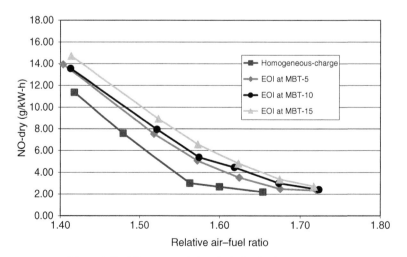

Figure 4.17 NO_x versus relative air–fuel ratio, λ, at 2500 rpm.

However, it will be seen in the next section that when using an approach that combines ultra-lean PSC combustion with some throttling, PSC operation provides significant reductions in NO_x emissions. The improved ignition, and formation of a strong initial flame kernel provided by the partial stratification, was also responsible for providing a reduction in both CO and tHC emission levels under very lean conditions.

4.4.2.2. Part-Load Performance at 2000 rpm

It was recognized that the 14 g/h pilot fuel flow rate used in the preliminary experimental work (which corresponded to only about 1% of total fuel flow) was not optimum, and so changes to the pilot fuel injection system were made to enable higher flow rates to be used. Preliminary measurements then indicated that with the higher flow rate of approximately 40 g/h (between 4% and 5% of total fuel flow), an injection pressure of 25 bar, and the beginning of pilot injection (BOI) set at 10° before MBT spark timing, significant improvements in operation could be obtained. Figure 4.18 shows both BSFC and brake mean effective pressure (BMEP) at WOT as a function of λ, with pure homogenous charge operation indicated by the open symbols, and PSC operation indicated by the solid red symbols. The overall air–fuel ratio has been made successively leaner, with a corresponding reduction of engine load (BMEP). It can be seen that operation with the PSC system provides a 10% extension of the lean limit of combustion.

If the data of Figure 4.18 is then re-plotted as BSFC against BMEP, as in Figure 4.19, it is clear that the extension of the lean limit provided by PSC operation provides an operating range that is also 10% greater, with minimal efficiency penalty. Wider control of the engine load without throttling has been made possible by running under the ultra-lean conditions enabled by the use of partial charge stratification.

A plot of NO_x emissions against BMEP is shown in Figure 4.20, with varying relative air–fuel ratios and varying throttle positions. Here, throttling has been used with the leanest possible air–fuel ratio for both PSC and homogeneous charge cases in order

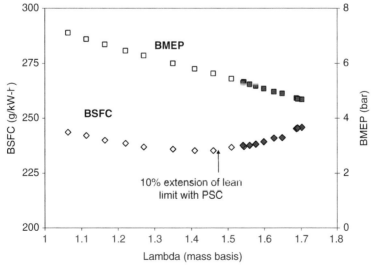

Figure 4.18 BSFC and BMEP versus relative air–fuel ratio, λ, at 2000 rpm.

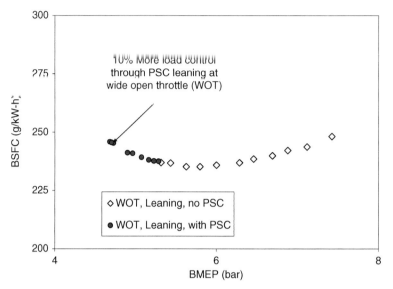

Figure 4.19   BSFC versus BMEP at 2000 rpm.

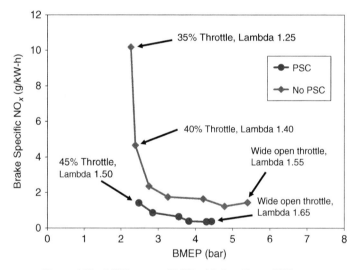

Figure 4.20   BSNO$_x$ versus BMEP with throttling at 2000 rpm.

to control engine load. It is clear that the PSC data shown is for values of λ significantly leaner than the non-PSC homogeneous charge data. These loads would be typical of much of the part-load operating range required for a motor vehicle engine. The highest load point in each case was obtained with a WOT, while the lowest load point corresponded to a throttle opening of 35% for the homogenous charge case, and 45% for the PSC case. The range of λ for homogeneous charge operation was 1.25–1.55, while for PSC operation it was 1.50–1.65. It can be seen that NO$_x$ levels are much lower with PSC operation compared to purely homogeneous charge operation over the complete range of engine loads during lean operation. At a given load requirement, the leaner operation made possible by the PSC system results in lower combustion temperatures, and this provides a substantial reduction in NO_x levels. Using this "ultra-lean" approach at an engine load of approximately 4 bar BMEP, the specific NO_x emissions have been reduced to less than one-third of the value measured without partial stratification. With PSC operation, the need for throttling at a given part-load operating condition is also reduced, leading to improvements in part-load thermal efficiency.

In order to investigate the effect of PSC on combustion stability, the in-cylinder pressure data for 100 consecutive combustion cycles were analyzed for both homogeneous charge operation, and operation with the PSC system. The results of this investigation are illustrated in the scatter plot shown in Figure 4.21. An operating point for each case was chosen which produced the same IMEP value of 5 bar, representative of part-load operation of a motor vehicle. The part-load IMEP was set by varying the air–fuel ratio at WOT. Without PSC operation the relative air–fuel ratio for this point was $\lambda = 1.61$, while with PSC operation it occurred at $\lambda = 1.74$. At $\lambda = 1.74$, the PSC approach had a COV IMEP that was close to the 5% limit normally chosen for stable operation, indicating that PSC operation continued to ensure reliable ignition and combustion. This is demonstrated in the lack of misfires evident in Figure 4.21 for PSC operation. However, misfires are readily observed in the pressure data for the homogeneous charge operation at an equivalent IMEP (at $\lambda = 1.61$). Homogeneous charge operation also had a much greater COV IMEP (over 21%) even though it was considerably less lean than with PSC operation. In the PSC case, the limiting factor to even leaner combustion is the poor flame propagation through the ultra-lean homogeneous charge. These results make it clear that reliability of flame initiation is dramatically improved with PSC operation, enabling the "lean-burn" engine to run leaner than is possible with a conventional spark plug.

The results described above show that a PSC system can extend the lean limit of combustion of homogeneous charge spark-ignition engines, including a substantial reduction in NO_x exhaust emissions. An additional benefit that may be realized with

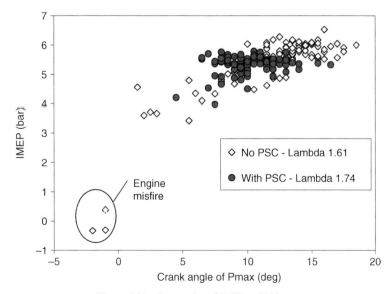

Figure 4.21 Scatter plot of IMEP at 2000 rpm.

the ultra-lean combustion made possible by PSC is an increase in thermal efficiency under part-load conditions. This is due to the improved ignition and flame propagation near to the lean limit of combustion and also due to the reduced amount of throttling required to achieve a given engine load. The PSC system has been successful in extending the lean limit by approximately 10%, and this extension enabled the NO_x emissions during part-load operation to be reduced dramatically. During the final set of experiments designed to optimize the PSC system, NO_x emission levels at an engine load of 4 bar BMEP were found to be less than one-third of the value obtained with a conventional lean homogeneous charge configuration. This "low-NO_x" property of ultra-lean engine operation, coupled with improved thermal efficiency, was the driving force behind the investigation.

Another benefit of PSC operation during part-load operation near the lean limit of combustion is increased engine stability, with a reduction in the number of misfiring cycles. This was demonstrated by examining the IMEP calculated from 100 individual cycles, both with and without PSC operation. Even though the PSC configuration was operated under leaner air–fuel ratios, there was a substantial reduction in the cycle-to-cycle variability of IMEP, which should result in improved drivability in a motor vehicle application. The spark-plug/pilot fuel injector used in this study provided a convenient way to examine the concept of partial stratification. The experimental design proved to be very reliable in operation, but was not designed to be a production-ready item. Further work is required to design a reliable, and cost effective, spark-plug/pilot fuel injector which is suitable for use in production engines.

4.5.   SUMMARY

This chapter concentrated on the kinds of engine design modifications that can be made to enable practical "lean-burn" operation in the context of a traditional spark-ignited engine. Lean combustion in a spark-ignition engine is advantageous because of the benefit associated with the increase in the ratio of specific heats, γ, which lean mixtures of air and fuel provide. A reduction in the equivalence ratio from stoichiometric conditions with $\Phi = 1.0$ to near the lean limit with $\Phi = 0.7$, provides an increase of nearly 10% in the theoretical thermal efficiency. Although, actual engine efficiency is only approximately one-half the theoretical air-standard value, there is nevertheless a significant increase in thermal efficiency with decreasing equivalence ratio. Another major benefit of lean operation is the accompanying reduction in combustion temperatures provided by the excess air, which leads directly to a significant reduction in the generation of NO, one of the most problematic exhaust emissions. Lean operation can therefore be used both to increase thermal efficiency, and reduce exhaust emissions. Ultimately, however, there is a lean limit of operation, beyond which it is impossible to maintain reliable ignition and combustion, resulting in an increased cyclic variation in combustion pressure, and misfires. An extension of the lean limit of operation through improved combustion chamber design is one way to further improve efficiency and reduce emissions. This chapter has summarized the results of two different techniques that have been proposed to extend the lean limit of operation of spark-ignited internal combustion engines.

Combustion chamber geometry has been shown to be an important factor in determining the performance and exhaust emissions of a lean-burn spark-ignition natural gas engine. The optimum spark advance, specific fuel consumption and specific emission levels are greatly affected by the burning rate in the combustion chamber, which in turn is controlled by both the intensity and scale of the mixture turbulence just prior to ignition and during the early combustion process. The "squish-jet" turbulence-generating chamber design was tested in a spark-ignited research engine, and was compared to a conventional bowl-in-piston design. The squish-jet chamber exhibited a significantly faster burning rate than the conventional chamber, as indicated by the reduction in MBT spark advance required. The faster burning rates led to a 5% reduction in BSFC, compared to the case with a conventional diesel type of quiescent combustion chamber. A comparison of two versions of the squish-jet combustion chamber and a conventional chamber design in a Cummins L-10 engine, both with and without enhanced swirl motion, has shown the squish-jet configuration to be more effective in increasing thermal efficiency than the use of high swirl levels.

A new design of PSC combustion system was also compared to a homogeneous charge configuration in single-cylinder research engine. Preliminary measurements of engine performance have shown that the stratified-charge design is effective in extending the lean limit of combustion and in significantly reducing the BSFC during lean operation. The use of the stratified-charge design should enable a spark-ignition engine to operate with reduced pumping losses and significantly higher thermal efficiency over a complete engine driving cycle. Further research is required to fully determine the effect of the new design in reducing exhaust emissions.

REFERENCES

Andrews, G.E., Bradley, D., and Lwakabamba, S.B. (1976). Turbulence and Turbulent Flame Propagation – A Critical Appraisal. *Combust. Flame* **24**, 285–304.

Benson, R. and Whitehouse, N. (1979). *Internal Combustion Engines*. Pergamon Press, London.

Blaszczyk, J. (1990). UBC Ricardo Hydra Engine Test Facility, UBC Alternative Fuels Laboratory Report AFL-90-15.

Campbell, A. (1979). *Thermodynamic Analysis of Combustion Engines*. John Wiley & Sons, New York.

Dymala-Dolesky, R. (1986). The effects of turbulence enhancements on the performance of a spark-ignition engine. M.A.Sc. thesis, University of British Columbia.

Evans, R.L. (1986). Internal Combustion Engine Squish Jet Combustion Chamber, US Patent No. 4,572,123.

Evans, R.L. (1991). Improved Squish-Jet Combustion Chamber, US Patent No. 5,065,715.

Evans, R.L. (2000). A Control Method for Spark-Ignition Engines, US Patent No. 6,032,640.

Evans, R.L. and Blaszczyk, J. (1993). The effects of combustion chamber design on exhaust emissions from spark-ignition engines. *Proceedings of the IMechE Conference on Worldwide Engine Emission Standards and How to Meet Them*, London.

Evans, R.L. and Cameron, C. (1986). A New Combustion Chamber for Fast Burn Applications, SAE Paper 860319.

Evans, R.L. and Tippett, E.C. (1990). The effects of squish motion on the burn-rate and performance of a spark-ignition engine. *SAE Technical Paper Series, Future Transportation Technology Conference and Exposition*, San Diego, CA.

Evans, R.L., Blaszczyk, J., Gambino, M., Iannaccone, S., and Unich, A. (1996). High efficiency natural gas engines. *Conference on Energy and the Environment*, Capri, Italy.

Goetz, W., Evans, R.L., and Duggal, V. (1993). Fast-burn combustion chamber development for natural gas engines. *Proceedings of the Windsor Workshop on Alternative Fuels*, pp. 577–601.

Gruden, D.O. (1981). Combustion Chamber Layout for Modern Otto Engines, SAE Paper 811231.

Heywood, J.B. (1988). *Internal Combustion Engine Fundamentals*. McGraw-hill, New York.

Mawle, C.D. (1989). The Effects of Turbulence and Combustion Chamber Geometry on Combustion in a Spark Ignition Engine, UBC Alternative Fuels Laboratory Report AFL-89-02.

Nakamura, H. *et al.* (1979) Development of a New Combustion System (MCA-Jet) in Gasoline Engine, SAE Paper 790016.

Overington, M.T. and Thring, R.H. (1981). Gasoline Engine Combustion Turbulence and the Combustion Chamber, SAE Paper 810017.

Reynolds, C. and Evans, R.L. (2004). Improving emissions and performance characteristics of lean burn natural gas engines through partial-stratification. *Int. J. Engine Res.* **5**(1), 105–114.

Semenov, E.S. (1963). Studies of turbulent gas flow in piston engines. *NASA Tech. Trans.* **F97**.

Young, M.B. (1980). Cyclic Dispersion – Some Quantitative Cause-and-Effect Relationships, SAE Paper 800459.

Chapter 5

Lean Combustion in Gas Turbines

Vince McDonell

Nomenclature

APU	Auxiliary power unit
B_g	Blockage ratio of bluff body to open flow channel
CO	Carbon monoxide
D_c	Bluff body characteristic length
Dp	Mass of NO_x in grams
DF_2	Diesel Fuel #2
Foo	Engine maximum takeoff thrust in kN
GE	General Electric
HSCT	High-speed civil transport
ICAO	International civil aviation organization
LDI	Lean direct injection
LPP	Lean premixed prevaporized
LTO	Landing-TakeOff
NASA	National Aeronautics and Space Administration
OEM	Original Equipment Manufacturer
P	Pressure
RQL	Rich burn quick mix, lean burn
s	Entropy
SCR	Selective catalytic reduction
T	Temperature
T_o	Initial temperature
Tu'	Turbulence intensity
U	Free stream velocity
v	specific volume
τ	Autoignition time
τ'_{eb}	Characteristic evaporation time
τ'_{fi}	Characteristics time associated with fuel injection
τ'_{hc}	Characteristic burning time for the fuel
τ_{sl}	Characteristic time associated with flame holder length scales
ϕ_{LBO}	Lean blow out equivalence ratio

Lean Combustion: Technology and Control

5.1. INTRODUCTION

5.1.1 ROLE OF COMBUSTION IN THE GAS TURBINE

In the gas turbine, the combustor serves as the source of heat addition to drive the turbine, thereby allowing work to be performed. Thermodynamically, the importance of the combustion to the overall work produced by the gas turbine is evident by examining the pressure–specific volume and temperature–entropy diagrams associated with the gas turbine cycle as demonstrated in Figure 5.1. Figure 5.1 illustrates the basic components of the gas turbine engine in the context of either power generation or aviation.

In both cases, air is compressed using a mechanical compressor (Points 2–3 in Figure 5.1) which elevates its pressure and temperature. It is noted that, in aviation applications, the ram compression (Points 1–2 in Figure 5.1) achievable by the "stationary" air impacting the engine moving at aircraft speeds is not capable of resulting in sufficient pressure rise to allow efficient operation (this is the case for sub- and supersonic flight up to Mach 2.5 or so).

Following the compression, which may take place in a number of stages depending upon the desired pressure ratio of the cycle and/or desired cycle enhancements such as intercooling, the air (or more appropriately, a portion of the air) enters the combustor. At this point, fuel is injected and the chemical energy contained in the fuel is converted to thermal energy which is then extracted from the gases in the turbine section. This temperature rise occurs from Points 3 to 4 in Figure 5.1. This moves the state point to the right on the P–v diagram and to the upper right on the T–s diagram. The hot gases are then expanded through the turbine (Points 4–5) which provides the work necessary to drive the compressor. Sufficient enthalpy remains within the gases to do additional work. In the aviation engine, the exhaust is sent through a nozzle which produces thrust.

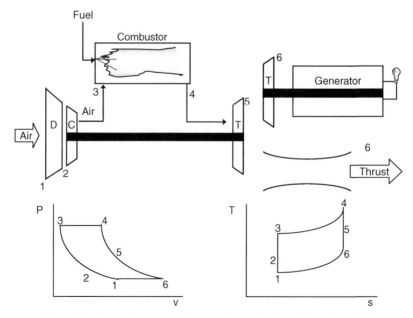

Figure 5.1   Role of combustion in gas turbine applications (Samuelsen, 2005).

In the power generation engine, the exhaust is further expanded through a power turbine which drives a generator or other mechanical device (e.g., pump). Note that in the absence of the state change from 3 to 4, the subsequent expansion through the turbine (Points 4–5–6) would simply return the system to Point 1. Since the work generated by the engine is represented by the area bound by the pathway on the P–v or T–s diagram, traveling from Point 3 back to Point 1 without reaching Point 4 or 6 would result in no work being accomplished. Hence without combustion, no work would result from these engines.

Two other factors are noteworthy which directly relate to the combustion process. First, higher pressures (i.e., Point 3) will result in potentially more area bound by the pathway on the P–v or T–s diagrams. Also, the higher temperature that can be reached at Point 4 will also result in a potentially larger bound area (i.e., more work output). This is important because it illustrates two other underlying principles related to emissions, namely higher pressure ratios and higher temperatures (for the simple cycles shown in Figure 5.1) should be targeted in order to achieve the highest efficiency systems possible. As a result, one of the critical aspects of the combustion science associated with gas turbines is the ability to achieve low emissions while at the same time operating at higher pressures and temperatures. This also illustrates a key difference in aviation and power generation applications. In power generation, the possibility of more advanced relatives of the simple cycle (e.g., recuperated, inter-cooled, reheat) can be considered as a means to achieve high efficiency. However, because these cycle modifications generally add weight (i.e., they reduce the specific power generation), they are not often viable for aviation applications where any added weight is difficult to tolerate. Hence, for aviation applications, the pathway to higher efficiency generally relies upon temperature and pressure increases. One notable exception is the consideration of reheat cycles for aviation gas turbines which is being explored for application to long range military aircraft (Sirignano and Liu, 1999; Sturgess, 2003).

5.1.1.1. Stationary Power

Stationary power gas turbines range in capacity from tens of kilowatts to hundreds of megawatts. They are manufactured by a large number of vendors worldwide. In the past decade, 30–120 GW worth of gas turbines have been ordered each year worldwide (DGTW, 2005). Figure 5.2 shows examples of commercial gas turbines representative

Figure 5.2 Examples of commercial power generation gas turbines: (A) Capstone Turbine 60 kW gas turbine and (B) GE 9H gas turbine (>400 MW in combined cycle).

at the two extremes of the power generation spectrum. Because gas turbines producing tens of kilowatts (order of kW) are contrasted to gigawatt class generation, they are often referred to as "microturbines." This is an unfortunate convention in the labeling of these devices because the prefix associated with the subject in this contraction generally refers to the physical dimension. In the case of microturbine, the physical scales are not on the order of microns, but rather their output is on the order of six orders of magnitude less than the largest systems.

The larger unit shown in Figure 5.2b is designed for power generation at a central station power plant, typically in a combined cycle. The most advanced of these combined cycle systems can achieve fuel to electricity efficiencies approaching 60% while producing sub-9 ppm of NO_x emissions using a complex combustion system discussed below (Myers *et al.*, 2003; Pritchard, 2003; Feigl *et al.*, 2005).

On the other hand, the type of engine shown in Figure 5.2a are expressly designed for distributed generation (Borbeley and Kreider, 2001). Versions with as much as 40–50 MW are available in "trailerable" configurations that would also be considered distributed generation. Deployment of gas turbines throughout a region can be an effective means of deferring transmission and distribution lines. Furthermore, if the engine is sited at a location that can make use of the heat from the exhaust, overall fuel to heat-and-power efficiencies exceeding 70% can be regularly attained. Of course, dispersing engines throughout a region must also be contemplated relative to air and acoustic emissions. As a result, the pollutant emissions from these devices must be extremely low. In the extreme case (California Air Research Board (CARB), 2000), these devices are being required to emit less than 0.07 lb/MW hr of NO_x which is equivalent to about 1 ppm NO_x at 15% O_2 for a 25% efficient engine. The regulation is based on power plants featuring engines like the one shown in Figure 5.2b, but *with* exhaust post treatment via selective catalytic reduction (SCR).

5.1.1.2. Aviation Gas Turbines

In aviation applications, gas turbine engines are used for both propulsion and auxiliary power. The more critical role is played by the main engine, though emissions and efficiency of auxiliary power engines also require consideration. Aviation engines are made in a wide range of thrust ratings which service aircraft from single-seaters to 747s. Unlike stationary engines, aviation engines are started and stopped frequently, and also require much more stringent safety consideration.

For aircraft propulsion, the gas turbine represents the principal source of thrust for military and commercial applications. Examples of engines are shown in Figure 5.3. Figure 5.3a shows a cross section of a turbo-fan engine. In this case, much of the air compressed by the fan bypasses the combustor altogether which results in overall improvement in cycle efficiency. A typical engine is shown in Figure 5.3b.

In addition to main engines, auxiliary power units (APUs) are needed to provide power to the aircraft when power cannot be obtained from the main engines. Figure 5.4 presents a schematic of an APU. Note that this engine is operated to produce shaft power, not thrust. As a result, it can be considered as a power generation engine, but one that must operate over an extremely wide range of conditions and on Jet-A fuel. These devices will not include air bypass. Compared to the main engines, APUs produce relatively little power; however, when the aircraft is on the ground, the APU can be an important source of acoustic and pollutant emissions. As a result, increased attention is

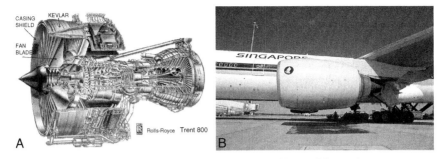

Figure 5.3 Propulsion gas turbine (Rolls-Royce Trent 800): (A) propulsion engine cross section and (B) Trent 800 main engine.

Figure 5.4 Hamilton Sunstrand lean-staged Pyrospin APU (Chen, 2004).

being given to these devices as aircraft manufacturers continue to reduce their overall emissions signature (which includes gate and taxi time).

5.1.2. DRIVERS AND EVOLUTION OF CONDITIONS FOR COMBUSTORS

The need for lean combustion in gas turbines is driven by a number of factors, including market need, regulatory pressure, performance, and reliability. The relative importance of these factors differs for stationary and aviation gas turbines. In this section, the major drivers for each application are discussed. Although the principal motivation for using lean combustion in gas turbines is generally associated with NO_x emissions reduction, it is helpful to summarize the current emissions issues for gas turbines in general. Table 5.1 highlights the current emissions drivers for aircraft and power generation gas turbines.

5.1.2.1. Stationary Power

A serious need for additional energy resources worldwide is apparent in the future. The International Energy Agency has indicated that 1750 billion kW h of additional energy will be required in the next 20 years (United States Department of Energy (US DOE), 2006). This is especially influenced by high population countries such as India and China which are rapidly developing infrastructure for energy, but it is also a factor in developed countries. Gas turbines play a preeminent role in the stationary power generation market place and should remain a critical part of the market mix for at least the next several decades. This is the case despite competition from reciprocating

Table 5.1

Primary emissions drivers for gas turbine engines

Species	Aircraft Engine		Power Generation	
	LTO	Cruise	Distillate	Gas
Soot	x	☒	x	
HC (VOC, ROG, NMHC)	x		x	x
CO	x		x	x
NO_x (NO, NO_2)	☒	☒	☒	☒
SO_x $(SO_2, SO_3,$ sulfates)		☒	x	
CO_2	☒	☒	☒	☒
H_2O	x	☒	x	x

x An issue.
☒ Current focus.

engines and newer technologies such as fuel cells. Alternative technologies compete with gas turbines in certain size classes, but at power generation levels above 5 MW, gas turbines offer the most attractive option due to their relatively low capital, operating, and maintenance costs. These engines are being looked to by the US DOE and the major Original Equipment Manufacturers (OEMs) (e.g., Siemens, GE) for clean power production from coal and other feedstocks. The configurations for these systems involve high efficiencies as well. As a result, the market will continue to demand gas turbines.

The more specific demand for lean combustion gas turbines is another consideration. Of the tens of gigawatts of stationary power engines sold each year, a relatively small number have "low emissions" systems and only a handful are equipped with the lowest emission combustion systems. These systems are required in only a relatively small number of areas, mostly in developed countries in regions with air quality issues such as the United States. Nonetheless, as needs for efficient, clean power generation grow, the number of applications featuring these low emissions systems is increasing. Furthermore, the engine OEMs view low emissions systems as a competitive and technological edge in the marketplace, and they also have an increasing interesting in being considered "green," which places emissions and efficiency at the forefront of their research and development. As a result, the market for lean combustion based gas turbines should increase substantially in the next two decades.

In terms of regulatory pressure, legislation involving criteria pollutants continue to bring additional challenges to the gas turbine industry. While post engine treatment is capable of providing regulated levels of criteria pollutants, some regions continue to lower the emissions levels that must be achieved. Conventional wisdom suggests that mitigating pollutant formation is preferred over post engine exhaust cleanup. Real drivers such as capital cost and operating and maintenance costs support this wisdom. As a result, great interest in further reducing NO_x emissions exists. As mentioned above, those OEMs that can offer the lowest emissions systems will have an edge not only in markets with tight regulations, but also in their image as environmentally friendly (i.e., "green").

5.1.2.2. Aviation

For main engines in aviation applications, the primary drivers for the combustion system do not include emissions. This is associated with the human safety factor that is

Table 5.2

Prioritized combustor design considerations for aviation engines

Design Consideration	Criteria	Comment
Safety		Prime factor
Operability	Takeoff to 45 000 ft	Lean blow off/altitude relight, rain/hail
Efficiency	99.9%	
Durability	6 000 cycles/36 000 hours	
Emissions	Best available emissions at 7%, 30%, 85% and 100% power ("Landing and Takeoff Operation" – LTO) and Cruise	Current regulations based only on "LTO" cycle, though pressure is being made to regulate "cruise" emissions
Downstream Turbomachinery Thermal and Life Integrity	Minimize impact of combustion output	Tailor temperature profiles to optimize life

Adapted from Rohde (2002).

involved in aviation. Table 5.2 summarizes the priorities associated with design of aviation gas turbine combustion systems. As shown, safety and operability are the most critical factors, followed by cost factors associated with operation and maintenance (e.g., efficiency and durability). Emissions rank near the bottom of the list. In reality, International Civil Aviation Organization (ICAO) regulations do drive combustion system development and, although emissions are relatively low on the list, they have driven the investment directions of the engine companies for the past few decades. For example, some air space authorities have imposed landing taxes based on engine emission levels. Hence to the extent lean combustion can impact emissions, it is an important option for aviation engines.

Regarding military applications, the concept of emissions reduction drops even further in priority. However, for public relations reasons and to reflect the overall leadership role of the government, some consideration is given to emissions reduction. A recent review discusses the status of emissions issues with respect to military engines and touches on the role lean combustion has played in this effort (Sturgess *et al.*, 2005).

For auxiliary power, cost is a principal driver. Since passenger safety is not as seriously affected by APU operation, the list shown in Table 5.2 for propulsion engines is somewhat different. Of particular note is the need to reduce noise emissions from the APU. This is a consideration while idling and moving about the airport. In this circumstance, the APU impacts local air quality, and also can provide a nuisance in terms of acoustic emissions. Table 5.3 summarizes these requirements for APU design. Note that the APU, like the stationary generator, is designed to do work, not produce thrust. However, it is still subject to aircraft safety requirements and must be operable over the very wide range of ambient conditions associated with sea level to flight altitudes.

5.2. RATIONALE FOR LEAN COMBUSTION IN GAS TURBINES

Very similar to the chart shown in Chapters 2 and 4 for internal combustion engines, the motivation for operating gas turbine combustors under lean conditions is illustrated

Table 5.3

Combustor design considerations for auxiliary power units

Design Consideration	Criteria	Comment
Operability	Stable operation	Lean blow off/altitude relight, rain/hail
	Relight at up to 43 100 ft	
Efficiency	99.9%	
Durability	Long hot section life	
Acoustic Emissions	9 dbA reduction at 20 m	Relatively new requirement
Air Emissions	40 % NO$_x$ reduction	Relatively new requirements
	No change in CO and UHC	
	No visible smoke	

Adapted from Chen (2004).

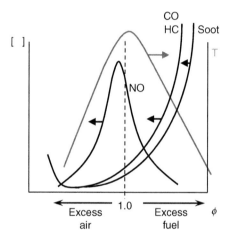

Figure 5.5   Motivation for operating combustion systems lean.

in Figure 5.5. As shown, by operating lean, the combustion temperatures at which substantial NO formation occurs can be avoided. Furthermore, the conditions desirable to oxidize unburned hydrocarbons as well as CO are also attainable. As a result, operating lean provides some obvious benefits from an emissions standpoint. It is worth pointing out that the definition for equivalence ratio here is taken at the local level. For gas turbine combustion, it is insufficient to consider an overall lean combustion situation (such that the total fuel and total air entering the combustor are in lean proportions) as being "lean combustion" because all engines must use excess air to cool the combustion products to within material limits. These material limits in the combustor and the turbine require in nearly all cases that the exhaust products exiting the combustor carry excess air or are "lean." Following the combustion process, additional air is introduced into the exhaust to tailor the temperature profile and to reduce the overall temperature of the exhaust gases entering the turbine.

What is meant by "lean" herein is operation at locally lean conditions in the combustion zone. This is accomplished only when the fuel and air are premixed. The level of premixing begins to take on a critical role in the performance of the combustor.

It has been observed that both spatial and temporal variation in the fuel/air mixing will impact emissions of NO_x (Lyons, 1982; Fric, 1993). As a result, the advent of lean combustion has led to auxiliary efforts in strategies to rapidly and completely mix the fuel and air prior to combustion. The need to premix also leads to additional operability issues which are discussed below.

Given the approach suggested in Figure 5.5, the next question generally raised is: How lean can the engine operate? Or, alternatively, What are the minimum NO_x levels expected for a given situation? This is a question of interest to researchers, manufacturers, and regulators. One approach to answering this question is illustrated in Figure 5.6, where researchers at General Electric (GE) summarized the NO_x emissions for a series of combustors with various premixers and developed what can be referred to as the "NO_x entitlement." In other words, the emissions levels for a given adiabatic reaction temperature and premixer performance should approach the level shown in Figure 5.6. This is helpful in that an assessment of relative NO_x emissions is available for a given system. This result is for stationary power generation.

For liquid fuel (Diesel #2 in this case) operation, a similar "entitlement" summary is shown in Figure 5.7. Liquid fuel has inherently higher emissions levels due to the activation of non-thermal NO_x pathways which are enhanced by the presence of the higher hydrocarbons at a given temperature (Lee *et al.*, 2001).

For aviation engines, another form of this entitlement plot has been produced (Tacina, 1990; Shaffar and Samuelsen, 1998). As shown in Figure 5.8, the ordinate value in this case is expressed in the form of an emissions index, namely grams of NO_x per kilogram of fuel consumed. It is noteworthy that the units associated with the emissions for stationary power and aircraft are different. The convention for aircraft engines is to use an emissions index which is mass of emissions per mass of fuel consumed. The convention for power generation is parts per million corrected to 15% O_2. Neither of these emission performance measures offers any "reward" for efficiency. Output based emissions (e.g., lb/MWh) do reward efficiency. Recent regulations

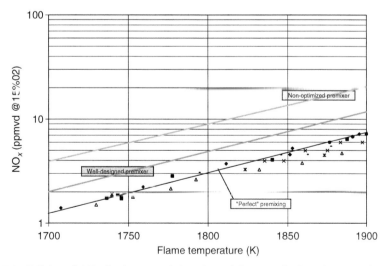

Figure 5.6 "Minimum" NO_x levels expected for advanced lean premixed combustors (adapted from Leonard and Stegmaier (1994)).

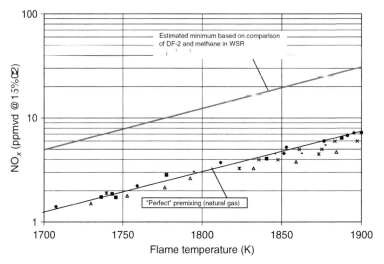

Figure 5.7 "Minimum" NO$_x$ levels expected for prevaporized premixed combustors.

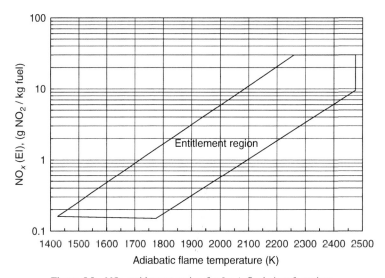

Figure 5.8 NO$_x$ entitlement region for Jet-A fired aircraft engines.

associated with stationary power are beginning to incorporate output based approaches (CARB, 2000). Aircraft engine certification values are in mass of pollutant per unit mass of thrust.

The implications for lean operation include a number of geometry modification considerations. To illustrate this, Figure 5.9 compares cross sections of conventional and lean premixed combustion strategies. In the conventional approach, the fuel is injected directly into the combustion chamber along with approximately 30% of the total air entering the combustor. In the lean combustion strategy shown, the fuel is premixed with about 60% of the combustor air flow. Hence, much more air enters the primary zone in the lean premixed case. This also reduces the amount of airflow

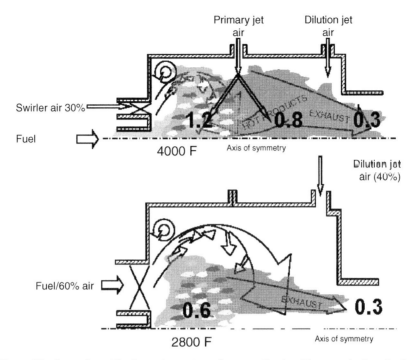

Figure 5.9 Comparison of basic reaction anatomy for conventional and lean premixed combustion.

available for cooling which can be a concern for maintaining liner integrity and can also impact the opportunity to properly tailor temperature profiles to meet the requirements of turbine materials at the turbine inlet.

5.2.1. BARRIERS FOR LEAN COMBUSTION IN GAS TURBINES

The barriers that must be overcome for success with lean combustion in gas turbines are associated with operability and emissions. Since emission is a principal driver motivating the use of lean combustion, the challenges are really associated with achieving low emissions while maintaining stability, avoiding autoignition and flash-back, and achieving sufficient turndown to cover the range of conditions needed to fulfill the operating map of the engine. In addition to the content herein, the reader is referred to other recent reviews of this subject (Richards *et al.*, 2001; Lieuwen *et al.*, 2006).

The challenges associated with lean combustion are illustrated in Figure 5.10 in the context of a typical combustor "stability loop." For a given inlet pressure and tempera-ture, the fuel/air ratio for a given mass flow through the combustor can be increased or decreased to a point where the combustor can no longer sustain the reaction. Limits in fuel/air ratio can be found on both rich and lean sides of stoichiometric at which the reaction will no longer be stable. This locus of conditions at which the reaction is no longer sustained is shown as the static stability limit in Figure 5.10. This loop may change as temperature and pressure change. Figure 5.10 also illustrates the presence of

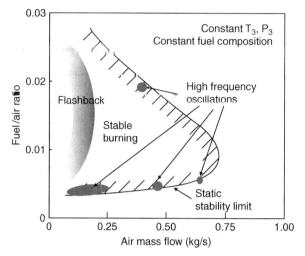

Figure 5.10 Illustration of combustion operability issues for gas turbine combustor at fixed inlet temperature and pressure (Lieuwen *et al.*, 2006).

flashback, which can be an issue as the velocities in the fuel injector/premixer become relatively low.

Finally, discrete points usually along the static stability limits are shown which correspond to operating conditions where combustion oscillations are problematic. Each of these operability issues is discussed in the next sections.

5.2.2. STABILITY

Stability in gas turbines is a major design consideration. As suggested in Figure 5.10, the stability limit effectively defines the operability of a given design for a given inlet condition. As shown, any combustion system will have a characteristic "stability loop" for a given fuel, temperature, and pressure. As the amount of air pushed through the system increases, the range of fuel/air ratios for which stable burning can be maintained will change. Figure 5.10 also shows a number of regions that are of particular interest. The lower region defining the separation of stable and unstable combustion is the lean stability limit, sometimes referred to as the "static" stability limit. It is desirable to operate the system at as low a fuel/air ratio as possible because the lower fuel/air ratio areas will produce beneficial low temperatures. Similarly, the upper line demarks the rich stability limit. Along the stability loop generally lie seemingly random isolation areas of high frequency oscillations. These acoustic oscillations are referred to as "dynamic stability" regions. The conditions giving rise to these relatively high frequency oscillations are difficult if not impossible to predict. Finally, as the air flow through the system reduces, the possibility of flashback occurs as the reaction has enough velocity to propagate upstream into injectors or premixers. Note that the most stable situation relative to airflow is the situation where the system is operated at stoichiometric fuel/air ratios (e.g., around fuel/air ratio of 0.01 in Figure 5.10). In the next sections, various aspects of stability and operability are discussed.

5.2.2.1. Static Stability (Lean Blow-Off)

In many combustion systems, the static stability of the reaction is a key parameter relative to the overall operability and performance of the system. Over the years, significant effort (e.g., McDonell *et al.*, 2001) has been directed at establishing relationships between stability and conditions/geometry of the flame-holding device. As a result, basic design tools have been developed which can give insight into how changes in operating conditions and/or flame stabilizer geometry may impact the stability. These are relevant for both stationary and aviation type combustors.

For example, specific work by Lefebvre (Lefebvre and Ballal, 1980; Lefebvre and Baxter, 1992) has provided insight into the functional relationships among stability, equivalence ratio, free stream velocity, pressure, temperature, and bluff body geometry. In that work, a cylindrical cone-shaped bluff body was introduced at the centerline of a larger diameter pipe within which either pure air or a mixture of gaseous fuel and air flowed. In order to study the stability of liquid fueled systems, pure air was used and liquid was introduced by a nozzle at the centerline in the wake formed downstream of the bluff body. The conclusion from the work conducted by Lefebvre and co-workers is summarized by the correlation expressed in Equation (5.1), which relates the weak extinction limit to the various parameters studied:

$$\phi_{\mathrm{LBO}} = \left\{ \frac{2.25[1 + 0.4U(1 + Tu')]}{P^{0.25}T_0 \mathrm{e}^{(T_0/150)}D_{\mathrm{c}}(1 - B_{\mathrm{g}})} \right\}^{0.16}. \tag{5.1}$$

The attractiveness of Equation (5.1) is that it provides a relationship of how the parameters studied influence the stability limits and also provides some guidance on how to design effective reaction holders for a given situation. It also provides a mechanism to scale results obtained at one condition to other conditions of interest. The physical mechanism leading to Equation (5.1) is the ratio of reaction time to residence time, otherwise referred to as the combustor loading parameter. This time ratio is also captured through the non-dimensional group, the Damköhler number.

Characteristic time models are based on the concept of the reaction zone acting like a well-stirred reactor, which may be a reasonable representation in some cases. Other examples of this type of approach include the work of Mellor and co-workers (Plee and Mellor, 1979; Leonard and Mellor, 1981) which is summarized by the equation,

$$\tau_{\mathrm{sl}} + 0.12\tau_{\mathrm{fi}}' = 2.21(\tau_{\mathrm{ho}}' + 0.011\tau_{\mathrm{eb}}') + 0.095 \tag{5.2}$$

In these characteristic time models, characteristic values are needed for velocity as well as the reaction zone size. A characteristic dimension divided by a characteristic velocity provides a typical time scale associated with the reaction. As a result, some thought is needed to identify appropriate values.

Another class of model involves a comparison of reaction propagation to the flow speed. This type of model may be more physically representative of a reaction stabilized by a vortex breakdown. In this class of flow, the ability of the reaction to anchor itself does seem related to the reaction speed propagating toward the flame holder versus the axial velocity of the flow approaching the stagnation point. This type of model has been typically implemented through the use of Peclet numbers (Putnam and Jensen, 1949) and has been utilized recently to correlate stability of swirl stabilized flames

(Hoffman *et al.*, 1994; Kroner *et al.*, 2002). A challenge with existing Peclet number approaches is the reliance upon turbulent flame speeds. As has been illustrated in Chapter 2 and in the literature, the typical dependency of turbulent flame speed on laminar flame speed is not intuitive and can be sometimes misleading (Kido *et al.*, 2002). Furthermore, recent work has suggested that neither the characteristic time nor propagation based models are sufficient for describing more subtle effects such as fuel composition due in part to their dependency upon the turbulent flame speed (Zhang *et al.*, 2005).

In summary, although extensive work has been done on developing tools for static stability limits, some inconsistencies have been observed which may or may not be dependent upon the type of stabilizing geometry used or upon the fuel composition. This is an area which needs further refinement.

5.2.2.2. Dynamic Oscillations

Dynamic oscillations are potentially significant problems for both stationary and propulsion engines. A review of these issues is available in both recent (Lieuwen and McManus, 2003) and earlier reviews (McManus *et al.*, 1993), and is discussed further in Chapter 7. The manifestation of the dynamic oscillations is tied to circumstances where perturbations in the heat release and pressure field couple together in a resonant manner. If the perturbation in heat release is in phase with an acoustic wave, the results can be destructive. Figure 5.11 illustrates an example of the type of destructive force combustion oscillations can cause to gas turbine engines.

Oscillations have been studied extensively for the past decade relative to lean gas turbines due to the problems that have been encountered (Candel, 2002). In most cases, solutions have been found through trial and error approaches. Considerable effort has been directed at predicting whether or not oscillations will occur for a given system, but a comprehensive model has yet to be fully developed (Ducruix *et al.*, 2003; Dowling and Stow, 2003; Lieuwen, 2003). Despite this, progress has been tremendous, and solutions ranging from altering fuel schedules or physical injection locations to the inclusion of damper tubes or hole patterns have been attempted successfully (Richards *et al.*, 2003). These so-called passive solutions are attractive because they are less complicated compared to more exotic closed loop active control approaches, but there

Figure 5.11   Transition failure in GE-F gas turbine due to dynamic stability issues (Tratham, 2001).

has also been effort directed at actively monitoring and controlling combustion oscillations (Docquier and Candel, 2002; Cohen and Banaszuk, 2003).

Stationary gas turbines seem to suffer more from dynamic stability issues than do aero engines. This is because stationary turbines are designed to operate very near the static stability limit. At some points near the static limit (see Figure 5.10), a situation may arise in terms of fuel loading, aerodynamics, and heat release that lead to the onset of an oscillatory behavior. Operating near the static stability limit nearly guarantees the presence of finite perturbations in the combustor.

Stationary combustion systems have a few features that make them especially susceptible to these problems (Lieuwen and McManus, 2003). For example, to minimize CO emissions, the use of cooling air, especially along walls ("hot wall strategies"), is minimized. Elimination of extensive cooling passages and jets results in "stiff" boundaries containing the acoustic field which provide minimal damping. Also, to achieve a high degree of premixing, the combustion system is set up to have the reaction sit downstream of some "dump plane" which potentially situates the reaction at an acoustic pressure maximum point. Finally, stationary combustors have some freedom relative to their length, and to ensure CO burnout, long burnout zones are common which leads to a relatively long dimension relative to the heat release zone and thereby the possibility of exciting organ pipe modes.

Nearly every gas turbine OEM involved in dry low emissions combustion has had to deal with acoustic oscillations. Both passive (e.g., moving fuel injection location, adding damper tubes, modifying "hard" boundaries in the combustion chamber, adjusting fuel split between a non-premixed pilot and a fully premixed main) and active (e.g., pulsed fuel) approaches have been demonstrated on fielded engines. Some very interesting experiences are described in the literature that illustrate the difficulty that the industry has faced with this issue (Hermann *et al.*, 2001; Lieuwen and Yang, 2006).

Stationary engines have been a viable platform for the evaluation of closed loop control systems. Examples of full scale implementation of closed loop control on a 240 MW central station power plant can be found (Hermann *et al.*, 2001).

In aviation engines, due to the focus on safety, engines will not operate as close to the lean stability limit, thereby resulting in some margin relative to the onset of oscillations. However, acoustic oscillations can occur in situations where the system is operated "conventionally" and even at stoichiometric combustion zones (Janus *et al.*, 1997; Mongia *et al.*, 2003; Bernier *et al.*, 2004). For aviation applications, where reliability and safety are key, passive approaches are highly preferred. Despite this, examples of implementing closed loop control on liquid fueled systems have appeared (Coker *et al.*, 2006). Applying active control on aviation applications could be possible if reliability and safety can be demonstrated conclusively. In advanced engines today, extensive sensor arrays are already in place, but comprehensive testing of any active control system would be needed before applying it in practice. Unmanned aircraft represent a viable test bed for such an application.

5.2.3. IGNITION/AUTOIGNITION

There are two ignition issues associated with gas turbines: the first is ignition of the engine for startup, and the second is autoignition of the fuel–air mixture. Both affect the design of systems and are described briefly below.

5.2.3.1. Stationary

For stationary engines, ignition at startup is not really a critical design issue. However, the occurrence of ignition within the premixer is a concern. Figure 5.12 illustrates the situation that exists in typical lean premixed combustion approaches. If reaction occurs within the premixer, it will result in high NO_x emissions and can also potentially damage the injector/combustor. As a result, if the ignition delay time is shorter than the premixer residence time, the system will have operability issues.

For gaseous fuels, one approach to quantifying the potential for this problem is carrying out a chemical kinetic calculation using an appropriate mechanism. Such mechanisms are commonly available for natural gas type fuels (http://www.me.berkeley.edu/gri_mech/; Ribaucour et al., 2000), although there is some debate as to the appropriateness of these mechanisms for ignition calculations.

Alternatively, global expressions for ignition delay have been developed which are convenient to apply. One example is shown in Equation (5.3) (Li and Williams, 2002):

$$\tau = \frac{2.6 \times 10^{-15}[O_2]_o^{-4/3}[CH_4]_o^{1/3}}{T_o^{-0.92} \exp\left(-13180/T_o\right)} . \tag{5.3}$$

Recent work has shown that the few global expressions for low temperature ignition delay times have a wide range (2–3 orders of magnitude) of predicted values (Chen et al., 2004), so they should be applied and verified with care.

Experience has shown that autoignition for natural gas is not a major issue for lean premixed combustion. However, with increasing interest in alternative fuels, this issue has again been raised. Compounding this matter, while expressions for ignition delay for non-methane gaseous fuels are available, far less evaluation of their accuracy has been completed. Of increasing recent interest is operation on hydrogen containing gases. In this case, although extensive work has been done on hydrogen/oxygen reaction systems, only limited results are available for lower temperature regimes that are relevant to gas turbine premixers. To illustrate, Figure 5.13 presents a comparison of measured ignition delay times with calculated delay times using a well-accepted hydrogen/air reaction mechanism. As shown, a significant lack of agreement in the low temperature regions (e.g., <1000°F) is evident. As a result, this is another area that is in need of further work. Chapter 8 includes some additional discussion of hydrogen use in gas turbines.

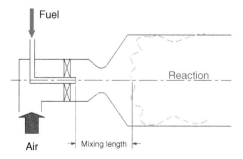

Figure 5.12 Lean premixed combustion strategy.

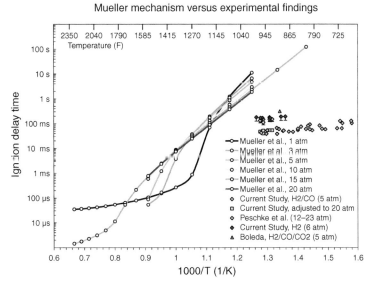

Figure 5.13 Comparison of ignition delay times for detailed mechanisms and measurements (Beerer *et al.*, 2006). (See color insert.)

5.2.3.2. Aviation

Ignition is a particularly critical issue for aviation engines, and altitude relight is always an important performance criterion. The ignitability of a mixture depends strongly upon the equivalence ratio in the vicinity of the igniter. Like static stability, a "loop" can be created for a given combustor which describes its ignition range. The ignition loop always falls within the static stability loop. This is because once the system is operating, energy from hot walls and radiation from the reaction help to widen the range of fuel/air ratios which will sustain the reaction. For ignition, cold fuel is on the walls and the entire combustion system is relatively cold. The differences between the static stability loop and the ignition loop give some characteristics of the combustion system. Ideally, these loops would be close to each other, but both must achieve requirements associated with overall stability requirements for a given combustor air flow.

Autoignition is a concern for lean premixed prevaporized (LPP) systems. For liquid fuels, the ignition delay time is generally shorter than it is for gaseous fuels. Because aviation fuels are so complex, it is difficult to identify a suitable kinetic mechanism which can be used for estimating ignition delay. However, global expressions for ignition delay are available in the literature for Jet-A fuel, and an example (Guin, 1998) is shown in Equation (5.4):

$$\tau = 0.508e^{(3377/T)}P^{-0.9}. \tag{5.4}$$

This expression suggests that at 30 bar and 700 K inlet temperature, ignition delay times are around 3 ms. This is on the order of the typical premixer residence times. This is consistent with other work done on ignition delay for liquid fuels (Lefebvre *et al.*, 1986).

5.2.4. FLASHBACK

Flashback in premixers is an issue that manifests itself much like autoignition. In the event of a sudden drop in premixer velocity or if conditions occur that lead to a spike in the reaction velocity, the possibility of reaction propagating back up into the premixer exists. The issues associated with flashback are similar for stationary and aviation gas turbines, only different fuels are used – so they need not be treated differently. Traditionally, flashback has been correlated with some form of laminar burning velocity. Unfortunately, as shown in Chapter 2, knowing the laminar flame speed has little to do with the turbulent flame speed found in practical devices.

Flashback often occurs in the flow boundary layer since this is the location of lowest flow velocity. Proximity to physical surfaces can also quench the reaction, however, so consideration must also be given to the thickness of the boundary layer and to how the properties change within the boundary layer (Lefebvre *et al.*, 1986). Regardless, another mechanism for flashback can exist in swirl stabilized flows. In this case, an interaction of the reaction and the vortex breakdown responsible for the reverse flow can lead to a sudden propagation of the reaction upstream (Kroner *et al.*, 2002). In this case, even if the flow velocity exceeds the flame speed everywhere, flashback can still occur. In these two mechanisms of flashback, the turbulent burning velocity appears to be the controlling factor. Unfortunately, establishing the turbulent burning velocity is not straightforward and although a number of correlations exist that relate the turbulent velocity to the laminar velocity, examples can be found where fuels with the same laminar burning velocity have significantly varying turbulent burning velocities even for the same flow turbulence intensities (Kido *et al.*, 2002; Lieuwen *et al.*, 2006). Finally, in the case when combustion oscillations are present, the local variation in flow velocity due to the oscillation can give rise to reactions propagating upstream (Davu *et al.*, 2005).

Recent work has been conducted to systematically characterize flashback in lean premixed gas turbines, including consideration of fuel composition effects (Thibaut and Candel, 1998). This study suggested a need for continued work in this area to elucidate the details of the mechanisms responsible for flashback.

5.2.5. FUEL FLEXIBILITY

Due to uncertainty about fuel reserves, gas turbine engine developers must consider using fuels different from those in the current fuel stream. Because aviation engines face more restrictions due to safety and must also be able to refuel at a large number of points, considerable effort has been taken to reduce the fuel flexibility requirements. On the other hand, for power generation, gas turbine engines are being required to operate on a fuel stock with widening properties. Unfortunately, while conventional combustion systems for gas turbines are relatively fuel tolerant, low emissions systems, and lean premixed systems in particular, face some challenges when fuel content is changed.

5.2.5.1. Stationary

As in the lean premixed burners described in Chapter 6, for stationary power generation, the Wobbe Index is typically used to classify the ability of a given gas

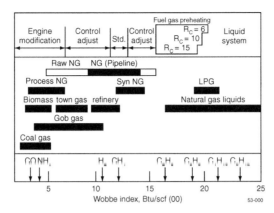

Figure 5.14 Wobbe Index for various fuels (Courtesy of Solar Turbines).

turbine to operate on a given fuel. Figure 5.14 illustrates how the Wobbe Index relates to fuel interchangability for a given system. As shown, a narrow range of Wobbe Index values can be tolerated for a given engine combustion system. At low Wobbe Index values, the volumetric flow requirement is high, requiring larger passages and control valves. At high Wobbe Index values, penetration may be insufficient, mixing may be reduced, and other issues may be encountered. Unfortunately, while the Wobbe Index describes some characteristics of the fuel, it does little to describe the combustion issues associated with these fuels. In fact, all the issues described in the previous sections on operability are more germane from a combustion view point.

What is apparent is that gas turbines, while traditionally quite fuel flexible, are now becoming more sensitive to changes in fuel composition. This is because lean premixed systems are expected to operate on the edge of stability. This leads to a need to carefully fine tune fuel injection strategies to minimize operability issues. The fuel composition can impact stability (both static and dynamic) and emissions. With natural gas, the presence of higher hydrocarbons facilitates non-thermal NO_x mechanisms which can lead to higher NO_x emissions (Meier *et al.*, 1999; Flores *et al.*, 2003) than for methane. As mentioned earlier, this same sensitivity to fuel composition near the boundary of stability is described for lean premixed burners in Chapter 6.

5.2.5.2. Aviation

For aviation gas turbines, the fuels of current interest have been standardized to a large extent and are produced per specification or contract. Originally, it was thought that jet engines could operate on nearly any fuel source. However, as engines have evolved in sophistication and are subject to emissions regulations, this viewpoint has changed as well. Current aviation fuels are generally straight-run distillates which, therefore, depend on crude oil type for their main characteristics. Detailed information on the fuels that are available throughout the world and a compilation of their properties is available (CRC, 2004).

Aviation fuels include aviation Gasoline (AVGAS) and turbine fuels are synthetic and crude oil based, respectively. In the United States, commercial turbine fuels include Jet-A and Jet-A1. These are nearly identical in specification with exception of the lower freezing point temperature for Jet-A1 (–47°C versus –40°C). Specifications in the

United Kingdom, Canada, PRC, and Europe are based on Jet A-1. In Russia, TS-1 is also produced. The TS-1 specifications call for lower freezing points (–50°C) and lower viscosities compared to Jet-A in order to facilitate use in the extremely cold temperatures of Russia. Jet A-1 is available at most major airports.

Jet-B is another fuel specification that has largely been phased out. However, Jet-B features a wider cut of the distilled product and therefore has a higher yield per barrel of crude oil. As the danger of fuel shortages increases, it is possible that a less demanding refining strategy like for Jet-B may again be considered.

For military applications, JP-4, JP-5, and JP-8 are used in the United States, although nearly all development is focused on use of JP-8 which is very similar to Jet-A. JP-8 has some additional additives to improve lubricity and other features of interest for military use. JP-8 can be "made" from Jet-A with an additive package. JP-4 has a much lower temperature distillation range (<100–270°C) than does Jet-A and is therefore more volatile. In contrast, JP-5 is closer to Jet-A in that its boiling range is 200–300°C. JP-4 was the traditional fuel of the Air Force, while the Navy used JP-5 which was safer to handle on aircraft carriers. However, due to experiences with incompatibilities of engines designed for operation on JP-4 when operated on JP-5 and vice versa, the decision was made to essentially phase out JP-4. The loss of JP-4 required reworking engines to handle extremely cold environments.

Again, as fuel prices become more volatile and oil supplies become questionable, it is probable that fuel flexibility may again become an issue for aviation gas turbines. In the 1970s, for example, in response to oil shortages, the US military considered liquid fuels derived from shale oil, which is plentiful in the central United States. Currently, renewed interest in coal-derived liquid via Fischer-Tropsche processes is occurring in the US as well.

One final note relative to aviation engine fuel flexibility is that hydrogen has been a fuel consideration for many years for gas turbines (Pratt *et al.*, 1974; Svennson *et al.*, 2004). In a liquefied or "slush" form, hydrogen may be viable. As crude oil is depleted or becomes too expensive, hydrogen may gain popularity as a fuel choice. The hydrogen option is discussed further in Chapter 8. For the near future, however, aviation gas turbines will operate on a single fuel: Jet-A. While this helps regarding implementation of lean combustion strategies, it also gives rise to complacency regarding consideration of fuel flexibility. Those engine manufacturers who can respond effectively to the potential need to provide fuel-flexible gas turbines may have an advantage as future markets evolve.

5.2.6. TURN DOWN

Up to this point, the issues described are applicable for any given operating point for the engine. However, when consideration is given to the entire operating range of the engine (i.e., "turn-down"), the situation becomes much more complex. Since the gas turbine thermodynamic cycle dictates that pressure ratio change as load changes, the fuel/air ratio will vary over the load range. Due to the different relationships between output power and gas flow (~squared) and output power and fuel flow (~linear), operating lean over the full load range while maintaining stable combustion requires "staging" of some form. To illustrate this, Figure 5.15 presents the relative fuel and air flows and the associated fuel/air ratio for a small gas turbine engine. As shown, the overall fuel/air ratio drops as load decreases. Consequently, if the system is optimized to

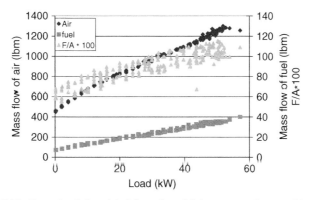

Figure 5.15 Example of air and fuel flows for a 3.5:1 pressure ratio gas turbine engine.

operate at minimum fuel/air ratio at full load (e.g., to minimize reaction temperatures), it would not be possible to reduce the load of the system since the flame would reach its lean blow-off limit. In response to this problem, lean premixed combustion systems are often staged, with multiple fuel injection points that can be operated sequentially. This allows the local fuel/air ratio for each point to be tailored while allowing the overall fuel/air ratio for the engine to vary as needed to accomplish the turndown desired.

To illustrate this in the context of emission performance, Figure 5.16 compares how staging allows the combustion system to stay within a lean combustion regime (locally for each fuel injection point), whereas the conventional non-staged engine must operate with combustion taking place over a wider range of equivalence ratios and inevitably ending up having to operate at conditions producing high NO_x emissions for at least some part of the load range. How the staging is accomplished differs substantially for stationary and aviation engines due to the differing drivers and constraints. Each is briefly reviewed here.

5.2.6.1. Stationary

Staging in stationary engines has evolved into a highly sophisticated engineering task. Consider the two engines originally shown as examples in Figure 5.2. Figure 5.17

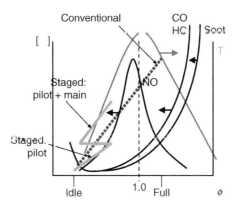

Figure 5.16 Comparison of staged and conventional combustion strategies – idle and full power points shown for conventional strategy.

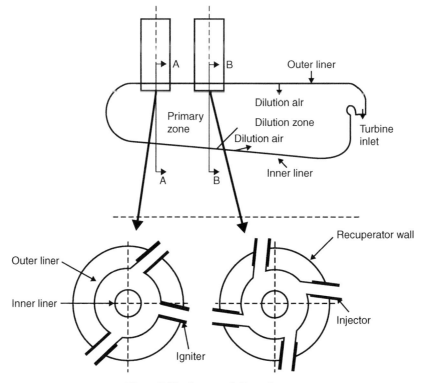

Figure 5.17   Capstone C-60 combustor.

shows the combustion strategy used in the engine in Figure 5.2a. Even though this is only a 60 kW engine, it makes use of a five-stage fuel injection strategy. In the first plane of injection (two injectors), both injectors are operated full time and are used to cover the startup and low power conditions. However, in the second plane, anywhere from 1 to 4 injectors are operated as the load ramps up. Within each stage, the equivalence ratio increases as load increases, but with staging, the overall equivalence ratio stays within the low emissions window. The first plane is not operated at extreme lean conditions since it is serving to ensure the reaction remains stable. The resulting NO_x emission profile is shown in Figure 5.18 and illustrates clearly where the stage points occur. It also shows that the NO_x emissions can be "tuned" for a given load point by altering the staging strategy. This staging point flexibility also provides a powerful "knob" for avoiding combustion oscillations.

The staging strategy taken in the engine shown in Figure 5.2b is illustrated in Figure 5.19. The GE DLN (Dry Low NO_x) strategy features more of a "can" style combustion chamber with traditional swirl, but the use of multiple discrete fuel injection points remain. In this system, four different fuel circuits are used featuring differing combinations of primary, secondary, premixed primary, and quaternary circuits. These are operated in different combinations as load changes. The resulting impact on NO_x emissions is shown in Figure 5.20. This engine can achieve very low emissions from 35% to 100% load using lean combustion.

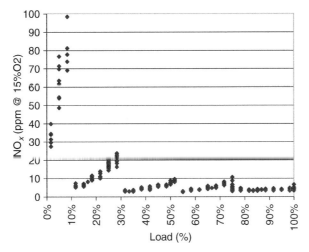

Figure 5.18 Staging influence on NO$_x$ emissions for Capstone C-60 (Phi *et al.*, 2004).

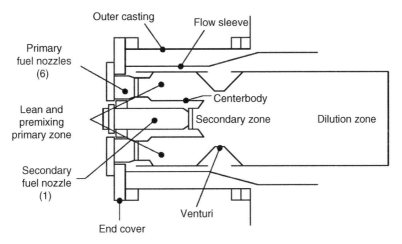

Figure 5.19 GE DLN fuel injection cross section.

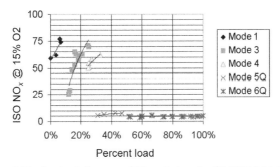

Figure 5.20 Staging influence on NO$_x$ emissions for GE DLN 2.6 injector.

5.2.6.2. Aviation

Stationary engines are not the only devices that need to use staging. Aviation engines are also capable of being staged. For example, Figure 5.21 illustrates the GE concept of using "dual annular" combustors for emissions reduction. The details regarding geometry modifications to accommodate the shape can be noted, although examples have been implemented in which the staged combustor can be retrofitted into an engine with a single annular design. However, this approach is not optimum and greater flexibility in design is generally preferred. The manner in which utilization of the dual annular concept impacts NO_x emissions is shown in Figures 5.22 and 5.23. The staging approach is beneficial from an operability standpoint as well. For example, the pilot stage can be optimized for ignition and low power operation while the main stage can be optimized for low emissions at high power. It is worth noting that the fuel injection approach used in both single and dual annular configurations features some degree of premixing prior to the reaction zone. In order to operate with low emissions, good mixing is essential for lean operation.

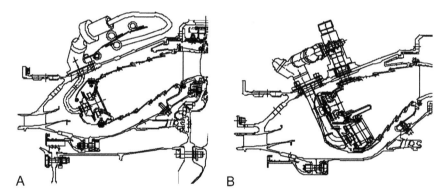

Figure 5.21   Comparison of (A) a single and (B) a dual annular combustor (GE CFM-56).

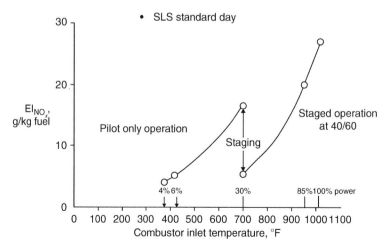

Figure 5.22   Example of dual annular staging influence on NO_x emissions.

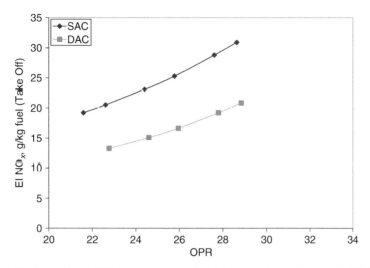

Figure 5.23 Comparison of NO$_x$ emissions for dual and single annular combustor (GE CFM-56-5).

5.3. LEAN GAS TURBINE COMBUSTION STRATEGIES: STATUS AND NEEDS

5.3.1. STATIONARY

For stationary power generation, it stands to reason that higher risk can be tolerated compared to aviation engines. As a result, in recent years, stationary power generation has received extensive focus as a test bed for study of a wide range of advanced combustion technologies. These can be classified as advanced homogeneous strategies and heterogeneous strategies. Each type is discussed in this section.

5.3.1.1. Homogeneous Strategies

"Homogeneous" refers to the state of the system involved in the combustion process. In the case of homogeneous combustion, gaseous fuel is reacted with gaseous oxidizer (air). Homogeneous lean combustion strategies basically involve mixing the fuel and air far upstream of the reaction zone. Various approaches are used in swirl stabilized combustion systems including injection through the swirl vanes or through "spokes" sitting in the air flow passage. With gaseous fuels, hundreds of injection points can be used to distribute the fuel over the injector exit plane. Some engine OEMs use relatively high pressure drops across the fuel injection passage ("stiff injection"), whereas others use relatively low pressures ("soft injection"). The approach taken generally depends upon the experience at each OEM.

The use of strong swirl to stabilize the reaction is taken from many years of design practice from combustors designed to operate stoichiometrically. As such, the recirculation zone serves to mix the fuel and air as well as provide a low velocity region that allows the reaction to anchor. Such strong swirl also causes high strain on the reaction. Indeed, with very high swirl levels, the reaction can be blown off as opposed to stabilized.

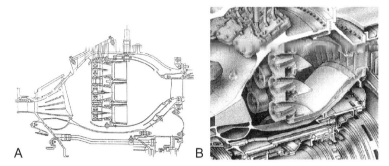

Figure 5.24 Example of large number of injection points for ultralow emissions lean premixed combustion (GE LM6000): (A) dual annular counter rotating swirler (DACRS) mixer and (B) triple annular implementation.

Through the use of staging and "million injection point" approaches, flexibility in overcoming combustion dynamics has been realized in many cases for swirl stabilized premixed systems. Figure 5.24 illustrates one example of such an approach. Even with the inherent flexibility of this design, damping tubes have been implemented into this system to overcome oscillation issues.

With premixing of fuel and air achieved through the use of multipoint injection strategies, injection from vane trailing edges, and other means, the absolute need for strong swirl is now being questioned. Similar to the low swirl burner described in Chapter 6, recent work on low swirl injectors is now showing some very promising emissions performance for gas turbine applications (Phi *et al.*, 2004). Furthermore, the use of fuel injection without significant swirl has also been demonstrated to yield low emissions (Fable and Cheng, 2001; Phi *et al.*, 2004; Neumeier *et al.*, 2005; Kalb *et al.*, 2006). Both of these strategies are reviewed in separate chapters in this book and are not discussed in detail here; however, a few sample results are presented in following sections.

In order to provide operability for lean premixed gas turbines, it is very common to rely on a "pilot" injector of some sort. The pilot often consists of a discrete injection of either pure fuel or of a relatively rich fuel/air mixture. Strategically targeting the pilot fuel allows the reaction to continue even if the reaction produced by the main fuel/air mixture starts to suffer some degree of instability. The pilot is generally directed into a strategic location in the combustor to enrich a region that will help sustain the reaction. A common location is along the centerline into a recirculation zone as illustrated in Figure 5.25.

The pilot typically generates a relatively high amount of NO_x due to the diffusion flame nature of the reaction it generates. As a result, efforts have been directed at alternative piloting approaches including alternative fuels and "energetic" piloting using catalysts or hydrogen as a means to produce energy to sustain the reaction (Nguyen and Samuelsen, 1999; Scheffer *et al.*, 2002; Karim *et al.*, 2003; Yoshimura *et al.*, 2005). As engines are developed for operation on hydrogen in support of the US DOE's 20-year vision for future power plants, opportunities may arise that can take advantage of energetic piloting.

Rather than relying upon long times and distances to fully premix fuel and air, the consideration for a "lean direct" injection approach has also been considered. In this

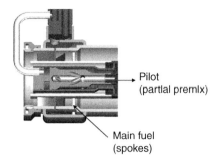

Pilot
(partial premlx)

Main fuel
(spokes)

Figure 5.25 Typical piloted lean premixed injector.

case, the fuel is injected into the main combustion chamber, and rapid mixing must occur with air inside the chamber as well. As long as space is available within which this high mixing is achievable, this approach can mitigate some of the concerns found with traditional premixers. A common way to accomplish this rapid mixing is with multi-point injection within which the scale of injection is small enough to preclude combustion within the basic flow passages (Tacina *et al.*, 2004; Lee *et al.*, 2006). An example is shown in Figure 5.26 which has demonstrated very low emissions for aviation applications as discussed in Section 5.3.2.2.

5.3.1.2. Heterogeneous Strategies

Heterogeneous strategies involve the physical placement of a solid material that is either reactive in nature or that serves as a source of thermal energy to sustain or promote the reaction while remaining unchanged in the process. These strategies provide an alternative means to stabilizing the reaction because they help overcome the activation energy needed to sustain the reaction. Figure 5.27 illustrates a conceptualization of the strategies that reduce the activation energy needed to sustain the reaction in contrast to conventional combustion and homogeneous lean combustion strategies.

Catalytic combustion is one of the most promising low NO_x technologies currently under development. By using a catalyst to promote fuel oxidation, catalytic combustion can sustain stable combustion at equivalence ratios far lower than conventional flame combustors. Ideally, this extremely lean combustion avoids the high reaction temperatures where large levels of NO_x are produced. Figure 5.27 illustrates how catalytic combustion can achieve very low levels of NO_x emission. In order to extract the chemical energy stored in fuel molecules, they must be energized beyond an activation energy. Normally, the needed energy comes from an arc produced by an electrical igniter or the re-circulation of hot combusted gas. As with other vortex breakdown stabilized systems, conventional gas turbine combustors recirculate hot products to heat and ignite the incoming fresh reactants, resulting in a highly stable reaction with high heat release. The penalty associated with this stability is high temperatures which can cause excessive NO_x formation.

Homogeneous lean pre-mixed systems approach the minimum limit of recirculated enthalpy and generally operate in a very narrow window adjacent to the temperatures at which high levels of NO_x form rapidly. However, through the use of a third body intermediary, it is possible to sustain stable combustion below this temperature, resulting in ultra-low NO_x levels.

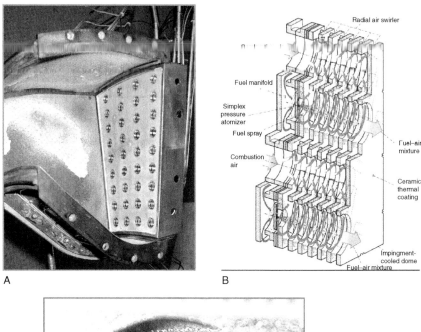

Figure 5.26 Illustration of multipoint injection for lean direct injection (Tacina *et al.*, 2004): (A) sector test rig, (B) schematic of flows, and (C) individual injection point.

Since catalytic combustion operates at much lower temperatures than conventional flame systems, a misconception exists that this type of combustion might be less efficient. Thermal efficiency increases with turbine inlet temperature. To maintain a given amount of power at a set pressure ratio, a lower turbine inlet temperature requires more airflow. Since compressors, combustors, and turbines all have losses, forcing these turbo-machineries to drive more mass would equate to having a lower overall system efficiency. However, as mentioned in Section 5.1.1, material limitations require dilution of the combustion products with a large amount of compressor discharge air. Since catalytic combustion can sustain stable combustion at conventional turbine inlet temperatures,

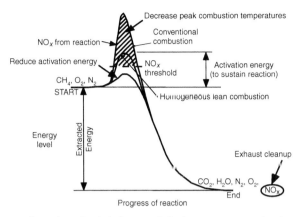

Figure 5.27 Illustration of technical approach for heterogeneous combustion strategies.

no dilution air is necessary. Because both systems have the same turbine inlet temperature, the overall system efficiency of the catalytic system could be higher since less air must be compressed. Figure 5.28 illustrates this point. Note that the turbine inlet temperature is the same because a conventional system required dilution.

Like all combustion systems, maintaining high efficiency and low emissions at various loads and ambient conditions is challenging due to the varying factors involved (pressure, velocity, mixing, residence time, inlet temperature, fuel/air ratio, etc.). This requirement drastically increases in complexity with catalytic combustion because of the operational requirements of catalytic reaction. Similar to conventional systems, catalytic combustion needs to implement advanced staging and control systems to operate over these different ranges. Some common catalytic combustor designs include single full

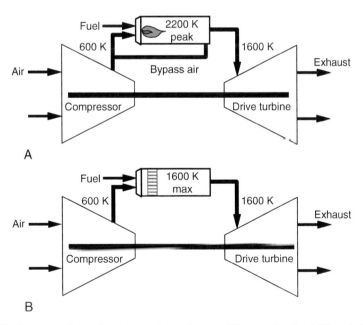

Figure 5.28 Comparing flow splits in the combustor between (A) conventional and (B) catalytic systems.

reaction catalysts, multistage different temperature catalysts, fuel/air staging before and after catalysts, and catalysts with burn out zone. One possible strategy is displayed in Figure 5.29, which is referred to as a staged catalytic approach and one that has been pursued in recent years as a commercial development for gas turbine combustion.

In this design, a preheated fuel and air mixture is fed into the catalyst bed where heterogeneous combustion begins. The first stage provides low light off temperatures while the second stage brings the mixture up to temperatures sufficient to allow the homogeneous reactions downstream to proceed to completion in a reasonable time. The second stage is usually a different type of catalyst more suited to higher temperatures. A gas phase burnout section lies downstream of the catalyst to allow time for any remaining CO and hydrocarbon species to fully oxidize. Tests on full-scale combustors of this type have demonstrated emissions of 5 ppm NO_x and 10 ppm CO (Cutrone *et al.*, 1999). Combustors using this approach have been commercialized to some degree, but still suffer from limitations related to the operating window of the catalytic material and the durability of the substrate on which it is mounted.

While the NO_x benefit is obvious for catalytic combustion, limitations and hurdles still need to be addressed before the technology can be widely adopted. One of these limitations is the light off temperature of the catalyst. If the temperature of the incoming fuel and air mixture is not sufficient (less than 400°C for PdO, Paladium Oxide), then no catalytic activity can occur. This temperature requirement is usually not a problem at full power because of the high compression ratio or recuperation. However, during engine startup and partial loads, the compressor or recuperator discharge temperature can drop below the catalyst light off temperature. Even a PdO catalyst, which possesses the lowest light off temperature of all the candidate catalysts for lean methane/air mixtures, still requires a temperature around 400°C. As a result, the engine may require a preburner upstream of the catalyst for these off-design conditions. However, a preburner increases the complexity of the engine and contributes to pollutant emissions counteracting the benefit of catalytic combustion.

The other principal limitation of catalytic combustion is the durability of the catalyst and its support structure. In conventional combustion systems, the maximum temperature occurs in an aerodynamic flow away from any structures. However, in a catalytic system, the catalyst and its supporting structure have to withstand the reaction temperature. Some possible failure modes in catalytic systems are thermal stresses leading to mechanical failure, oxidation and/or phase change of metallic supports, and sintering.

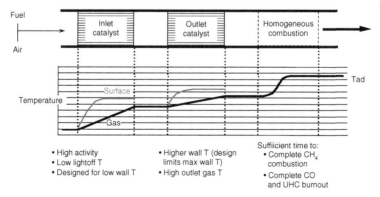

Figure 5.29 Two-stage catalytic combustor with temperature profile along the centerline.

Sintering is a process where the metal substrate starts combining with the catalyst wash coat resulting in a reduction in the available reaction surface area.

Combining the material limitations and light off requirements, a catalytic combustion system must meet the following operating requirements:

- The fuel and air must be highly premixed with a uniform velocity profile to prevent local overheating (Lefebvre, 1999).
- The inlet temperature must be sufficiently high to ensure catalyst light off (Cutrone *et al.*, 1999).
- The temperature on the catalyst monolith must be kept below the limits where the failure mechanisms listed above can destroy it, but must be high enough that gas phase reactions downstream will proceed to completion (Lefebvre, 1999).

Figure 5.30 illustrates these limitations. As shown, the catalyst bed will only function in a small operating window. Nevertheless, at least one commercial implementation of the approach shown in Figure 3.29 has occurred, with Kawasaki offering it as an ultra-low emissions (<3 ppm NO_x) option for their 1.4 MW M1A-13X stationary gas turbine.

It is worth noting that an alternative to operating the catalytic combustor lean is to operate it fuel rich and then introduce additional air downstream of the catalytic reactor to complete the oxidation of fuel and to achieve the final temperature desired. While this concept is beyond the scope of this chapter, it is an approach that has been successfully demonstrated in gas turbine applications and appears to overcome some of the catalytic combustion limitations outlined above (Smith *et al.*, 2003). In particular, a lower light off temperature is possible, and non-uniformities in the fuel/air mixture entering the catalytic reactor will be less likely to cause thermal runaway because the reaction will be oxygen limited at some point in the bed. Of course, gross non-uniformities will still cause problems.

Similar to catalytic combustion, surface stabilized combustion uses an external object to sustain a stable flame. However, unlike catalytic combustion, the external object does not play a part in the reaction. Instead, it is used as both a thermal mass and a flow separator/guide. One surface stabilizer currently under development utilizes a "checker shape" design with alternating regions of porous and perforated metal (Weakley *et al.*, 2002). The porous zone is where stable surface reactions occur. Premixed fuel and air flows through the porous fiber where it is heated. The stabilizer is designed so that the mixture will absorb just enough heat going through the fiber that combustion will take place on the exiting surface. Combustion continues into the burnout zone with the maximum temperature occurring slightly beyond the face. Figure 5.31 illustrates this concept.

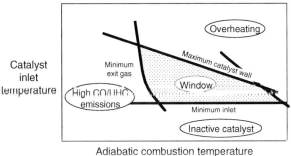

Figure 5.30 The operation window for catalytic combustion.

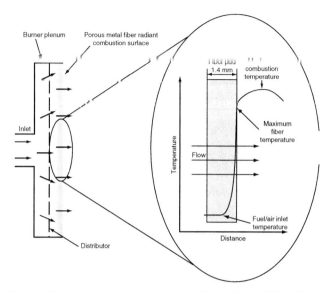

Figure 5.31   Principle of operation for the porous section of the surface stabilizer (Weakley *et al.*, 2002).

The heat transfer and diffusion of combustion products back to the surface sustains stable combustion even at extremely low combustion temperatures. Under higher loads, the porous surface would not be able to sustain a stable flame because of the flame detachment or "lift off" due to the large flow rate. For this reason, the perforated sections were installed between the porous surfaces. This divides the large flow rate into localized regions of high (perforated) and low (porous) mass flux. A stretched laminar flame is anchored on the perforated section due to the high flow velocity (stretching) and the hot porous surface supplying a continuous pool of radicals (anchored) (Weakley *et al.*, 2002). Figure 5.32 presents an illustration of one burner based on this approach. The variation in the surface seen in Figure 5.32b reflects the porous and perforated region. Figure 5.33 illustrates a prototype design by Alzeta Corporation which has been demonstrated in rig tests and is being considered for commercialization (Weakley *et al.*, 2002). The advantages of applying a surface stabilized system over a conventional lean premixed one are as follows:

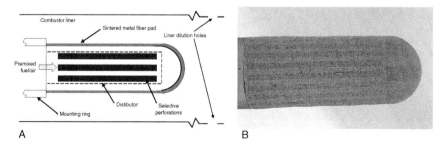

Figure 5.32   Surface stabilized gas turbine burner: (A) schematic and (B) as fabricated (Greenberg *et al.*, 2005).

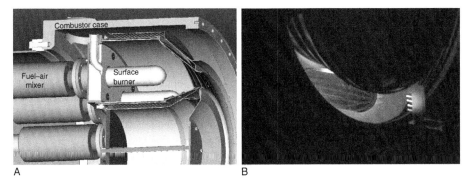

Figure 5.33 Illustration of the porous/perforated design of a surface stabilized combustion system (Images Courtesy of Alzeta Corporation): (A) schematic illustrating installation and (B) rig test. (See color insert.)

- The increase in surface area due to the stretching of the flame allows rapid lowering of the high flame temperature in the burnout stage.
- The thinning of the flame brought about by the stretching reduces the residence time in the high flame temperature zone.

Combining these two features with a large burnout zone results in low NO_x and CO emissions.

5.3.1.3. Current Performance

Based on the above discussion of advanced low emissions combustion technology for natural gas, the question of how low emissions can be taken, posed originally in Section 5.2, is worth revisiting. The results in Figure 5.34 compare recently reported results for a number of advanced lean combustion technologies along with the rich burn catalytic lean

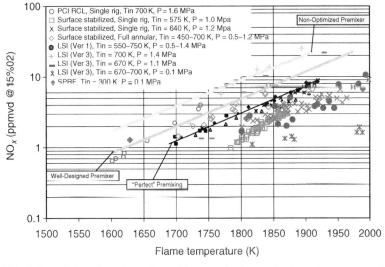

Figure 5.34 NO_x Emissions for various advanced lean premixed strategies for stationary power. (See color insert.)

burn strategy. These results suggest that the entitlement for NO_x levels needs to be adjusted, and also reveal how much progress has been made in lean combustion for gas turbines.

5.3.2. AVIATION

As mentioned previously, aviation engines require great care relative to safety, so emissions issues are secondary. Nonetheless, emissions reduction has been pursued due to increasing concerns for air quality. National Aeronautics and Space Administration (NASA) has engaged in numerous low emissions developmental efforts for aviation engines. In the 1980s, the race to develop the next generation High Speed Civil Transport (HSCT) resulted in considerable effort on low emissions engines. Many approaches were considered, including rich burn quick mix, lean burn (RQL), LPP, and lean direct injection (LDI). In the early 1990s, NASA decided to direct its attention to LDI concepts. However, Pratt & Whitney has continued to evolve rich burn dome strategies for its commercial engines. This is attractive because the stability of this type of combustion system is inherently high.

On the other hand, GE and Rolls-Royce have focused on lean combustion approaches, although they generally stop short of going to LPP operation. It is generally accepted that LPP combustion should provide the lowest possible emissions levels, which is why developmental work continues in Europe (Bittlinger and Brehmn, 1999; Ripplinger *et al.*, 1999; Behre *et al.*, 2003; Charest *et al.*, 2006) and Japan (Hayashi *et al.*, 2005). However, because of the higher pressure ratios and temperatures found in advanced aeroengines as well as their wide-ranging operational requirements, concerns regarding flashback, autoignition, and acoustic instabilities are substantial. As a result, LDI has been the focus of work in the United States (Tacina, 1983).

Lean direct injection appears to have tremendous potential for emissions reductions. Work on test rigs has shown that LDI combustion can approach LPP performance for NO_x emissions (Lee *et al.*, 2001; Shaffar and Samuelsen, 1998), and more recent work by Tacina *et al.* (2004) has confirmed this result. These successes have been facilitated by recent breakthroughs in manufacturing strategies that have allowed the "million injection point" approach, which has been demonstrated for gaseous fuels to be applied to liquid fueled systems.

Early NASA programs (e.g., C^3) considered the use of catalytic combustion for aeroengines. Figure 5.35 shows an example of one of the concepts developed. At the time, the technology did not appear viable for commercialization. However, given all the developments in catalytic combustion in recent years, there may be an opportunity to revisit this strategy.

Similarly, to the situation with stationary combustion systems, a question concerning the lowest achievable NO_x levels has been raised for aviation systems. Figure 5.36 illustrates a comparison of the natural gas fired entitlement plot compared with data for a well-stirred reactor (Lee *et al.*, 2001), a Capstone C-30 microturbine (Nakamura *et al.*, 2006) that operates in an LPP manner, and the Parker Hannifin LDI multipoint injector (Mansour, 2005). As shown, this recent fuel injection technology is approaching the "entitlement" that would be expected in the case of "perfect premixing" (as suggested by a well-stirred reactor with 10–20 ms residence times).

For aeroengines, Figure 5.37 presents a summary of where current technology is today relative to current emissions regulations and targets in terms of the NO_x emission

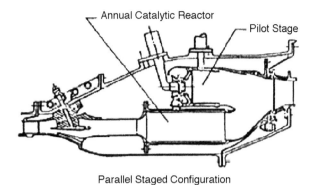

Parallel Staged Configuration

Figure 5.35 NASA C^3 program catalytic aircraft engine combustor configuration.

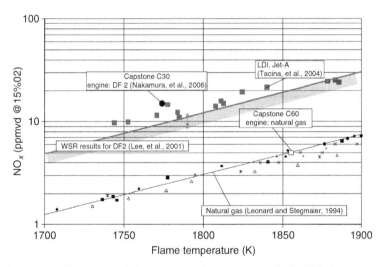

Figure 5.36 Comparison of advanced low emissions strategies for liquid fueled systems.

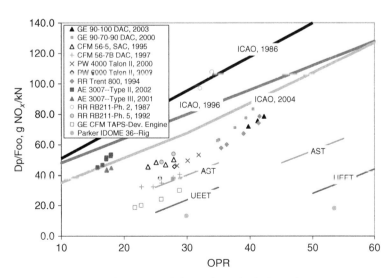

Figure 5.37 Comparison of advanced aviation engine technologies and NASA program targets.

characteristic, *Dp/Foo* (the ratio of mass of NO_x in grams to engine maximum takeoff thrust in kN), versus the engine overall pressure ratio (OPR). The most aggressive targets (NASA Ultra Efficient Engine Technology – UEET) are either approached or exceeded with the latest technology enhancements: either the GE "twin annular pre-mixing swirler—TAPS" or the Parker LDI technology. Note that the data for both of these systems is not engine certification data, but rather results obtained on test rigs. Nonetheless, it shows that technology improvements are continuing to achieve further reductions in NO_x emissions.

5.4. SUMMARY

Tremendous progress has been made in the area of lean combustion as applied to gas turbines. This has been a principal strategy for achieving ultra-low emissions performance for both stationary and aviation engines. With the shift away from traditional "diffusion" type combustion systems, a wide variety of technologies that attempt to lower the peak combustion temperatures found in the reaction zone have appeared. Relative to power generation, these approaches (several of which appear in other chapters and lean combustion technologies as well) have been included:

- Lean premixed combustion
  - Homogeneous
  - swirl stabilized (with vortex breakdown)
  - swirl stabilized (without vortex breakdown)
  - dilute combustion (aka MILD combustion, flameless oxidation, flameless combustion, etc.)
- Heterogeneous
  - Catalytic
  - surface stabilized

Relative to aviation engines, the need for reliability and safety has led to a focus on LDI of liquid fuels. However, the viability of LPP continues to be explored as a low emissions strategy.

Emissions reductions have been substantial and are now considered a basic design constraint that must be met regardless of all other aspects. To that end, emissions targets must always be met, but in doing so, the operability of the combustion system can become the limiting factor. As a result, improved stability (both static and dynamic), avoidance of flashback and autoignition, good ignition characteristics, and good turn down are all key design aspects. Relative to future needs, the shift from fossil fuels to alternative fuels coupled with low emissions requirements will call for increased understanding of the fundamental behavior of these potential fuels. Understanding how fuel composition affects flame speeds, reaction rates, and kinetics is one important requirement. Additionally, the desire for higher firing temperatures in order to achieve higher efficiencies means that further research is needed to fill the gaps in our comprehension of lean-burn gas turbine combustion.

REFERENCES

Beerer, D.J., Greene, M.U., McDonell, V.G., and Samuelsen, G.S. (2006). Correlation of Ignition Delay with IGCC and Natural Gas Fuels, Final Report, Contract 03-01-SR112, Prepared for South Carolina Institute for Energy Studies.

Behrendt, T., Heinze, J., and Hassa, C. (2003). Experimental investigation of a new LPP injector concept for aero engines and elevated pressures, Paper GT2003-38444. *Proceedings of ASME Turbo Expo 2003*, Atlanta.

Bernier, D., Lacas, F., and Candel, S. (2004). Instability mechanisms in a premixed prevaporized combustor. *J. Prop. Power* **20**(4), 648–657.

Bittlinger, G. and Brehmn, N. (1999). High Pressure Combustion Test of lean Premixed Prevaporized (LPP) Modules in an Axially Staged combustor using a MultiSector Rig, IGADE Paper 99-8008

Borbeley, A.M. and Kreider, J.F. (Eds.) (2001). *Distributed Generation: The Power Paradigm for the New Millennium*. CRC Press, Boca Raton.

California Air Resources Board (CARB) Small Generator Emissions Certification Program (2000). Senate Bill 1298.

Candel, S. (2002). Combustion dynamics and control: Progress and challenges. *Proc. Combust. Inst.* **29**, 1–28.

Charest, M.R.J., Gauthier, J.E.D., and Huang, X. (2006). Design of a lean premixed prevaporized can combustor, *ASME Turbo EXPO 2006*, Barcelona.

Chen, D.Y. (2004). Lean staged Pyrospin combustor for next generation APU. *Presented at 2nd Engineering Foundation Workshop on Lean Burn Combustion*, Tomar.

Chen, J.H., McDonell, V.G., and Samuelsen, G.S. (2004). Effects of ethane and propane additives on the autoignition behavior of natural gas fuels. *Spring Meeting of the Western States Section/The Combustion Institute*, Paper 04S-24, submitted to *J. Prop. Power*.

Cohen, J.M. and Banaszuk, A. (2003). Factors affecting the control of unstable combustors. *J. Prop. Power* **19**(5), 811–821.

Coker, A., Neumeier, Y., Zinn, B.T., Menon, S., and Lieuwen, T. (2006). Active instability control effectiveness in a liquid fueled combustor. *Combust. Sci. Technol.* **178**(7), 1251–1261.

Coordinating Research Council (CRC) (2004). *Handbook of Aviation Fuel Properties*, Report No. 635, 3rd edition.

Cutrone, M.B., Beebe, K.W., Dalla Betta, R.A., Schlatter, J.C., Nickolas, S.G., and Tsuchiya, T. (1999). Development of a catalytic combustor for a heavy duty utility gas turbine. *Catal. Today* **47**, 391–398.

Davu, D., Franco, R., Choudhuri, A., and Lewis, R. (2005). Investigation on Flashback Propensity of Syngas Premixed Flames, Paper AIAA 2005–3585, Reno.

Diesel and Gas Turbine Worldwide (DGTW) (2005). 29th Annual Power Generation Survey, www.dieselpub.com.

Docquier, N. and Candel, S. (2002). Combustion control and sensors: A review. *Prog. Energy Combust. Sci.* **28**(2), 107–150.

Dowling, A.P. and Stow, S.R. (2003). Acoustical analysis of gas turbine combustors. *J. Prop. Power* **19**(5), 751–764.

Ducruix, S., Thierry, S., Durox, D., and Candel, S. (2003). Combustion dynamics and instabilities: Elementary coupling and driving mechanisms. *J. Prop. Power* **19**(5), 122–134.

Fable, S.E. and Cheng, R.K. (2001). Optimization of a premixed low-swirl burner for industrial applications. *2nd Joint Meeting of the US Sections of the Combustion Institute*, Paper 109.

Feigl, M., Setzer, F, Feigl-Varela, R., Myers, G., and Sweet, B. (2005). Field test validation of the DLN2.5H combustion system on the 9H gas turbine at Baglan Bay power station, Paper GT2005–68843. *ASME Turbo Expo 2005*, Reno.

Flores, R.M., McDonell, V.G., and Samuelsen, G.S. (2003). Impact of ethane and propane variation in natural gas on the performance of a model gas turbine combustor. *ASME J. Eng Gas Turb. Power* **125**(3), 701–708.

Fric, T.F. (1993). Effects of fuel–air unmixedness on NO_x emissions. *J. Prop. Power* **9**(5), 708–713.

Greenberg, S., McDougald, N.K., Weakley, C.K., Kendall, R.M., and Arellano, L.O. (2005). Surface stabilized fuel injectors with sub-three PPM NO_x emissions for a 5.5 MW gas turbine engine. *J. Eng. Gas Turb. Power* **127**(2), 276–285.

Guin, C. (1998). Characterization of autoignition and flashback in premixed injection systems. *AVT Symposium on Gas Turbine Engine Combustion, Emissions, and Alternative Fuels*, Lisbon.

Hayashi, S., Yamada, H., and Makida, M. (2005). Extending low NO_x operating range of a lean premixed-prevaporized gas turbine combustor by reaction of secondary mixtures injected into primary stage burned gas. *Proc. Combust. Inst.* **30**, 2903–2911.

Hermann, J., Orthmann, A., Hoffmann, S., and Berenbrink, P. (2001), *Proceedings, NATO RTO MP 051, Active Control Technology for Enhanced Performance Operational Capabilities of Military, Aircraft, Land Vehicles, and Sea Vehicles*. Neuilly-Sur-Seine Cedex, France.

Hoffman, S., Habisreuther, P., and Lenze, B. (1994). Development and assessment of correlations for predicting stability limits of swirling flames. *Chem. Eng. Proc.* **33**, 393–400.

Janus, M.C., Richards, G.A, Yip, M.J., and Robey, E.H. (1997). Effects of ambient conditions and fuel composition on combustion stability, Paper ASME 97-GT-266. *Turbo Expo 1997*, Orlando, FL.

Kalb, J.R., Hirsch, C., and Sattelmayer, T. (2006). Operational characteristics of a premixed sub-ppm NO_x burner with periodic recirculation of combustion products, Paper GT2006–90072. *ASME Turbo Expo 2006*, Barcelona.

Karim, H., Lyle, K., Etemad, S., Smith, L.L., Pfefferle, W.C., Dutta, P., and Smith, K.O. (2003). Advanced catalytic pilot for low NO_x industrial gas turbines. *J. Eng. Gas Turb. Power* **125**, 879–884.

Kido, H., Nakahara, M., Nakashima, K., and Hashimoto, J. (2002). Influence of local flame displacement velocity on turbulent burning velocity. *Proc. Combust. Inst.* **29**, 1855–1861.

Kroner, M., Fritze, J., and Sattelmayer (2002). Flashback limits for combustion induced vortex breakdown in a swirl burner, Paper GT-2002–30075. *Proceedings of Turbo Expo 2002*, Amsterdam.

Lee, J.C.Y., Malte, P.C., and Benjamin, M.A. (2001). Low NO_x combustion for liquid fuels: Atmospheric pressure experiments using a staged prevoprizer–premixer (2001), Paper 01-GT-0081. *Presented at the 46th ASME IGTI Conference*, Louisiana.

Lee, S., Svrcek, M., Edwards, C.F., and Bowman, C.T. (2006). Mesoscale burner arrays for gas-turbine reheat applications. *J. Prop. Power* **22**(2), 417–424.

Lefebvre, A.H. (1999). *Gas Turbine Combustion*. Taylor and Francis, Philadelphia.

Lefebvre, A.H. and Ballal, D.R. (1980). Weak extinction limits of turbulent heterogeneous fuel/air mixtures. *J. Eng. Power, Trans. ASME* **102**, 416.

Lefebvre, A.H. and Baxter, M.R. (1992). Weak extinction limits of large-scale flameholders. *J. Eng. Power, Trans. ASME* **114**, 777.

Lefebvre, A.H., Freeman, W., and Cowell, L. (1986). Spontaneous Ignition Delay Characteristics of Hydrocarbon Fuel/Air Mixtures, NASA Contractor Report 175064.

Leonard, G. and Stegmaier, J. (1994). Development of an aeroderivative gas turbine dry low emission combustion system. *J. Eng. Gas Turb. Power* **116**, 542–546.

Leonard, P.A. and Mellor, A.M. (1981). Lean blowoff in high-intensity combustion with dominant fuel spray effects. *Combust. Flame* **42**, 93–100.

Li, S-C. and Williams, F.A. (2002). Reaction Mechanism for Methane Ignition. *J. Eng. Gas Turb. Power* **124**(3), 471–480.

Lieuwen, T. (2003). Modeling premixed combustion – acoustic wave interactions: A review. *J. Prop. Power* **19**(5), 765–781.

Lieuwen, T. and McManus, K. (2003). Combustion dynamics in lean premixed prevaporized (LPP) gas turbines. *J. Prop. Power* **19**(5), 721–846.

Lieuwen, T. and Yang, V. (2006). Combustion instabilities in gas turbine engines: Operational experience, fundamental mechanisms, and modeling. *Prog. Astro. Aero.* **210**.

Lieuwen, T., McDonell, V., Petersen, E., and Santavicca, D. (2006). Fuel flexibility influences on premixed combustor blowout, flashback, autoignition, and stability, Paper GT2006-90770. *ASME Turbo EXPO 2006*, Barcelona. *J. Eng. Gas Turb. Power*, to be published.

Lyons, V.J. (1982). Fuel/air nonuniformity – effect on nitric oxide emissions. *AIAA J.* **20**(5), 660–665.

Mansour, A. (2005). Gas turbine fuel injection technology, Paper GT2005-68173. *ASME Turbo EXPO 2005*, Reno.

McDonell, V.G., Couch, P.M., Samuelsen, G.S., Corr, R., and Weakley, C.K. (2001). Characterization of flameholding tendencies in premixer passages for gas turbine applications. *2nd Joint Meeting of the US Sections of the Combustion Institute*, Oakland.

McManus, K., Poinsot, T., and Candel, S.M. (1993). A review of active control of combustion instabilities. *Prog. Energy Combust. Sci.* **19**, 1–29.

Meier, J.G., Hung, W.S.Y., and Sood, V.M. (1999). Development and application of industrial gas turbines for medium-BTU gaseous fuels. *J. Eng. Gas Turb. Power* **108**, 182–190.

Mongia, H.C., Held, T.J., Hsiao, G.C., and Pandalai, R.P. (2003). Challenges and progress in controlling dynamics in gas turbine combustors. *AIAA J. Prop. Power* **19**, 822–829.

Myers, G.D., Tegel, D., Feigl, M., and Setzer, F. (2003). Dry low emissions for the "H" heavy-duty industrial gas turbine: Full-scale combustion and system rig tests, Paper GT2003-38192. *ASME Turbo Expo 2003*, Atlanta.

Nakamura, S., McDonell, V.G., and Samuelsen, G.S. (2006). The effect of liquid-fuel preparation on gas turbine emissions. *ASME J. Eng. Gas Turb. Power*, to be published.

Neumeier, Y., Waksler, Y., Zinn, B.T., Seitzman, J.M., Jagoda, J., and Kenny, J. (2005). Ultralow emissions combustor with non-premixed reactants injection, Paper AIAA 2005-3775. *41st Joint Propulsion Conference*, Tucson.

Nguyen, O.M. and Samuelsen, G.S. (1999). The effect of discrete pilot hydrogen Dopant injection on the lean blowout performance of a model gas turbine combustor, Paper 99-GT-359. *Presented at the 44th ASME International Gas Turbine and Aeroengine Congress and Exposition*, Indianapolis.

Noble, D.R., Zhang, Q., Shareef, A., Tootle, J., Meyers, A., and Lieuwen, T. (2006). Syngas mixture composition effects upon flashback and blowout, Paper GT2006-90470. *ASME Turbo Expo 2006*, Barcelona.

Phi, V.M., Mauzey, J.L., McDonell, V.G., and Samuelsen, G.S. (2004). Fuel injection and emissions characteristics of a commercial microturbine generator, Paper GT-2004-54039. *ASME TurboExpo 2004*, Vienna.

Plee, S.L. and Mellor, A.M. (1979). Characteristic time correlation for lean blowoff of bluff-body stabilized flames. *Combust. Flame* **35**, 61–80.

Pratt, D.T., Allwine, K.J., and Malte, P.C. (1974). Hydrogen as a turbojet engine fuel – technical, economical, and environmental impact. *2nd International Symposium on Air Breathing Engines (ISABE)*, Sheffield.

Pritchard, J. (2003). H-system technology update, Paper GT2003-38711. *ASME Turbo Expo 2003*, Atlanta.

Putnam, A.A. and Jensen, R.A. (1949). Application of dimensionless numbers to flashback and other combustion phenomena. *3rd International Symposium on Combustion*, pp. 89–98.

Ribaucour, M., Minetti, R., Sochet, L.R., Curran, H.J., Pitz, W.J., and Westbrook, C.K. (2000). Ignition of isomers of pentane: An experimental and kinetic modeling study. *Proc. Combust. Inst.* **28**, 1671–1678.

Richards, G.A., McMillian, M.M., Gemmen, R.S., Rogers, W.A., and Cully, S.R. (2001). Issues for low-emission, fuel-flexible power system. *Prog. Energy Comb. Sci.* **27**, 141–169.

Richards, G.A., Straub, D., and Robey, E.H. (2003). Passive control of combustion dynamics in stationary gas turbines. *J. Prop and Power* **19**(5), 795–810.

Ripplinger, T., Zarzalis, N., Meikis, G., Hassa, C., and Brandt, M. (1999). NO_x reduction by lean premixed prevaporized combustion. *Proceedings on Gas Turbine Engine Combustion, Emissions, and Alternative Fuels*, RTO-MP-14, pp. 7.1–7.12, ISBN 92-837-0009-0, Lisbon.

Rohde, J. (2002). Overview of the NASA AST and UEET emissions reduction projects. *UC Tech Transfer Symposium*, San Diego.

Samuelsen, G.S. (2005). UCI Gas Turbine Combustion Short Course Lecture Notes.

Scheffer, R.W., Wicksall, D.M., and Agrawal, A.K. (2002). Combustion of hydrogen-enriched methane in a lean premixed swirl-stabilized burner. *Proc. Combust. Inst.* **29**, 843–851.

Shaffar, S.W. and Samuelsen, G.S. (1998). A liquid fueled, lean burn, gas turbine combustor injector. *Combust. Sci. Technol.* **139**, 41–57.

Sirignano, W.A. and Liu, F. (1999). Performance increases for gas-turbine engines through combustion inside the turbine. *J. Prop. Power* **15**(1), 111–118.

Smith, G. P. *et al.*, http://www.me.berkeley.edu/gri_mech/.

Smith, L.L., Karim, H., Castaldi, M.J., Etemad, S., Pfefferle, W.C., Khanna, V.K., and Smith, K.O. (2003). Rich-catalytic lean-burn combustion for low-single digit NO_x gas turbines, Paper GT-2003-38129. *Presented at ASME Turbo Expo 2003*, Atlanta.

Sturgess, G.J. (2003). Turbine burner for near-constant temperature cycle gas turbine engines: Application to large commercial subsonic transport engine – a progress report. *Workshop on Inter-Turbine Burning Engines*, Cleveland.

Sturgess, G.J., Zelina, J., Shouse, D.T., and Roquemore, W.M. (2005). Emissions reduction technologies for military gas turbine engines. *J. Prop. Power* **21**(2), 193–217.

Svennson, F., Hasselrot, A., and Moldanova, J. (2004). Reduced environmental impact by lowered cruise altitude for liquid hydrogen-fuelled aircraft. *Aero. Sci. Technol.* **8**, 307–320.

Tacina, R. (1983). Autoignition in a premixing–prevaporizing fuel duct using three different fuel injection systems at inlet air temperatures to 1250 K. *NASA Technical Memorandum 049909320.*

Tacina, R. (1990). Low NO$_x$ potential of gas turbine engines, Paper AIAA-90-0550. *28th Aerospace Sciences Meeting,* Reno.

Tacina, R., Mansour, A., Partelow, L., and Wey, C. (2004) Experimental sector and flame tube evaluations of a multipoint integrated module concept for low emission combustors, Paper GT2004-53263. *Turbo Expo 2004,* Vienna.

Thibaut, D. and Candel, S. (1998). Numerical study of unsteady turbulent premixed combustion: Application to flashback simulation. *Combust. Flame* **113**, 53–65.

Tratham, S. (2001). *AGTSR Combustion Workshop 8,* Charleston.

United States Department of Energy (US DOE) (2006). Energy information administration. *Annual Energy Outlook 2006 Electricity Forecast.*

Weakley, C.K., Greenberg, S.J., Kendall, R.M., and McDougald, N.K. (2002). Development of surface-stabilized fuel injectors with sub-three PPM NO$_x$ emissions, *International Joint Power Generation Conference,* IJPGC2002-26088.

Yoshimura, T., McDonell, V.G., and Samuelsen, G.S. (2005). Evaluation of hydrogen addition to natural gas on the stability and emissions behavior of a model gas turbine combustor, Paper GT2005-68785. *ASME Turbo Expo 2005,* Reno.

Zhang, Q., Noble, D.R., Meyers, A., Xu, K., and Lieuwen, T. (2005). Characterization of fuel composition effects in H$_2$/CO/CH$_4$ mixtures upon lean blowout, Paper GT2005-68907. *Proceedings of Turbo EXPO 2005.*

Chapter 6

Lean Premixed Burners

Robert K. Cheng and Howard Levinsky

Nomenclature

A	Stoichiometric air factor for a gas
B	Thermal load of a burner
F	Volumetric flow rate
G_{ang}	Axial flux of angular momentum
G_{x}	Axial thrust
H	Higher heating value
K	Empirical correlation constant
L	Swirler recess distance
m	Mass flux ratio, $\dot{m}_{\mathrm{c}}/\dot{m}_{\mathrm{a}}$
$\dot{m}_{\mathrm{a}}$	Mass flux through the annulus
$\dot{m}_{\mathrm{c}}$	Mass flux through the centerbody
q'	Two-component turbulent kinetic energy, $[(u'^2 + v'^2)/2]^{1/2}$
r	Radius
R	Ratio of centerbody-to-burner radii, $R_{\mathrm{c}}/R_{\mathrm{b}}$
R_{b}	Burner radius
R_{c}	Center channel radius
Re	Reynolds number
S	Swirl number
S_{L}	Laminar flame speed
S_{T}	Turbulent flame speed
u'	Turbulent rms velocity in axial direction
U	Axial velocity
U_{a}	Mean axial velocity through swirl annulus
U_{c}	Mean axial velocity through the center core
U_{o}	Bulk axial flow velocity
v'	Turbulent rms velocity in radial direction
V	Volumetric flow of air and radial velocity
W	Wobbe Index and velocity in radial direction
x_{f}	Leading edge position of the flame brush
x_{o}	Virtual origin of the linearly divergent portions of the axial profiles
Y	Blockage of center channel screen
α	Blade angle

ρ Gas density
ϕ Equivalence ratio
ϕ_{LBO} Equivalence ratio at lean blow off

6.1. INTRODUCTION

Interest in lean premixed burners as an attractive alternative for "state-of-the-art" practical combustion systems began in the early 1980s. In addition to their common application in NO_x control strategies (Bowman, 1992), these burners were independently introduced in high-efficiency condensing boilers in domestic central heating appliances as a method for the efficient use of the fan, which is necessary to overcome the pressure drop in the condensing heat exchanger. In this sense, these appliances were "low emission" before NO_x emission standards ever existed. In this chapter, we describe a number of developments in the design of lean premixed burners for use in heating applications. As will be discussed, lean premixing presents two challenges to burner design: achieving acceptable flame stability (flashback/blow-off) over the desired range of turndown ratio and maintaining stability with varying fuel composition. Although the effects of fuel variability are well known in gas engines and premixed gas turbines, the analogous principles governing the operation of burner systems are often neglected. Chapter 2 discusses the foundations of some of these principles in general. We shall review them briefly in the particular context germane to lean premixed burners.

6.2. PRINCIPLES OF FUEL VARIABILITY

For the purposes of this chapter, we will only consider the effects of possible variations in fuel composition using natural gas as a fuel because it is by far the most prevalent. For other gaseous fuels, the extension of the basic principles is straightforward. Most of the principles can be traced to older literature (*AGA Bulletin* #10, 1940; von Elbe and Grumer, 1948; Harris and South, 1978) and are general for all burners having some degree of premixing.

Although natural gas is predominately methane with lesser amounts of higher hydrocarbons and inerts, the natural variation in gas composition can have large effects on burner performance. The two largest consequences of this variability pertain to the thermal input and primary aeration of a burner. At a fixed thermal load, fuel gas is supplied in the vast majority of burners at constant pressure drop across a restriction. Since the thermal load of a burner is, by definition, $B = HF$, where H is the calorific or "higher heating" value of the gas (say in MJ/m^3) and F is the volumetric flow rate, substituting two gases (1 and 2) at constant pressure changes the thermal load according to:

$$\frac{B_1}{B_2} = \frac{H_1}{H_2}\frac{F_1}{F_2}. \tag{6.1}$$

Recognizing that at constant pressure F is inversely proportional to the square root of the density,[1] this suggests that $B \propto H/\sqrt{\rho}$, where ρ is the density of the gas. This relation is used to define the so-called Wobbe Index (Wobbe, 1926) or Wobbe Number,

---

[1] Here, we recall that $F \propto \sqrt{2\Delta P/\rho}$, where ΔP is the pressure drop over the restriction.

$W = H/\sqrt{\rho}$, where in practice the relative density of the gas (relative to air) is used, giving the Wobbe Index units of energy/volume, similar to the calorific value. Thus, the thermal input to a burner is directly proportional to the Wobbe Index of the gas, not its higher heating value.

When characterizing the system in terms of volumetric flows of fuel and air, we define the primary equivalence ratio, ϕ, for a given gas as $\phi = AF/V$, where A is the stoichiometric air factor for the gas (m³ air/m³ gas) and V is the volumetric flow of air. Whereas $F \propto 1/\sqrt{\rho}$, for V one can make the distinction between naturally aspirated burners (using a gas nozzle and venturi to entrain combustion air) and fan burners. For naturally aspirated systems, which are also used in some commercial systems to obtain lean equivalence ratios, the amount of air entrained (and thus the volumetric flow rate of primary air, V) is solely determined by the momentum flux of the gas jet (von Elbe and Grumer, 1948). The momentum flux of the gas is, to a good approximation, solely determined by the pressure drop over the nozzle. Since the gas pressure is usually not changed when substituting one gas for another, the amount of air entrained, and thus V, is independent of the gas composition (Harris and South, 1978). Therefore, since V is constant, for our two gases 1 and 2

$$\frac{\phi_1}{\phi_2} = \frac{A_1 F_1}{A_2 F_2}. \tag{6.2}$$

The calculation of variations in equivalence ratio with Wobbe Index for many natural gases shows that $\phi_1/\phi_2 = W_1/W_2$. This rather surprising result (although known in the older literature, see *AGA Bulletin* #10 (1940)) can be accounted for by recalling one of the peculiar properties of alkanes, that is, that the air factor divided by the calorific value is constant to within 0.7% for the major alkanes in natural gas. Thus, since $F \propto 1/\sqrt{\rho}$ and $A \propto H$, the right hand side of Equation (6.2) reduces to the ratio of the Wobbe Indices.

The vast majority of practical lean premixed burners use fans to supply the air to the burner. For those systems that do not regulate the air supply by means of an oxygen sensor, the air supply is also independent of the fuel composition, and the relation between Wobbe Index and equivalence ratio is valid. Further, even in industrial combustion equipment that does not employ premixed flames, the *overall* equivalence ratio still varies with Wobbe number, as described above.

Depending on the location, the Wobbe Index can vary by as much as 15% (Marcogaz Ad Hoc Working Group on Gas Quality, 2002). Thus, both the thermal input and the equivalence ratio can also vary by this amount. Obviously, variations of this magnitude can have significant consequences for the efficacy of heating processes, as well as for the stability and pollutant emissions from lean premixed systems (see, e.g., Chapter 2). For stability, the changes in burning velocity caused by the changes in equivalence ratio can create serious challenges to the design of the system in terms of blow-off and flashback.

6.3. STABILIZATION METHODS

6.3.1. HIGH-SWIRL FLAME STABILIZATION FOR INDUSTRIAL BURNERS

For industrial burners, the primary function of a flame holder is to maintain a stable flame that has high combustion efficiency and intensity throughout the range of turndown for load following. Turndown is the ratio of the peak-to-bottom power outputs of a given system. Except for some small domestic and commercial appliances where turndown can be achieved by turning the burner on and off intermittently, large industrial and utility systems modulate the power output of the burner (or burners) by adjusting the fuel and the air flows. Therefore, the critical role of the flame holder is to maintain a stable lean flame within a range of flow velocities as well as through the course of rapid changes in flow velocity during transitions from one load point to the next. Typically, industrial systems require minimum turndown of 5:1 (i.e., 100–25% load range), but many advanced processes need turndown approaching 20:1 to enhance system and process efficiencies.

Swirl is the predominant flow mechanism found in premixed and non-premixed combustion systems because it provides an effective means to control flame stability as well as the combustion intensity (Chigier and Beer, 1964; Davies and Beer, 1971; Beer and Chigier, 1972; Syred and Beer, 1974; Lilley, 1977; Gupta *et al.*, 1978). Until now, all practical systems utilized high swirl in which the swirling motion is sufficiently intense (attained beyond a critical swirl number, *S*, and Reynolds number, *Re*) to generate a large and stable central internal recirculation zone that is also known as the toroidal vortex core. Non-premixed burners use the recirculation zone for mixing the fuel and oxidizers to ensure stable and complete combustion. For premixed combustion, the role of the recirculation zone is similar to the wake region of a bluff body flame stabilizer where it retains a steady supply of radicals and hot combustion products to continuously ignite the fresh reactants. Another benefit of using swirling flows for flame stabilization is the generation of a compact flame with much higher combustion intensity than in a non-swirling situation. Syred and Beer (1974) gave an extensive review of the basic processes and practical implementation of two types of swirl combustors. For the sake of clarity, we shall refer to the type of swirling flows that produce strong and well-developed recirculation zones as high-swirl flows, and the burners designed or configured to produce high-swirl flows as high-swirl burners (HSB).

Numerous books, review articles, and a vast number of research papers in both the scientific and engineering literatures attest to the prominent role of high-swirl flows in combustion systems. The overwhelming majority of these studies concentrate on characterizing the size and strength of the recirculation zone as functions of Reynolds and swirl numbers to determine their effects on flame stability and combustion intensity. Flow structures within the recirculation bubble are shown to be highly complex with steep instantaneous velocity gradients producing high intensity non-isotropic shear turbulence. The recirculation bubble also exhibits spatial fluctuations and can swell and shrink from one instance to the next.

The complexity of the recirculation zone structure is a primary cause for the non-linear characteristics of flames generated by HSB. These non-linearities have lesser effects on non-premixed burners and partially premixed burners, but can be detrimental to premixed burners that operate at very fuel-lean conditions for NO_x control. In a typical lean premixed HSB, the flame becomes unstable when ϕ is decreased to a limiting value and eventually blows off with further reduction in ϕ. The instability limit and the lean blow-off limit (expressed in terms of the equivalence ratio at lean blow-off, ϕ_{LBO}) are non-linear functions of the Reynolds number. Therefore, maintaining low emissions at certain load points can require the burner to operate close to these limits where vortex shedding, mixture homogeneity, and acoustic feedback to the fuel or air systems can trigger larger flame instabilities and premature blow-off. The onset of flame instabilities when coupled with the acoustics mode of the combustion chamber often produces large

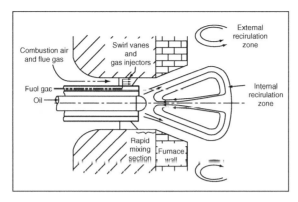

Figure 6.1   Schematic of RMB.

amplitude oscillations. If unmitigated or uncontrolled, combustion oscillations can lead to rapid hardware degradation and even catastrophic system failure. This sensitivity of lean premixed HSB designs near their lean limit has given rise to the expectation that lean combustion is inherently unstable. As a result, elucidating the basic processes responsible for initiating and sustaining combustion instabilities as well as the development of passive and active control methodologies for combustion oscillations are significant current topics of combustion research. Chapter 7 discusses such oscillations in the context of recirculation-stabilized flames. We will see, however, that appropriately designed fluid mechanics within the burner can substantially reduce sensitivity to lean operation.

Despite the challenges in stabilizing lean premixed turbulent flames, combustion equipment manufacturers have succeeded in developing "fully premixed" HSB that attain <9 ppm NO_x (at 3% O_2) emissions. One example is the Rapid-Mix Burner (RMB) manufactured by Todd Combustion, a schematic of which is shown in Figure 6.1. As its name implies, the key feature of the RMB is to inject a homogeneous mixture of reactants in the shortest distance possible. This is to address the major concern of flashback in fully premixed burners. The RMB approach is to integrate the fuel injection port into the swirl vanes. To mitigate NO_x below 9 ppm NO_x, this burner relies on introducing flue gas to dilute and slightly preheat the air stream. Other designs such as the Coen Micro-NO_x burners and Net-Com's Hyper-Mix burners utilize multistaged combustion. The basic premise is to create a fuel-rich central combustion zone to serve as a stable pilot to ignite the fuel-lean premixture injected into the surroundings. Several sets of air and fuel injectors are needed to produce and control the fuel-rich/fuel-lean sequence. Additionally, a recirculation zone in the central region is necessary to entrain flue gases into the fuel-rich combustion core to reduce the peak temperatures and in turn provide a means for NO_x control.

6.3.2. SURFACE-STABILIZED LEAN PREMIXED INDUSTRIAL BURNER CONCEPTS

Surface-stabilized, or surface-radiant, burners use perforated ceramic tiles, ceramic foam, or porous metal fiber materials to support lean premixed combustion. The operating principle of the surface-stabilized burner is essentially identical to that of the water-cooled sintered burner used in fundamental combustion research (Mokhov and Levinsky, 2000). By reducing the exit velocity of the premixed fuel–air mixture below that of the free-flame burning velocity, the flame transfers heat to the burner (Botha and Spaulding, 1954), lowering the flame temperature and thus the NO_x emissions. The scale of the porous medium is generally the same as, or smaller than, the laminar flame thickness (of <1 mm), preventing the flame from flashing back inside the material. This method is distinct from submerged radiant combustion (Rumminger *et al.*, 1996), in which the flame is located within the open porous structure. In the submerged case, there is also substantial heat transfer to the burner surface downstream of the primary flame front. For surface radiant burners, the heat transferred to the burner surface is subsequently radiated to the surroundings. Increasing the exit velocity of the combustible mixture reduces the heat transfer to the burner surface (and increases the NO_x emission). When the exit velocity exceeds the free-flame burning velocity, the flame must be stabilized aerodynamically to prevent it from becoming unstable. Because of this relation to the laminar burning velocity (Mokhov and Levinsky, 2000), in the "radiant mode" (exit velocity lower than the free-flame burning velocity), the specific power is limited to $<1\,MW/m^2$. At low exit velocities, the flame temperature becomes so low that the flames extinguish (Mokhov and Levinsky, 1996). This limits the lowest achievable specific power input to $\sim 100 - 200\,MW/m^2$. Interestingly, in this mode the flame properties (temperature, NO_x emissions) are independent of the nature of the surface (Bouma and de Goey, 1999; Mokhov and Levinsky, 2000).

Depending on the design for stabilization, a few megawatts per meter square is an average maximum before the flames become unstable. Stabilization can be enhanced by perforating the porous surface to allow some of the mixture to burn as jet flames, which are stabilized by low velocity hot gases from the surface combustion. Scaling is usually accomplished by increasing the surface area (i.e., constant velocity scaling) and consequently the physical size of surface-stabilized burners is directly proportional to the power output. Clearly, for commercial and industrial systems of even a few megawatts, this requires an inconveniently large burner surface. This characteristic is quite unlike that of typical burner designs where constant velocity and constant area scaling (i.e., increasing throughput of reactants to attain high power output) are both applicable. The high cost of some of the porous materials and their fragile nature had limited their use in the past to high-end residential, commercial, and industrial applications. With the development of tough and inexpensive ceramic foams and moderately priced metal fabrics that are pliable and less fragile, more affordable surface-stabilized burners are being marketed to meet a wide variety of utilization needs. Currently, ceramic foam burners are a standard feature in European domestic condensing boilers, where a turndown ratio of 6:1 is reliably achieved over the range of distributed gas compositions.

In addition to emitting very low NO_x concentrations and enhancing radiative heat transfer, the main advantage of surface-stabilized burners is that their shapes can be customized to meet the specific (radiant) efficiency and system requirements of various industrial processes. As mentioned above, the main drawbacks, however, are their relatively small turndown capacity and large sizes.

6.3.3. LOW-SWIRL STABILIZATION CONCEPT

The low-swirl burner is a recent development originally conceived for laboratory studies on flame/turbulent interactions (Chan *et al.*, 1992; Bedat and Cheng, 1995;

Cheng, 1995; Plessing *et al.*, 2000; Cheng *et al.*, 2002; Shepherd *et al.*, 2002; Kortschik *et al.*, 2004; de Goey *et al.*, 2005). Its operating principle exploits the propagating nature of turbulent premixed flames, and its basic premise is to stabilize a turbulent premixed flame as a stationary "standing wave" unattached to any physical surfaces. This is accomplished by generating a divergent flow that is formed when the swirl intensity is below the vortex breakdown point. Therefore, the low-swirl burner concept is fundamentally different from the high-swirl concept where vortex breakdown is a vital prerequisite to producing a strong and robust toroidal recirculation zone.

The original low-swirl burner for laboratory studies known as the jet-LSB is shown in Figure 6.2 (Bedat and Cheng, 1995). This burner consists of a cylindrical tube of 5.08 cm ID fitted with a tangential air swirler section located 7 cm upstream of the exit. The swirler consists of four small jets of 0.63 cm ID inclined at 70° to the central axis. Reactants supplied to the bottom of the tube through a turbulence generating plate interact with the tangential air jets. The size of the air jets is kept small so that the swirling fluid motion clings to the inner wall and does not penetrate deep into the reactant core. When the flow exits the burner, centrifugal force due to the swirling motion causes the central non-swirling flow to expand and diverge. The divergent center region is characterized by a linear velocity decay and this feature provides a very stable flow mechanism for a premixed turbulent flame to freely propagate and settle at a position where the local velocity is equal and opposite to the turbulent flame speed (see Chapter 2 for discussion of some of the elements governing turbulent flame speed). Varying the stoichiometry of the reactants (without changing the mean velocity and swirl rate) simply shifts the flame brush to a different position within the divergent flow. This behavior is similar to the behavior of flames stabilized in the divergent regions produced in stagnating flows, but the absence of a downstream stagnation plate or plane in the low-swirl burner allows it to support flames that are much leaner than is possible in stagnation burners.

For practical applications, the LSB approach has many desirable attributes. Its capability to support very lean premixed flames is of course the primary motivation. The features of the flame stabilization mechanism also address the important operational and safety concerns regarding the use of fully premixed burners for industrial processes. The flame cannot flashback into the burner when the velocity at the exit is

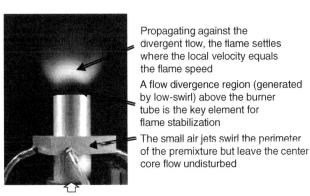

Propagating against the divergent flow, the flame settles where the local velocity equals the flame speed

A flow divergence region (generated by low-swirl) above the burner tube is the key element for flame stabilization

The small air jets swirl the perimeter of the premixture but leave the center core flow undisturbed

Fuel/Air
mixture supply

Figure 6.2 A jet-LSB demonstrates the principle of low-swirl flame stabilization.

higher than the turbulent flame speed. Blow-off is also mitigated because the flame retreats to a lower velocity region in the divergent flow when any sudden decrease in stoichiometry occurs. Additionally, the consequence of changes in mixture inhomogeneity or slight flow transients is a slight shift in the flame position. Therefore, the likelihood of catastrophic flameout is reduced substantially. The LSB flowfield provides automatically a robust self-adjusting mechanism for the flame to withstand transients and changes in mixture and flow conditions.

One of the first issues for practical adaptation of the LSB is to determine the effect of an enclosure on the flame stabilization mechanism and flame behavior. Figure 6.3 displays centerline velocity profiles from LDV measurements of CH_4/air flames at $\phi = 0.8$ and a bulk flow velocity of $U_o = 3.0$ m/s (18.5 kW) generated by the jet-LSB of Figure 6.2 inside quartz cylinders of 7.62 cm diameter and 20 and 30 cm lengths, with or without an exit constriction of 5.4 cm (Yegian and Cheng, 1998). The features of the mean axial velocity profiles (top) in the nearfield region ($x < 50$ mm) are similar to those measured in stagnation burners (Cho *et al.*, 1988; Kostiuk *et al.*, 1993) where flow divergence away from the burner exit is illustrated by a linear decay of U with increasing x followed by an abrupt upturn due to combustion-generated flow acceleration within the flame brush. The inflection point marks the leading edge of the flame brush and provides a convenient means to determine the turbulent flame speed S_T. Differences in the enclosed and open LSB profiles are shown in the farfield region where the open flame sustains a higher mean velocity while velocities in the enclosures are lower. This is due to the open flame generating a jet-like plume and the plumes of the enclosed flames expanding to fill the enclosure volume. These data show that the LSB flames are not sensitive down to at least this 3:1 diameter ratio enclosure.

Technology transfer and commercial implementation of low-swirl combustion began with adaptation to residential pool heaters of 15–90 kW (50–300 K Btu/h). These small domestic appliances are commodities and can only afford to include very simple and low cost technologies controlled by rudimentary electronics. The jet-LSB requiring two flow controls was too elaborate, which motivated the development of a simpler burner

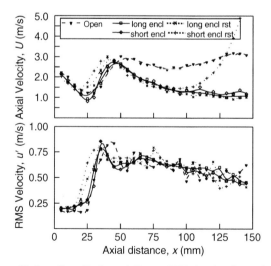

Figure 6.3 Velocity profile from (from Yegian and Cheng, 1998) showing that enclosures have little effect on the flame stabilization mechanism of a jet-LSB for a CH_4/air flame with $\phi = 0.8$ at 18.5 kW.

that is easy to manufacture and requires few controls. The outcome was a patented vane swirler (Cheng and Yegian, 1999) that has since been adapted for industrial burners and gas turbines.

The main challenge in developing a swirler for LSB was lack of knowledge on the fluid mechanics of low-swirl flows because all prior research focused on high swirl. Laboratory experimentation with LDV led to the design of Figure 6.4. This LSB is sized for domestic heaters of 18 kW and has a radius R_b of 2.54 cm. It has an annular swirler that consists of eight blades inclined at 37.5° with respect to the burner axis. Its unique feature is an open center channel ($R_c = 2$ cm) to allow a portion of the reactants to bypass the swirl annulus. The center channel is fitted with a perforated screen (3 mm holes arranged in a rectangular grid to give 81% blockage) to balance the flow split between the swirled and the unswirled flow passages. The screen also produces turbulence in the center unswirled core. The swirler assembly is recessed from the exit at a distance L of 6.3 cm. This vane LSB produces the same key flowfield features as the jet-LSB and as shown in Figure 6.5, the flame generated by the vane LSB is lifted with a bowl shape slightly different from the one produced by the jet-LSB. The vane LSB was

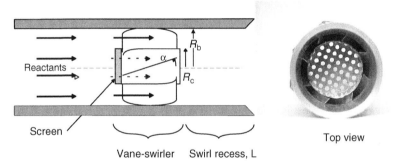

Figure 6.4 Schematic and photograph of a vane swirler developed for the low-swirl burner.

Figure 6.5 A vane LSB firing at 18 kW.

found, however, to have very high turndown (approaching 60:1) and very low emissions. More importantly, the design is scalable (Cheng *et al.*, 2000) and the development of cooling rules is crucial to its smooth transition to commercialization in larger industrial systems.

6.3.3.1. Scaling of LSB to Industrial Capacities

Central to the scaling rule for low-swirl vane burners is a new definition of swirl number, S, that includes the flow through the center channel. It is derived from the formal definition of S based on the geometry of the device (Syred and Beer, 1974):

$$S = \frac{\text{axial flux of angular momentum}}{\text{axial thrust} \times R} = \frac{G_{\text{ang}}}{G_{\text{x}}R}. \tag{6.3}$$

Assuming that the distribution of the axial flow remains flat, and that U and W at the burner exit are kinematically related to the blade angle as $\tan \alpha = U/W$, the axial flux of angular momentum in the annular section can then be written as follows:

$$G_{\text{ang}} = 2\pi\rho \int_{R_c}^{R_b} U_a(U_a \tan \alpha)r^2 dr = 2\pi\rho U_a^2 \tan \alpha \left(\frac{R_b^3 - R_c^3}{3}\right). \tag{6.4}$$

Here, U_a is a mean axial velocity supplied through the swirl annulus. By assuming a flat axial velocity distribution, the linear momentum flux from the two regions of the burner is then calculated as follows:

$$G_{\text{x}} = 2\pi\rho \int_{R_c}^{R_b} U_a^2 r\,dr + 2\pi\rho \int_{0}^{R_c} U_c^2 r\,dr = \pi[\rho U_a^2(R_b^2 - R_c^2) + \rho U_c^2 R_c^2], \tag{6.5}$$

where U_c is the mean axial velocity through the center core. With Equation (6.3) as defined, the geometric swirl number for the vane swirl burner is then

$$S = \frac{\frac{2}{3}\tan\alpha(1 - R^3)}{\left(1 - R^2 + \frac{U_c^2}{U_a^2}R^2\right)} = \frac{2}{3}\tan\alpha \frac{1 - R^3}{1 - R^2 + [m^2(1/R^2 - 1)^2]R^2}. \tag{6.6}$$

Here, R is the ratio of the centerbody-to-burner radii $R = R_c/R_b$. Equation (6.6) is simplified further when U_c/U_a is expressed in terms of m, the mass flux ratio (flow split) $m = \dot{m}_c/\dot{m}_a$ through the centerbody ($\dot{m}_c$) and annulus ($\dot{m}_a$); m is the same as the ratio of the effective areas of the center core and the swirl annulus, and can be determined simply by the use of standard flow pressure drop procedures. Obviously, Equation (6.6) is a more convenient form of the swirl equation for engineering design than is provided by Equation (6.3).

The scaling rules were determined from studying the influences of S, L, and R_b on burner performance using lean blow-off, flame stability, and emissions as the criteria. To start, the LSB prototype shown in Figure 6.4 was used as a benchmark with its swirl number varied by using four different screens with 65–75% blockage. The swirl numbers were 0.4 < S < 0.44 corresponding to m of 0.8–1, meaning that 44–50% of the reactants bypassed the swirl annulus. These swirlers were fitted with recess distance L from 4 to 12 cm. The 16 LSBs with various S and L combinations were tested with

CH_4/air flames at $5 < U_o < 25\,\text{m/s}$ covering a thermal input range of 18–90 kW. All burners were found to be operable. Increasing S pulled the flame closer to the burner, but the lean blow-off remained relatively unaffected, indicating that the performance of the LSB is not highly sensitive to small variations in S. The differences were mainly in flame positions and fuel/air equivalence ratio at lean blow-off, ϕ_{LBO}. Large swirler recesses resulted in a highly lifted flame, but the overall flame stability remained relatively unchanged. A short recess distance produced higher ϕ_{LBO}, indicating a compromise in the capability to support ultra-lean flames.

Additional studies were performed to explore the effects of varying radius R_b as well as R (0.5–0.8) and α (30–45°). The swirl numbers of the burners with various combinations of R_b, R, and α were varied by fitting them with screens of different blockages. The most significant finding was that the LSBs with larger R_b operated at the same range of S (around 0.4–0.5) as the smaller burner. Their performances in terms of flame stability and ϕ_{LBO} were also identical. Moreover, decreasing R had no effect on emissions or performance but brought about a significant benefit in lowering the pressure drop of the LSB. This can be explained by the fact that reducing R enlarges the swirl annulus and lowers its drag. To maintain a swirl number of 0.4–0.5, a screen with lower blockage is required. For example, the screen used for $R_b = 6.35\,\text{cm}$ with $R = 0.5$ has a 60% blockage compared to 65–81% needed for $R = 0.8$. This combination effectively lowers the overall pressure drop of the burner. The drag coefficients determined for the different LSBs show them to depend only on R and to be independent of R_b. This knowledge is very important for engineering LSBs to meet various system requirements and efficiency targets.

The scaling rules for the LSB were established from the above results. They are independent of burner radius (up to $R_b = 25.4\,\text{cm}$). For stable and reliable operation, the S of an LSB should be between 0.4 and 0.55. The swirler can have straight or curved vanes with angle α from 37° to 45°. The center channel-to-burner radius ratios R can range between 0.5 and 0.8. Once α and R are defined, the blockage of the center channel screen can be varied to give Y within the desired range of 0.4–0.5. In addition, the swirl recess distance L can be two to three times the burner radius. To determine the appropriate burner size, R_b, guidelines have been developed to optimize for the desired thermal input range, turndown, fuel pressure, fan power (pressure drop), combustion chamber size, and other physical constraints. The criterion for minimum thermal input is a bulk flow velocity of $U_o = 3\,\text{m/s}$. This is simply the flashback point for natural gas ($U_o \approx 1.7\,\text{m/s}$) with a built-in safety factor. There is no restriction on the maximum thermal input owing to the high turndown (at least 20) available. To optimize for the fuel pressure and fan power, the drag coefficient for different R can be used. The optimum enclosure radius for the LSB is between three and four times R_b. Smaller enclosures restrict flow divergence and force the flame to move inside the burner. Larger enclosures allow the flame to over-expand and generate internal flow patterns that affect emissions. We have found these rules and guidelines easy to apply, as they are quite lenient in providing many design options to build simple and low-cost LSBs for integration into existing or new systems.

6.3.3.2. Development of a Commercial LSB

Subsequent to the development of LSBs for pool heaters, several projects were pursued to adapt the LSB to industrial and commercial heaters. These studies proved that the LSB design is robust with regards to application. To investigate turndown, the smallest LSB with $R_b = 2.54$ cm was fired in the open (no enclosure). It generated stable flames from 10 to 600 kW that remained stationary despite the 60:1 change in input rate. At the lowest thermal input of 10 kW, the bulk flow velocity U_o corresponded to 1.7 m/s. This is the minimum allowable operating point for natural gas. Flashback becomes likely if U_o is reduced further because the velocity at the burner exit would be too close to S_T. The minimum U_o criterion to prevent flashback also applies to larger burners because the LSB has constant velocity scaling. This simply means that the thermal input of the LSB is directly proportional to U_o and R_b^2. The effects of enclosure geometry on LSB performance were also investigated by testing various versions of the $R_b = 6.35$ cm LSB in boilers and furnaces at 150 kW–2.3 MW. The results showed that vane shape and screen placement have little effect on flame noise, flame stability, and lean blow-off. Most significantly, emissions of NO_x depend primarily on ϕ. As shown in Figure 6.6 by the NO_x emissions from LSBs of various sizes, the trends with ϕ are similar despite differences in thermal inputs and combustor geometries. Additional tests of the 6.35 cm LSB were also performed with alternate fuels including natural gas diluted with up to 40% flue gases, and with refinery gases with hydrogen constituent up to 50% to show its capability to accept different fuels. Hydrogen can be a particularly challenging fuel to blend because of its dramatic effect on flame speed (see Chapter 8).

In 2003, Maxon Corporation demonstrated an industrial implementation of LSBs called M-PAKT burners. These products were developed for direct process heat applications of 0.3–1.8 MW (1–6 MM Btu/h) with a guarantee of 4–7 ppm NO_x and CO (both at 3% O_2) throughout its 10:1 turndown range. These ultra-low emissions meet the most stringent air-quality rules in the United States. As shown by the schematic in Figure 6.7, the M-PAKT burner has a very simple and compact design consisting of a swirler with air supplied by a blower through a plenum, and a multi-port natural gas injector delivering fuel just upstream of the swirler. The control system is also standard with conventional mechanical linkages and flow dampeners. The performance of these commercial LSBs demonstrated that the implementation of low-swirl combustion can

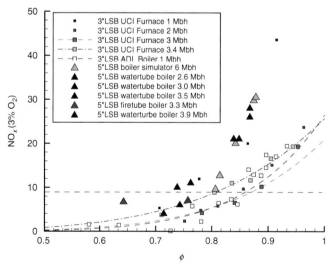

Figure 6.6 NO_x emissions of LSB in furnaces and boilers of 300 kW to 1.8 MW. (See color insert.)

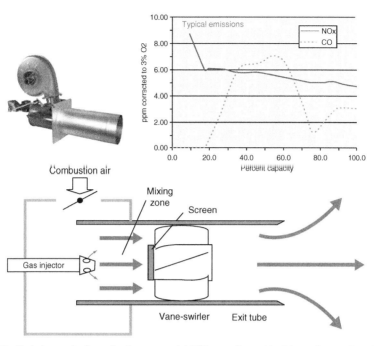

Figure 6.7 Emissions and schematic of a commercial LSB manufactured by Maxon Corporation of Muncie, Indiana.

not only provide very effective emissions control, but can also improve system performance and reliability by eliminating the need for elaborate controls and intricate auxiliary components. The economic and operational benefit of this approach is significant. Continuing efforts by Maxon to commercialize low-swirl combustion technology include the design of a new LSB product of 15 MW (50 MM Btu/h). The first installation was complete in February of 2005. This large burner has a radius R_b of 25.4 cm and has a 20:1 turndown. It also incorporates a liquid fuel injector for dual-fuel firing. Commercial demonstrations of LSB technology over a range of scales indicate that the fluid mechanics of these burners exhibits some form of self-similarity that is advantageous for burner design. Detailed studies of this similarity can provide insights for future designs.

6.3.3.3. Flowfield Characteristics, Turbulent Flame Speed and their Relevance to LSB Performance

The advent of PIV has greatly facilitated the characterization of LSB flowfields by providing the data needed to explain how the LSB delivers consistent performance over a large range of conditions. PIV measures velocity distributions in 2D over a relatively large region and is a much quicker method than LDV for investigating the overall flowfield features. In a recent study, a series of experiments were conducted to compare the flowfield and flame features with increasing U_o. Shown in Figure 6.8 is an example of the velocity vectors and turbulence stresses for an LSB with $S = 0.53$, $R_b = 2.54$ cm, $R = 0.6$, $\alpha = 37°$, and $L = 6.3$ cm burning a methane/air flame of $\phi = 0.8$ and $U_o = 5.0$ m/s. The leading edge of the flame brush is outlined by the

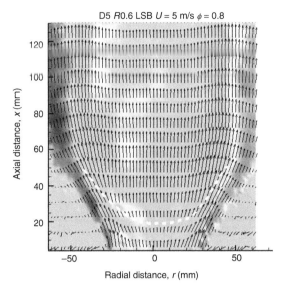

Figure 6.8 Velocity vectors and turbulence stresses for an LSB burning CH_4/air at $\phi = 0.8$ and $U_o = 5.0\,\text{m/s}$. (See color insert.)

broken line. These velocity vectors show that the LSB flowfield is relatively uniform and free of steep mean velocity gradients. The background contours of the positive (red) and negative (blue) shear stresses also show that high turbulence stresses are confined to the outer edges where the reactants mix with the ambient air. The center flow entering the flame brush is relatively free of large shear stresses. The absence of high stresses means that the flames are much less vulnerable to stress-induced non-uniform heat release and local quenching at ultra-lean conditions. This feature is different than in HSB where steep mean velocity gradients exist in the flame brush and the shear stresses can lead to premature flame blow-off.

As discussed above, S_T is defined in an LSB by the centerline velocity at the leading edge of the flame brush. Figure 6.9 displays values of S_T reported in Cheng *et al.* (2006) for several versions of the LSB with methane, ethylene, propane, and methane diluted with CO_2. Despite variations in burner configuration and in ranges of flow conditions, stoichiometry, and fuels, the turbulent flame speed data show an unequivocal linear correlation with u'. The only exceptions are three outlying points from the jet-LSB. These results clearly show that the two non-jet implementations of low-swirl combustion, that is, tangential injection swirler and vane swirler, are fully compatible. More importantly, they indicate that the linear behavior of the turbulent flame speed is a characteristic and unique combustion property central to the operation of low-swirl burners.

The evolution of the flowfield with U_o was investigated by applying PIV at $5 < U_o < 17.5\,\text{m/s}$. Figure 6.10 presents the normalized axial velocity profiles from the non-reacting cases. The fact that the profiles of the normalized axial velocity, U/U_o, and the normalized two-component turbulent kinetic energy, q'/U_o (where $q' = [(u'^2 + v'^2)/2]^{1/2}$), collapse to their respective trends shows that the LSB flowfield exhibits a similarity feature. Similarity is also shown by the radial profiles of U/U_o and V/U_o, meaning that the key flowfield features are preserved at different bulk flow

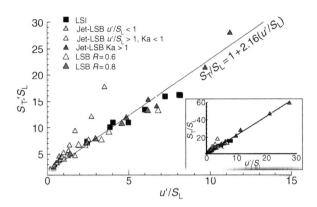

Figure 6.9   Turbulent flame speed correlation. (See color insert.)

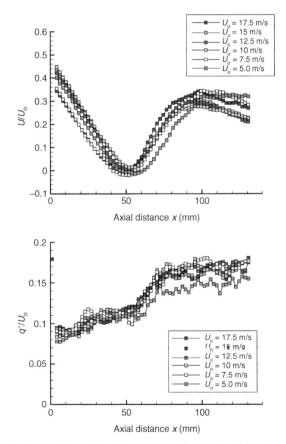

Figure 6.10   Normalized centerline profiles of non-reacting flow produced by a laboratory LSB showing self-similarity features. (See color insert.)

velocities. Recent PIV measurements in the reacting flowfields of a low-swirl injector developed for gas turbines show that self-similarity is also preserved in the nearfield divergent region (Cheng *et al.*, 2006). The effect of combustion heat release is to increase the aerodynamic stretch rate and decrease x_0.

Self-similarity explains why the flame maintains a relatively fixed position regardless of U_o. This is illustrated by invoking an equality at the leading edge position of the flame brush, x_f, (typically at $15 < x < 25$ mm for this LSB):

$$U_o - \frac{dU}{dx}(x_f - x_o) - S_1. \tag{6.7}$$

Here, x_o is the virtual origin of the linearly divergent portions of the axial profiles and has a negative value. As discussed earlier, S_T of the LSB is linearly dependent on the rms velocity of the turbulence u' such that $S_T = S_L(1 + Ku')$, where S_L is the laminar flame speed and K is an empirical correlation constant that is 2.16 for methane. Substituting this into Equation (6.4) and dividing both sides by U_o results in

$$1 - \frac{dU}{dx}\frac{(x_f - x_o)}{U_o} = \frac{S_T}{U_o} = \frac{S_L}{U_o} + \frac{Ku'}{U_o}. \tag{6.8}$$

The similarity feature of the U/U_o profiles means that the normalized axial divergence rate [i.e., $(dU/dx)/U_o$] has a constant value. On the right hand side (RHS), Ku'/U_o is a constant because turbulence produced by the perforated plate is nearly isotropic. The first term of the RHS tends to a very small value for large U_o because the laminar flame speeds for hydrocarbons are from 0.2 to 0.8 m/s. As long as flow similarity is preserved, the flame position at large U_o tends to an asymptotic value independent of the laminar flame speed. This provides an explanation why the lifted flames generated by LSBs do not show significant flame shift during turndown. Flame shift is found only when operating the burner at a U_o that is on the same order as S_L. Equation (6.8) also shows that the flame moves closer to the burner exit with decreasing U_o until flashback occurs when $x_f - x_o = 0$.

6.4. SUMMARY

Lean premixed burners can be deceptively simple devices that must nevertheless provide flame stability (i.e., resistance to flashback and blow-off) over the desired range of turndown ratio, and they must maintain this stability with varying fuel composition (even when the fuel is nominally "natural gas"). The reliance on blowers or fans to provide high volume flows of air economically can make these burners' performance very sensitive to fluctuations in fuel composition, particularly when operating fuel lean. High swirl and surface stabilization have two mechanisms to achieve reliable combustion behavior, but turndown can be limited. The relatively new low-swirl approach, on the other hand, has shown great promise in expanding the stability map of lean premixed burners.

REFERENCES

AGA Bulletin #10 (1940). Research in fundamentals of atmospheric gas burner design.
Bedat, B. and Cheng, R.K. (1995). *Combust. Flame* **100**, 485–494.
Beer, J.M. and Chigier, N.A. (1972). *Combustion Aerodynamics*. Applied Science Publishers Ltd., London.
Botha, J.P. and Spaulding, D.B. (1954). *Proc. R. Soc. London A* **225**, 71–96.
Bouma, P.H. and de Goey, L.P.H. (1999). *Combust. Flame* **119**, 133–143.

Bowman, C.T. (1992). *Proc. Combust. Inst.* **24**, 859–878.

Chan, C.K., Lau, K.S., Chin, W.K., and Cheng, R.K. (1992). *Proc. Combust. Inst.* **24**, 511–518.

Cheng, R.K. (1995). *Combust. Flame* **101**, 1–14.

Cheng, R.K. and Yegian, D.T. (1999). Mechanical Swirler for a Low-NO$_x$ Weak-Swirl Burner, US Patent #5879148.

Cheng, R.K., Yegian, D.T., Miyasato, M.M., Samuelsen, G.S., Pellizzari, R., Loftus, P., and Benson, C. (2000). *Proc. Combust. Inst.* **28**, 1305–1313.

Cheng, R.K., Shepherd, I.G., Bedat, B., and Talbot, L. (2002). *Combust. Sci. Technol.* **174**, 29–59.

Cheng, R.K., Littlejohn, D., Nazeer, W.A., and Smith, K.O. (2006). Power for land sea and air, ASME Paper GT2006-90878. *ASME Turbo Expo 2006*, Barcelona, Spain.

Chigier, N.A. and Beer, J.M. (1964). *Trans. ASME, J. Basic Eng.* **4**, 788–796.

Cho, P., Law, C.K., Cheng, R.K., and Shepherd, I.G. (1988). *Proc. Combust. Inst.* **22**, 739.

Davies, T.W. and Beer, J.M. (1971). *Proc. Combust. Inst.* **13**, 631–638.

de Goey, L.P.H., Plessing, T., Hermanns, R.T.E., and Peters, N. (2005). *Proc. Combust. Inst.* **30**, 859–866.

Gupta, A.K., Beer, J.M., and Swithenbank, J. (1978). *Combust. Sci. Technol.* **17**, 199–214.

Harris, J.A. and South, R. (1978). *Gas Eng. Manag.* **May**, 153.

Kortschik, C., Plessing, T., and Peters, N. (2004). *Combust. Flame* **136**, 43–50.

Kostiuk, L.W., Bray, K.N.C., and Cheng, R.K. (1993). *Combust. Flame* **92**, 396–409.

Lilley, D.G. (1977). *AIAA J.* **15**, 1063–1078.

Marcogaz Ad Hoc Working Group on Gas Quality (2002). *National Situations Regarding Gas Quality*, Report UTIL-GQ-02-19, http://marcogaz.org/information/index_info4.htm.

Mokhov, A.V. and Levinsky, H.B. (1996). *Proc. Combust. Inst.* **26**, 2147–2154.

Mokhov, A.V. and Levinsky, H.B. (2000). *Proc. Combust. Inst.* **28**, 2467–2474.

Plessing, T., Kortschik, C., Mansour, M.S., Peters, N., and Cheng, R.K. (2000). *Proc. Combust. Inst.* **28**, 359–366.

Rumminger, M.D., Dibble, R.W., Heberle, N.H., and Crosley, D.R. (1996). *Proc. Combust. Inst.* **26**, 1755–1762.

Shepherd, I.G., Cheng, R.K., Plessing, T., Kortschik, C., and Peters, N. (2002). *Proc. Combust. Inst.* **29**, 1833–1840.

Syred, N. and Beer, J.M. (1974). *Combust. Flame* **23**, 143–201.

von Elbe, G. and Grumer, J. (1948). *Ind. Eng. Chem.* **40**, 1123.

Wobbe, G. (1926). *L'Industria de Gas e degli Acquedotti* **30**, November.

Yegian, D.T. and Cheng, R.K. (1998). *Combust. Sci. Technol.* **139**, 207–227.

Chapter 7

Stability and Control

Sivanandam Sivasegaram

Nomenclature

DLE	Dry low emission
f_{er}	Extinction and relight cycle frequency
ϕ	Fuel-to-air ratio
ϕ_{lean}	Lean flammability limit
$\phi_{lean,\ overall}$	Overall equivalence ratio at the lean flammability limit
ϕ_{stable}	Stability limit
Re	Reynolds number
rms	Root mean square
Sw	Swirl number

7.1. INTRODUCTION

For many years, land-based gas turbines have operated with diffusion flames, often resulting in NO_x concentrations greater than 200 ppm with 15% oxygen in the products (Correa, 1993). Strategies for reducing NO_x emissions from engines have been developed over the past two decades in the context of growing environmental concerns and stringent government regulations relating to permissible emission levels, with target dates emphasizing the urgency of emission control (Bowman, 1992). Since the formation of thermal NO_x depends on residence time at high temperatures, methods for NO_x reduction have involved lowering the temperature of the combustion products by the addition of water or steam, staging, and operating with low equivalence ratios of premixed fuel and air (Allen and Kovacik, 1984; Touchton, 1985; Bowman, 1992; Norster and De Pietro, 1996; Li *et al.*, 1997; Moore, 1997; Gore, 2002;). Imposed oscillations have also been examined as a way of reducing NO_x emissions, with combustion taking place close to stoichiometric conditions (Keller and Hongo, 1990; Poppe *et al.*, 1998).

Premixed lean combustion combines the benefit of NO_x reduction with the prospect of achieving compact flames with complete combustion (Bradley *et al.*, 2006). As described in Chapter 5, a variety of lean burn gas turbines have been developed over the past two decades, driven by the demand for gas turbine engines with reduced emission levels and low specific fuel consumption. Pre-vaporized premixed gas turbine combustors for power generation need to be operated as lean as possible to secure sub-10 ppm concentrations of NO_x. However, achieving stable combustion in a combustor under fuel-lean conditions requires overcoming several inter-related problems associated with low intensity heat release, such as flame stabilization, flame stability and extinction, and combustion oscillations arising from thermo-acoustic coupling between time-dependent heat release rates and the natural frequencies of the combustion chamber.

Combustion oscillations in ducted flames have generally been quantified in terms of the amplitude of pressure fluctuations in the combustor cavity. Large amplitudes can compromise the structural integrity of the combustor, cause fluctuations in fluid flow rates to induce early extinction, limit the safe and stable operating range, and give rise to noise levels that threaten health and safety. Hence, they have been extensively investigated, and passive and active control strategies have been devised to overcome the problem, as is evident from the reviews of Culick (1989), McManus *et al.* (1993), and Schadow and Gutmark (1992), with the latter providing the possibility of operation over a wide range of operational conditions. However, the emphasis of these studies, until recently, has been on oscillations occurring close to stoichiometric conditions in ducted premixed flows.

Close to the lean limit, combustors can give rise to large amplitudes of oscillation that would restrict the operating envelope of the combustor and prevent the realization of the full potential of low emissions (Bradley *et al.*, 1998; De Zilwa *et al.*, 2001). The purpose of this chapter is to discuss the sources and mechanisms of the combustion instability that lead to oscillations of large and medium amplitude under lean burn conditions in premixed combustion, and examine possible ways to ameliorate, if not eliminate, such oscillations so as to ensure stable combustion even under conditions of very lean burning close to the flammability limit.

Early studies of combustion oscillations concerned high pitch, radial, and circumferential mode acoustic oscillations in aircraft engine combustors with frequencies on the order of several kHz. Most of these oscillations have been successfully overcome by geometric modifications, such as the attachment of baffles, acoustic liners, and Helmholtz resonators to the combustion chamber to hamper the propagation of the acoustic waves (Culick, 1989). Oscillations of large amplitude observed in ramjets, tunnel burners, and after-burners were generally associated with longitudinal acoustic and bulk-mode frequencies in the range between 50 and 500 Hz, of which one tends to be dominant. These frequencies, unlike the radial and circumferential frequencies, could not be easily averted with simple geometric modifications.

Subsequent research on combustion oscillations in round ducts, and in ducted premixed flows with flame stabilization behind a bluff body or a step formed by a sudden expansion in plane, showed that the oscillations were driven by coupling between oscillatory heat release and a natural frequency of the combustor duct, and that they occurred over a range of values of equivalence ratio, ϕ (the fuel-to-air ratio expressed as a fraction of that for stoichiometric burning) typically between 0.75 and 1.25 for saturated hydrocarbon fuels, such as methane and propane. The dominant frequency was identified as either an acoustic longitudinal frequency (Putnam, 1971; Heitor *et al.*, 1984; Crump *et al.*, 1985) or a bulk-mode frequency (Crump *et al.*, 1985; Sivasegaram and Whitelaw, 1988) of the combustor cavity. Acoustically closed ends of combustor ducts and flame-holding steps with a large area expansion ratio provided pressure antinodes of the longitudinal frequency modes, while acoustically open ends acted as pressure nodes (see Figure 7.1). Since the oscillations were driven by a

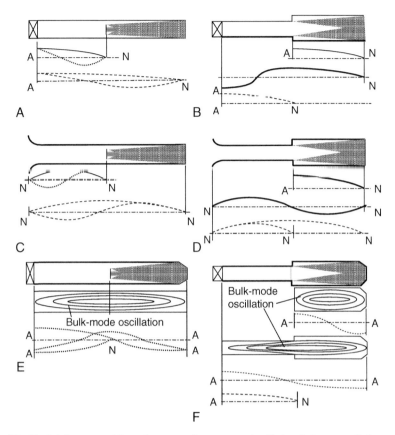

Figure 7.1 Ducted flow geometries and common frequency modes (fuel and air are supplied premixed at upstream, left hand end): (A), (C), and (E): bluff-body stabilized flames; (B), (D), and (F): sudden-expansion flows; (A) and (B): upstream end acoustically closed, downstream end acoustically open; (C) and (D): upstream end acoustically open, downstream end acoustically open; (E) and (F): upstream end acoustically closed, downstream end constricted by nozzle. Uppermost frequency modes are for large area expansion ratios and lower ones for moderate ratios. ········frequency mode commonly associated with large amplitudes; ----- frequency mode that could be associated with large amplitudes; ─·─·─ frequency mode usually associated with small amplitudes. A – pressure antinode; N – pressure node.

coupling between fluctuating heat release and an acoustic frequency of the duct, the location of intense heat release and hence the plane of the flame holder lay close to an acoustic boundary, a pressure node in the case of a bluff body of modest blockage, and a pressure antinode close to a step of large or moderate area ratio. Although severe constriction of the combustor exit by a nozzle could give rise to a longitudinal acoustic frequency with an antinode at the exit (Crump *et al.*, 1985), constriction by a nozzle often resulted in a bulk-mode (Helmholtz) frequency dominating the oscillations (Sivasegaram and Whitelaw, 1988).

Outside this range of equivalence ratios, oscillations are of smaller amplitude and are associated with one or several of the natural frequencies identified in Figure 7.1. Close to extinction, a broadband low frequency dominates oscillations, with amplitude increasing with proximity to the lean flammability limit (De Zilwa *et al.*, 2000; 2001) and accompanied by acoustic or bulk-mode frequencies of the combustor cavity.

Several recent designs of dry low emission (DLE) gas turbines (Moore, 1997; Hubbard and Dowling, 2003; Milosavljevic *et al.*, 2003) involve lean burning in sudden-expansion flow arrangements with a moderate area ratio (Figure 7.2). Although sudden expansions with area ratio around 2.5 or less were known to be associated with oscillations dominated by the acoustic frequency of the entire duct length, as opposed to that of the length downstream of the step for area ratio greater than 3, the step could sometimes serve as a pressure node for an acoustic wave in the upstream duct section (Sivasegaram and Whitelaw, 1987a). Detailed studies in sudden expansions with moderate area ratio were carried out only in later years, especially in view of their relevance to lean burn combustors, such as in land-based gas turbines (Yang and Schadow, 1999).

Attempts to control or, if possible, avoid such oscillations included a variety of strategies (see Figure 7.3), such as attaching a resonator to the combustor cavity (Putnam, 1971; Sivasegaram and Whitelaw, 1987a), the use of sound absorbent material (Heitor *et al.*, 1984), constriction of the upstream duct (Sivasegaram and Whitelaw, 1987b), modifications to the fueling arrangement (Whitelaw *et al.*, 1987) and to the flame holder (Gutmark *et al.*, 1989; Perez-Ortiz *et al.*, 1993), which have been used with varying degrees of success to suppress longitudinal acoustic frequencies, and the

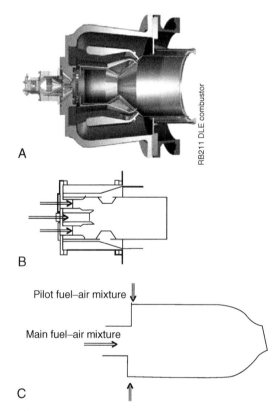

Figure 7.2 Selections of lean burn gas turbine combustors with sudden-expansion flow: (A) RB211 DLE combustor (Rolls Royce), (B) GE DLN-1 combustor (General Electric)(Moore, 1997), and (C) cross-section of combustion chamber sector of experimental axisymmetric premixed burner (ALSTOM Power) (Milosavljevic *et al.*, 2003).

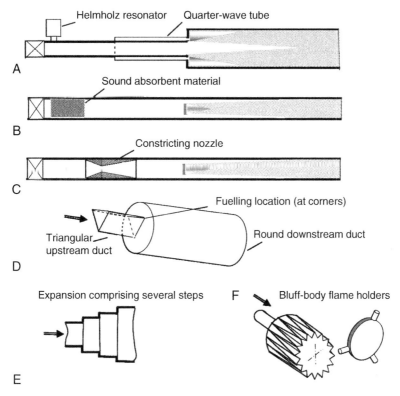

Figure 7.3 Modifications to suppress combustion oscillations: (A) use of resonators, (B) use of sound absorbent material, (C) constriction of duct, (D) modified fuelling arrangement, (E) modified sudden expansion, and (F) modified disk-type flame holder.

addition of swirl (Sivasegaram and Whitelaw, 1988; 1991) to control bulk-mode oscillations. It is also possible to avert combustion oscillations under conditions of lean combustion by ensuring good flame stabilization by modifications to the burner design (Milosavljevic, 2003).

Arrangements to avert combustion oscillations of large amplitude by modifications to the combustor, even when they are highly effective, tend to be specific to each combustor geometry and, sometimes, to a given range of operating conditions, and are often unacceptable to the designer. The notion of actively controlling combustion oscillations, stemming from the concept of anti-sound used to control discrete frequency noise by Sreenivasan *et al.* (1985), has led to the successful development of a variety of control techniques and strategies in the laboratory (Bloxsidge *et al.*, 1987; Poinsot *et al.*, 1989; Gulati and Mani, 1990; Langhorne *et al.*, 1990; Billoud *et al.*, 1992; Hendricks *et al.*, 1992; Sivasegaram and Whitelaw, 1992; Wilson *et al.*, 1992; Sivasegaram *et al.*, 1995a, b; Neumeir and Zinn, 1996; Lee *et al.*, 2000), with some of them applied to industrial scale combustors (Seume *et al.*, 1997; Pascherit *et al.*, 1998; Moran *et al.*, 2000). Again, the emphasis has been on oscillations observed at values of equivalence ratio close to stoichiometry.

Since the imposed oscillations (unlike in anti-sound) led to a weakening of the source of the controlled oscillation, the input power for control was much less than the energy of the oscillations to be controlled; and control essentially concerned the oscillation of the bulk flow (Bloxsidge *et al.*, 1987), the imposition of pressure oscillations on the flow field (Poinsot *et al.*, 1989; Gulati and Mani, 1990; Hendricks *et al.*, 1992; Sivasegaram and Whitelaw, 1992), or the oscillation of heat release (Langhorne *et al.*, 1990; Billoud *et al.*, 1992; Hendricks *et al.*, 1992; Wilson *et al.*, 1992; Sivasegaram *et al.*, 1995a, b; Neumeir and Zinn, 1996; Lee *et al.*, 2000) in a way that the imposed oscillations had an appropriate phase difference with the naturally occurring oscillations. A variety of actuators have been developed to provide the necessary oscillated input at the required frequency and phase difference, and will be discussed along with active control methods later in this chapter in the context of their usefulness to the control of oscillations under conditions of lean burn.

Recent investigations of combustion oscillations in premixed flows at conditions close to the lean flammability limit (Bradley *et al.*, 1998; De Zilwa *et al.*, 2001; Emiris and Whitelaw, 2003; Emiris *et al.*, 2003) have provided considerable insight into the nature and source of these oscillations, which relate to flame stability and local extinction, and are distinct from oscillations observed close to stoichiometry. The investigations also pointed to difficulties in applying conventional active control to these oscillations, and studies have been carried out to develop control strategies to suppress the oscillations and thereby lower the lean flammability limit and extend the operating range of the lean burn combustor. The study has since been extended to stratified flows (Emiris *et al.*, 2003; Luff *et al.*, 2004; Luff, 2005), since such flows occur in practical combustors as well as in view of the prospect of using stratification to extend the lean flammability limit.

The next section of this chapter deals with the nature of oscillations close to the lean flammability limit and the section following it presents an overview of active control and the application of suitable control strategies to the oscillations. The fourth and final section summarizes the main conclusions.

7.2.   OSCILLATIONS AND THEIR CHARACTERISTICS

As described above, the emphasis of early studies of combustion oscillations in ducted premixed flows was on large-amplitude oscillations occurring close to stoichiometry. Studies of oscillations close to the lean flammability limit became important in the context of lean burn combustion with a need to operate close to the lean flammability limit and to widen the operating range by lowering the lean flammability limit, thus including the implications of naturally occurring and imposed oscillations close to the lean flammability limit.

Since oscillations close to the lean flammability limit have been related to cycles of local extinction and relight, which in turn give rise to oscillations of low frequency, the process of extinction will be discussed in the subsection below, which will also deal with the importance of fuel and the influence of imposed oscillations and swirl to the lean flammability limit. The two subsections that follow will deal, respectively, with ducts without and with exit nozzles, in view of the differences in the amplitudes and frequencies associated with the oscillations. The final subsection will deal with stratified flows, since stratification exists in premixed gas turbine flows and has implications for the lean flammability limit and for oscillations close to the lean flammability limit.

7.2.1. THE PROCESS OF EXTINCTION

Extinction in stable laminar flames occurs when the local flow speed exceeds the laminar flame speed, which has been quantified as a function of the fuel (Warnatz, 1984; Egolfopoulos and Dimotakis, 2001), and of the equivalence ratio and temperature of the fuel–air mixture. Susceptibility to extinction will be less for an unsaturated hydrocarbon fuel such as ethylene than for a saturated hydrocarbon such as methane or propane, for near-stoichiometric fuel–air mixtures than for lean mixtures, and for higher mixture temperatures than for lower temperatures. As presented in Chapter 2, in turbulent flows, the speed of flame propagation is enhanced by turbulent diffusion and the effective flame speeds are higher. Thus the flammability range would be wider for ethylene than for methane, as for example in the results of Korusoy and Whitelaw (2004) shown in Figure 7.4 for a round sudden-expansion flow with an unconstricted exit. The flammability range narrows and the importance of fuel diminishes with an increase in Reynolds number, *Re*, based on bulk flow properties just upstream of the sudden expansion, owing to the increase in influence of turbulence.

Flame studies close to the lean extinction limit in turbulent premixed flames have shown that a sequence of cycles of extinction and relight precede global extinction of the flame (Bradley *et al.*, 1998; De Zilwa *et al.*, 2001). The study of extinction in flames stabilized in the stagnation flow formed by a pair of opposed-jets flame by Mastarakos *et al.* (1992), Kostuik *et al.* (1999), Sardi and Whitelaw (1999), and Sardi *et al.* (2000) has provided valuable insight into the mechanism of extinction in turbulent flames. The effect of strain rate on extinction time was quantified by Sardi and Whitelaw (1999) and Sardi *et al.* (2000) by imposing pressure oscillations on opposed-jet premixed methane–air flames. It was shown that global extinction occurred at the end of a process of continuous weakening of the flame through a series of cycles of local extinction and relight, and that large strain rates led to extinction in a fewer number of cycles. Extension of this work to propane and ethylene fuels as well as to the measurement of extinction time under unforced conditions and profiles of velocity by Korusoy and Whitelaw (2002) and Luff *et al.* (2003) confirmed that the flame was gradually weakened by high strain rates before total extinction.

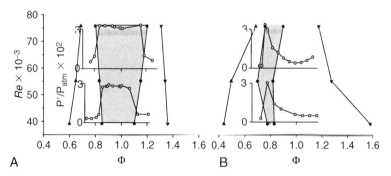

Figure 7.4 Importance of fuel to flammability and stability limits (Korusoy and Whitelaw, 2004). Round duct: 970 mm long, 51 mm diameter upstream duct with acoustically closed end; 840 mm long, 80 mm diameter downstream duct without exit nozzle: (A) methane–air and (B) ethylene–air. ▲ – lean limit, ▼ – rich limit, unstable region is shaded, ∘ – rms pressure amplitudes superimposed for $Re = 39\,500$ and $55\,400$.

Flame visualization using cycle-resolved CH images (Hardalupas *et al.*, 1999) was carried out in plane sudden expansion of De Zilwa *et al.* (2001) to understand the mechanism of extinction in sudden-expansion flows because of its relevance to annular sudden-expansion flows, and since the rectangular duct was more amenable than the round for imaging the flame. Global extinction in the plane sudden expansion occurred in two stages, with one of the two branches of the flame extinguishing much earlier than the other as ϕ was lowered (Figure 7.5). Stable combustion, characterized by small amplitudes of oscillation (peak-to-peak amplitude of the order of 0.1 kPa), prevailed over a range of values of ϕ between the lean flammability limit ϕ_{lean} and the stability limit ϕ_{stable} (defined as the lower limit of the range of equivalence ratios associated with large-amplitude oscillations close to stoichiometry). As ϕ was decreased from the stability limit, the two branches of the flame started a flapping movement, in which the branches moved together with increasing vigor until one of the branches extinguished. This led to less vigorous oscillation of the other branch due to the decrease in strain rate, but with the resumption of vigorous oscillation as ϕ was decreased towards the flammability limit as evident from the cycle-resolved images of the flame and the pressure signals in Figure 7.6. The pattern of flame movement clearly demonstrates that, close to global extinction, the flame undergoes repeated cycles of local extinction close to the step, followed by downstream translation, relight, and flashback leading to the re-establishment of the flame close to the step.

It is difficult to measure the velocity field under conditions of flow oscillation. However, the distribution of local time-averaged strain rates in the vicinity of the sudden expansion (Figure 7.7), calculated as the transverse velocity gradient of the axial velocity component measured in a plane sudden expansion in smooth combustion (Khezzar *et al.*, 1999), shows that the strain rates are largest close to the edges of the steps and decrease with downstream distance. This is consistent with the pattern of flame movement close to the lean flammability limit, with local extinction caused by

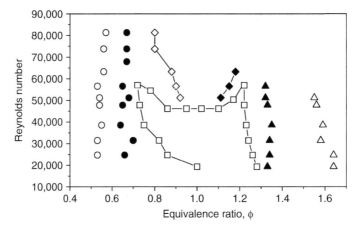

Figure 7.5 Flammability and stability limits (De Zilwa *et al.*, 2001): 160 mm wide plane sudden expansion: upstream duct with acoustically closed end, internal height 20 mm, length 880 mm; downstream duct of internal height 40 mm, length 680 mm. ○ – lean flammability limit; ● – lean extinction of one branch of flame; ◇ – lean stability limit (near-stoichiometric oscillations of large amplitude); ◆ – rich stability limit; ▲ – rich extinction of one branch of flame; △ – rich flammability limit; □ – boundary of moderate amplitude flow-dependent oscillations.

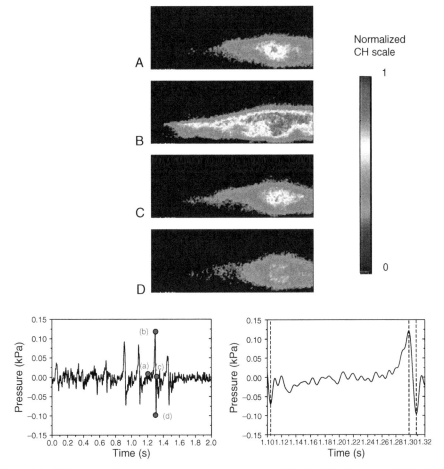

Figure 7.6 CH images of flame and pressure signal at lean flammability limit (De Zilwa *et al.*, 2001). Plane sudden expansion, duct geometry as in Figure 7.5; $Re = 57\,000$, $\phi = 0.55$. Observation window: 40×100 mm; exposure time $= 5$ ms; CH values normalized by maximum; average frequency of extinction and relight cycle $= 5.6$ Hz. (See color insert.)

large strain rates and re-establishment of the flame further downstream with a sufficient reduction in the strain rate.

Constriction of the combustor duct with an exit nozzle made the flammability range narrower, with an increase in the lean flammability limit and a decrease in the rich limit with increasing constriction, as is evident from the results of Figure 7.8 for the round duct of De Zilwa *et al.* (2001). Also, in plane sudden-expansion flows with an exit nozzle, the extinction of the two branches of the flame, unlike in ducts without an exit nozzle, was simultaneous except for rich extinction over a limited range of values of *Re*. The reason for extinction occurring earlier in ducts with a constricted exit as compared to unconstricted ducts seems to involve the larger amplitude of oscillations observed close to the flammability limit in ducts with an exit nozzle, giving rise to larger strain rates.

The effect of high swirl on flame stabilization has been studied in sudden-expansion flows (Sivasegaram and Whitelaw, 1988; Feikema *et al.*, 1991; De Zilwa *et al.*, 2001) where it has the effect of shortening the recirculation region behind the step and, as

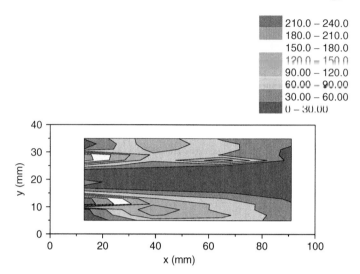

Figure 7.7 Time-averaged strain rates (De Zilwa *et al.*, 2001). Strain rate calculated as transverse gradient of axial velocity measurements of Khezzar *et al.* (1999); plane sudden expansion (duct geometry as in Figure 7.5), $Re = 76\,000$, $\phi = 0.72$ (See color insert.)

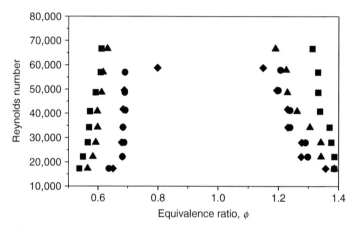

Figure 7.8 Effect of exit constriction on flammability range (De Zilwa *et al.*, 2001). Round duct, geometry as in Figure 7.4, without and with exit nozzle: methane-air ■ – without exit nozzle; ▲ – exit nozzle diameter, $D_E = 40$ mm; ● – $D_E = 25$ mm; ♦ – $D_E = 25$ mm.

observed by De Zilwa *et al.* (1999), improving flame stabilization by increasing the residence time. The addition of swirl, however, leads to an increase in axial velocity close to the step and the addition of a tangential component to the velocity so that the shear rate in the vicinity of the step is greater than in the corresponding unswirled flow. Figure 7.9A shows that the addition of a moderate amount of swirl with swirl number, Sw (the ratio of the tangential momentum of the flow upstream of the step and the product of the axial momentum with the radius of the upstream duct), on the order of 0.1, yielded a marginal reduction in ϕ_{lean} at low flow rates (Re less than 30 000). In contrast, larger flow rates and swirl numbers led to an increase in shear rate and in ϕ_{lean}, and the flame could not be stabilized on the step for Sw greater than 0.32 and Re

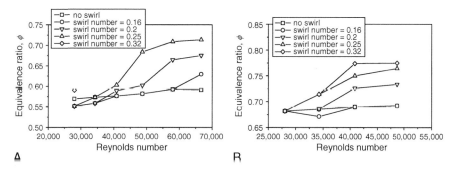

Figure 7.9 Influence of swirl on lean flammability limit (De Zilwa *et al.*, 2001). Round duct, geometry as in Figure 7.4, without exit nozzle, $Re = 49\,000$: (A) without exit nozzle and (B) with exit nozzle.

greater than 34 000. Note that this configuration corresponds to stabilization involving swirl and recirculation as opposed to the low swirl stabilization mechanism described in Chapter 6 that has no recirculation.

Laboratory studies of the effect of imposed pressure oscillations (De Zilwa *et al.*, 2001; Emiris and Whitelaw, 2003) on the lean flammability limit of round sudden-expansion flows without an exit nozzle showed that imposed oscillations led to an increase in ϕ_{lean}, owing to the increase in shear rate. However, in ducts with a constricted exit, where constriction leads to large amplitudes of oscillation dominated by the bulk-mode frequency of the combustor cavity over a range of values of ϕ close to ϕ_{lean}, the addition of swirl (Sivasegaram and Whitelaw, 1988), like the addition of pressure oscillations at a suitable frequency (Sivasegaram and Whitelaw, 1987a; Langhorne *et al.*, 1990), could lead to the amelioration of the bulk-mode oscillation frequency and hence to a decrease in ϕ_{lean}, over a limited range of flow condition as is evident from Figure 7.9B.

7.2.2. OSCILLATIONS IN DUCTS WITHOUT AN EXIT NOZZLE

Bradley *et al.* (1998) reported oscillations of large amplitude in a lean-burn swirl combustor close to the flammability limit. These studies were followed by an extensive investigation of oscillations and flame imaging in uniformly premixed methane–air flows in plane and round sudden expansions of moderate area expansion ratio (less than 3), with ϕ close to ϕ_{lean} (De Zilwa *et al.*, 2000, 2001). The results showed that the oscillations were dominated by a broadband low frequency of the order of 10 Hz, which was much smaller than any of the accompanying longitudinal acoustic frequencies of the combustor duct (~ 200 Hz) as well as the bulk-mode frequency (~ 50 Hz), thus confirming that the source of the low frequency was cycles of extinction and relight.

Flame behavior close to the lean flammability limit was similar to that close to the rich limit in plane as well as in round sudden-expansion flows (De Zilwa *et al.*, 2001). As the flame approached the step, the flame boundary moved closer to the center of the duct, in a manner similar to that observed during acoustic oscillations close to stoichiometry (De Zilwa *et al.*, 1999). However, the amplitude of oscillation, despite excursions of the flame up to 20 mm (two step heights) along the duct, was an order of magnitude less than that observed close to stoichiometry. Moreover, in the pressure signals of Figure 7.6 for a plane sudden expansion and Figure 7.10A for a round sudden expansion (De Zilwa *et al.*, 2001), the broadband low frequency is accompanied by the longitudinal ¼-wave frequency of the entire duct length, and not the ¾-wave frequency dominating oscillations close to stoichiometry. This is due to the heat release being distributed over a longer distance downstream of the step, unlike at near stoichiometry where heat release is intense close to the step.

The pressure signal of Figure 7.10A also shows that the period and the peak-to-peak amplitude of the cycles increased with time. The period of the extinction and relight cycle was found to be close to the value estimated as the sum of the time for forward movement at the average axial flow velocity of the shear layer, and the time for return at the burning velocity of the mixture. Figure 7.10B and C, respectively, shows the amplitude and period of the extinction and relight cycles as a function of Reynolds number, *Re* (based on the mean flow just upstream of the step) and time to global extinction. The growth in amplitude and in the period of the cycles is directly related to the increased travel of the flame preceding relight and return, resulting from the gradual weakening of the flame with each cycle of extinction and relight, and consistent with the findings of Sardi and Whitelaw (1999) that the opposed-jet flame weakened with successive imposed oscillations until extinction.

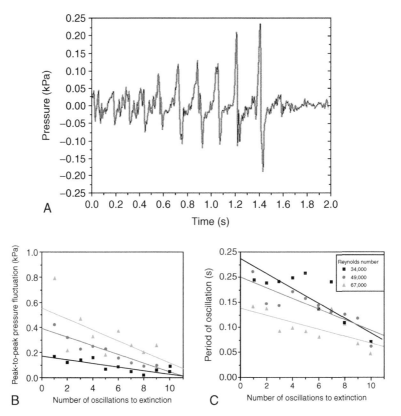

Figure 7.10 Pressure signal at lean limit, amplitude, and period of oscillation (De Zilwa *et al.*, 2001). Round duct, geometry as for Figure 7.4, *Re* = 67 000, ϕ = 0.61: (A) pressure signals preceding extinction, (B), and (C) amplitude and period of the 10 cycles preceding extinction.

The mean frequency of the relight and extinction cycles increased from 5 to 8.8 Hz as *Re* increased from 34 000 to 67 000 in the round sudden-expansion flow and from 4.4 to 7 Hz for *Re* from 39 000 to 81 000 in the plane duct. The higher frequency for the round duct was due to a shorter mean reattachment length than the plane duct (Gabruk and Roe, 1994). The amplitude of the cycle preceding extinction increased from 0.2 to 0.8 kPa as *Re* increased from 34 000 to 69 000, due to the larger intensity of heat release, while the increase in mean frequency with *Re* was due to the shorter downstream excursion time of the flame following local extinction.

The mean frequency of the extinction and relight cycles also increased with swirl (Figure 7.11) from around 7 to 10 and 20 Hz as *Sw* was increased from zero to 0.16 and 0.25, respectively, for *Re* = 49 000, and this is consistent with the shorter recirculation zone in flows with swirl (Ahmed and Nejad, 1992). Part of the increase in frequency at *Sw* = 0.25 is, however, due to the larger equivalence ratio. Also, with *Sw* = 0.25, the heat release close to the step became intense and led to the excitation of the 170 Hz,

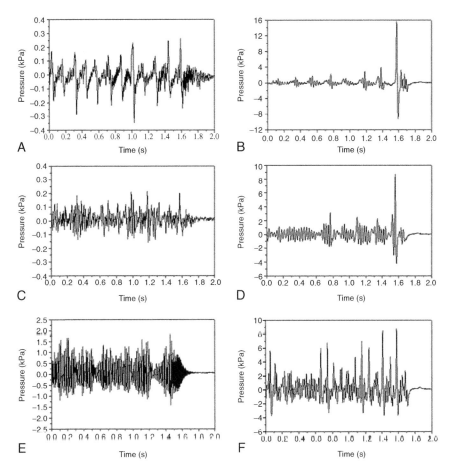

Figure 7.11 Influence of swirl on oscillations (De Zilwa *et al.*, 2001). Round duct, geometry as in Figure 7.4, *Re* = 49 000; (A), (C), and (E) signals in duct without exit nozzle; (B), (D), and (F) with 25 mm exit nozzle: (A) *Sw* = 0, ϕ = 0.58, average extinction and relight cycle frequency, f_{er} = 7 Hz, (B) *Sw* = 0, ϕ = 0.58, f_{er} = 4.6 Hz, (C) *Sw* = 0.16, ϕ = 0.58, f_{er} = 10 Hz, (D) *Sw* = 0.16, ϕ = 0.58, f_{er} = 5.5 Hz, (E) *Sw* = 0.25, ϕ = 0.68, f_{er} = 20 Hz, and (F) *Sw* = 0.25, ϕ = 0.68, f_{er} = 11.6 Hz.

¾-wave frequency of the whole duct with a pressure antinode near the step, compared with a 70 Hz, ¼-wave frequency for the smaller swirl number and without swirl.

With ϕ greater than ϕ_{stable}, the oscillations were of large amplitude and dominated by the acoustic ¾-wave frequency of the entire duct length. However, the larger flame speed of ethylene, which leads to a shorter flame and bigger amplitudes than for methane or propane, also implied a tendency for periodic flashback of the flame behind the step as ϕ was increased towards stoichiometry. This induced oscillations at the ¼-wave frequency of the upstream duct with a pressure node near the step, which were in competition with the dominant ¾-wave frequency of the whole duct with a pressure antinode near the step. The incompatibility of the boundary conditions near the location of flame stabilization for the competing frequencies led to a decline in amplitude with increase in equivalence ratio beyond 0.8, as may be seen from Figure 7.4B. Oscillations close to the lean flammability limit in ethylene flames were, however, similar to those described earlier for methane, with low frequency cycles of extinction and relight preceding extinction, but with somewhat higher frequency and amplitude because of the larger return speed of the flame following re-ignition and a greater intensity of heat release.

7.2.3.   DUCTS WITH AN EXIT NOZZLE

As was noted in Section 7.2.1, constriction of the duct had the general effect of narrowing the flammability range through an increase in the lean flammability limit and a decrease in the rich limit in the round duct of De Zilwa *et al.* (2001). Early flame extinction in plane sudden-expansion flows with an exit nozzle is often caused by the amplitude of the oscillations occurring close to the flammability limit being much larger than without an exit nozzle. The frequency dominating combustion oscillations close to stoichiometry depend on the degree of constriction of the duct exit: constriction by up to 80% of the duct sectional area favors the acoustic ¼-wave frequency of the upstream duct of around 100 Hz, while more severe constriction favors the bulk-mode frequency of the combustor cavity at around 50 Hz.

Flame behavior close to the flammability limit in ducts with an exit nozzle, however, resembles that in ducts without an exit nozzle, except for a much larger peak-to-peak amplitude of oscillations (order of 1 to 10 kPa compared with 0.1 to 1 kPa without an exit nozzle as may be seen from Figure 7.11A, B) and the presence of a bulk-mode frequency of around 50 Hz rather than a longitudinal acoustic frequency of the combustor alongside the broadband low frequency of the cycles of extinction and relight. The amplitude of pressure oscillations decreased with duct length as a result of the tendency for bulk-mode oscillations to weaken with cavity volume, as explained by Dowling and Williams (1983).

The larger amplitude with the exit nozzle appears to be a result of the cycles of extinction and relight inducing bulk-mode oscillations of the combustor cavity which in turn increase local strain rates and therefore proneness to local extinction. Another contributory factor to the large amplitudes is the higher value of ϕ_{lean} caused by the greater proneness to extinction, implying a larger heat release rate at the extinction limit than without an exit nozzle. As in ducts without an exit nozzle, the peak-to-peak amplitude and the frequency of the extinction and relight cycle also increased with flow rate owing to the larger heat release.

The mean value of the extinction and relight cycles was around 5 Hz, slightly higher than without the nozzle (5 Hz compared with around 4.5 Hz for the same bulk-flow rate) and has been explained by De Zilwa *et al.* (2001) in terms of a 25% higher flame speed based on the calculations of Andrews and Bradley (1972) for the higher equivalence ratio at the lean limit. In addition, the frequency of the extinction and relight cycle could, as discussed by De Zilwa *et al.* (2001), reach values as high as 100 Hz based on the time taken for the downward convection of the detached flame following local extinction, and the time for its return following relight. Thus, the prospect is high in practical combustors for coupling by resonance between the cycles of extinction and relight and a longitudinal acoustic frequency of the order of 100 Hz or a bulk mode frequency of the order of 20 Hz (Hubbard and Dowling, 2003), but not with radial and circumferential frequencies which are likely to be between 600 and 1000 Hz (Hubbard and Dowling, 2003). This could mean peak-to-peak amplitudes of the order of 0.1–1 bar compared with combustion chamber pressures of order 10 bar in practical combustors, with heat release rates of over a hundred times the 100 kW found in a laboratory combustor.

Mean concentrations of NO_x, measured close to the lean flammability limit in the exit plane of a round duct with an exit nozzle of 75% area blockage, were found to be around 50% higher than in a duct with an un-constricted exit for the same bulk-flow rate and equivalence ratio, but with a more stable flow (De Zilwa *et al.*, 2001). The increase in NO_x emissions appears to be due to the oscillations causing a skewing of the temperature distribution towards higher values than without oscillations, and is consistent with the opposite effect of a decrease in NO_x emissions caused by imposed oscillations in the vicinity of the adiabatic flame temperature, as in the gas turbine combustor of Poppe *et al.* (1998), where the oscillations skewed the temperature towards lower values. Large amplitudes of oscillation in practical lean burn combustors will therefore represent greater skewing of the temperature profile and higher levels of NO_x emission.

7.2.4. STRATIFIED FLOWS

Lean burn combustors comprising round sudden expansions as with Norster and De Pietro (1996) use a pilot stream at the core to stabilize an otherwise lean flow. Stratification is also known to exist in other gas turbine flows because of the way the air and fuel supply is arranged (Moore, 1997; Milosavljevic *et al.*, 2003). Experimental studies in stratified flows have been carried out in round ducts with fuel concentration decreasing with distance from the center of the duct (Emiris, 2003; Luff *et al.*, 2004; Luff, 2005) to explore the effect of stratification on combustion oscillations and examine the prospect of reducing the overall equivalence ratio at which a combustor could be stably operated. The flow arrangement of Figure 7.12 had premixed core and annular flows of different equivalence ratios, and the radial distribution of fuel concentration could be modified by varying the equivalence ratios and the distance between the exit of the core duct and the step. Results have been reported with the mass flow rates and equivalence ratios of the core and annular flows as variables, and the fuel (methane, propane, or ethylene) as an additional variable.

Figure 7.13 shows the variation in rms pressure fluctuation with equivalence ratio in the core and annulus of the round sudden-expansion flow of Luff (2005) with an area expansion ratio of 2.5 and without an exit nozzle, for methane and ethylene and two values of *Re* based on the bulk-mean velocity upstream of the step. While it is possible

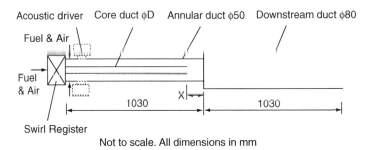

Figure 7.12 Flow arrangement for stratified flow (Emiris *et al.*, 2003; Luff *et al.*, 2004; Luff, 2005).

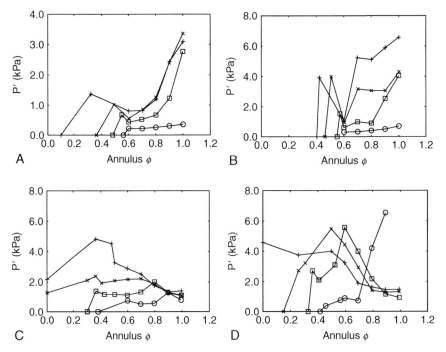

Figure 7.13 The rms pressure fluctuation as function of ϕ_{core} and $\phi_{annulus}$ for stratified flows with methane and ethylene as fuels (Luff, 2005). Duct geometry of Figure 7.12. Extinction limits are at $P' = 0$: (A) methane, $Re = 52\,000$, (B) methane, $Re = 62\,000$, (C) ethylene, $Re = 52\,000$, and (D) ethylene, $Re = 62\,000$. $\bigcirc\ \phi_{core} = 0.6$, $\square - 0.8$, $\times - 1.0$, $+ - 1.2$.

to achieve a stable flame with a very lean annular flow and a rich core, for comparable core and annular mass flow rates, the overall equivalence ratio at the lean flammability limit $\phi_{lean,\,overall}$ is often more than that in a uniformly premixed flow. This is due to the core flow carrying a larger mass flow than is realistic for a pilot stream. A modest reduction in $\phi_{lean,\,overall}$ was observed, however, with a smaller core pipe with bulk-flow rate nearly a third of that in the annulus (for example, $\phi_{lean,\,overall} = 0.55$ compared with $\phi = 0.58$ in a uniformly premixed flow, with methane as fuel and $Re = 62\,000$). It may, however, be possible to reduce $\phi_{lean,\,overall}$ with a richer core by employing a much smaller mass flow rate than in the annulus, and with swirl added to enhance entrainment by the annular flow.

Stable flames have, on the other hand, been achieved with methane, propane, and ethylene as fuels, with $\phi_{core} = 0$ and $\phi_{annulus}$ slightly above ϕ_{lean} for the uniformly premixed flow (Emiris *et al.*, 2003) so that $\phi_{lean, \, overall}$ was significantly lower than ϕ_{lean} ($\phi_{lean, \, overall} = 0.4$ or less with methane as fuel and $Re = 62\,000$ compared with $\phi_{lean} = 0.58$ in uniformly premixed flows). Although such stratification also meant smaller amplitudes of oscillation than in uniformly premixed flows (see Figure 7.14), it is of no benefit to practical gas turbine combustors to have the annular flow richer than in the core.

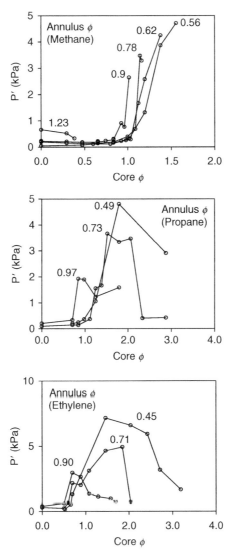

Figure 7.14 The rms pressure fluctuations for stratified flows with methane, propane and ethylene as fuel and ϕ_{core} varied from 0 to 3 (Emiris *et al.*, 2003). Duct dimensions as in Figure 7.10, core pipe diameter = 35 mm, length = 810 mm; $Re = 62\,000$.

With a rich core and a lean annular flow, the acoustic ¾-wave frequency of the entire duct dominated oscillations for ϕ_{overall} close to unity for methane and propane and over 0.6 for ethylene, almost as in uniformly premixed flows. With ethylene, again, as in uniformly premixed flows (Korusoy and Whitelaw, 2002; Emiris *et al.*, 2003), the amplitude decreased with ϕ_{overall} greater than around 0.8, as a consequence of the tendency for the flame to attach close to the step and flash back, leading to the excitation of the ¼-wave frequency of the upstream duct length in competition with the dominant ¾-wave in the whole duct.

Flows with fuel concentrations at the center of the duct that were higher than near the wall gave rise to cycles of extinction and relight with amplitudes comparable to those observed close to stoichiometry, and over a fairly wide range of values of ϕ_{overall} close to the lean flammability limit. For example, for *Re* around 62 000, large-amplitude oscillations dominated by the broadband low frequency of median value between 5 and 10 Hz prevailed for ϕ_{annulus} between 0.45 and 0.6 and ϕ_{core} around unity. With a richer core, such oscillations prevailed over an even wider range of ϕ_{annulus}. The amplitude of the low frequency increased with $\phi_{\text{core}} - \phi_{\text{annulus}}$ as well as with overall heat release rate, and the rich core flow ensured that extinction and relight cycles did not lead to a gradual weakening of the flame and global extinction, except very near the lean flammability limit.

The pressure signal in Figure 7.15 for the stratified flow comprised bursts of large-amplitude oscillations, interspersed with periods of smaller amplitude of the order of 0.05 s. With a very lean annular flow and a rich core, the duration of small amplitude close to extinction could be as long as 0.5 s, due to re-ignition occurring far from the step and a slow return of the flame. Although the rms pressure fluctuations shown in Figure 7.13 for a very lean annulus and a rich core are comparable to those for flows with a smaller difference between ϕ_{core} and ϕ_{annulus} and a similar ϕ_{overall}, the peak-to-peak fluctuations in pressure are much larger than what the rms values suggest. The bursts of large amplitude followed by spells of smaller amplitude in Figure 7.15A and the spectral information of Figure 7.15B point to cycles of extinction and relight at a median frequency of 7 Hz.

The reason for the larger peak-to-peak amplitude when the core is much richer than the annulus could have been explained based on flame visualizations of De Zilwa *et al.* (2004) in uniformly premixed methane–air flames, which showed that the instantaneous

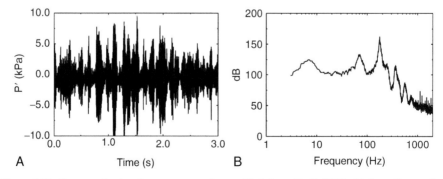

Figure 7.15 Pressure signal and power spectra for stratified flows (Luff, 2005). Methane flames, duct geometry of Figure 7.12: (A) and (B) $\phi_{\text{core}} = 1.2$, $\phi_{\text{annulus}} = 0.9$, *Re* = 62 000, (C), and (D) $\phi_{\text{core}} = 1.20$, $\phi_{\text{annulus}} = 0.43$, *Re* = 62 000.

pressure in the vicinity of the sudden-expansion rose as the flame boundary shifted towards the duct axis in the upstream direction during acoustic oscillations, and fell as the flame moved downstream. Thus, a rich core would enhance the intensity of heat release at the time of a pressure maximum, which coincides with the movement of the flame close to the center of the duct, leading to an increase in the amplitude of oscillation, as explained by Rayleigh (1896). On the other hand, a lean core and a rich annular flow would lead to amplitudes less than in uniformly premixed flows with the same overall equivalence ratio, as may be seen from the results of Figure 7.13.

Since large amplitudes of oscillation in the stratified flow with a rich core imply large strain rates, the flame would be susceptible to cycles of extinction and relight, as well as an increase in $\phi_{lean,\ overall}$. Thus, the need for control of the low frequency, thereby averting early extinction, is greater in flows with a rich premixed core and a lean annulus than in uniformly premixed flows.

While the results of Luff *et al.* (2004) and Luff (2005) involve mostly near-stoichiometric flow conditions, they identify a fairly wide range of flow conditions with a very lean annular flow and a near-stoichiometric core with cycles of extinction and relight, where the large amplitudes of the low frequency do not lead to blow-off. This offers an opportunity to explore the effectiveness of control strategies in flows where the broad-band low frequency coexists with an acoustic frequency (or a bulk-mode frequency in the case of a duct with an exit nozzle), without risk of blow-off during the process of varying control parameters.

7.3.   CONTROL STRATEGIES

Active control of combustion oscillations has been demonstrated in bluff-body stabilized flames as well as in sudden-expansion flows with a large area expansion ratio via the oscillation of the bulk-flow rate, pressure field, and local heat release rate with an appropriate phase difference in naturally occurring oscillations. Control strategies applied to heat release rates of up to a few hundred kilowatts in the laboratory have been successfully extended to industrial scale burners with heat release rates of several megawatts (Seume *et al.*, 1997; Pascherit *et al.*, 1998) and oscillations dominated by a natural frequency of the combustor duct. For active control to be effective, deviation from the dominant frequency needs to be small and predictable to provide the oscillated input at the correct frequency and phase. Alternative strategies are needed for flow conditions that are not amenable to active control by oscillated input with a phase difference.

Addition of pressure oscillations is the simplest to arrange in the laboratory, and impressive reductions in the amplitude of oscillations have been reported in laboratory combustors with modest rates of heat release, varying from attenuation of an rms pressure fluctuation of 1 kPa by 40 dB (a factor of 100) in a ducted premixed flame with a heat release rate of 500 W (Gulati and Mani, 1990) to attenuation of rms pressure fluctuations of the order of 5 kPa by up to 20 dB (a factor of 10) with heat release rates of up to 100 kW (Sivasegaram and Whitelaw, 1992). Oscillation of the bulk flow (Bloxsidge *et al.*, 1987), although successful, has not been followed up in view of practical difficulties. Oscillation of heat release proved necessary with larger heat release rates and rms pressure oscillations in excess of 10 kPa in laboratory combustors and in industrial scale combustors.

Since existing actuators and control strategies have been developed to control combustion oscillations in near-stoichiometric flows dominated by an acoustic or bulk-mode frequency of the combustor, they will not always be suitable for use in the control of oscillations observed under lean burn conditions. Some of the important types of actuators that have been developed and strategies for active control will be briefly outlined and commented upon in the next two subsections. The third subsection will deal with control strategies for oscillations under lean burn conditions, especially close to the flammability limit.

7.3.1. ACTUATORS

Pressure oscillations may be added using strategically located acoustic drivers (Billoud *et al.*, 1992) as well as loudspeakers (Sivasegaram and Whitelaw, 1992) as shown in Figure 7.16. Since acoustic power input decreases as the square of the frequency, acoustic drivers, although easy to install in laboratory combustors without significantly altering the flow and acoustic characteristics of the combustor, are not very effective with frequencies less than 50 Hz. While wide-range loudspeakers could provide a larger acoustic power input than acoustic drivers at low frequencies, they significantly alter the acoustic boundary conditions of the combustor cavity and are impractical in many situations. Thus, input at frequency values less than 50 Hz has generally required pulsing the fuel flow in laboratory combustors. However, adding pressure oscillations would not be suitable for practical flows with large heat release rates or large amplitudes of oscillation.

Many actuator designs have been successfully developed to implement the oscillation of heat release using arrangements to pulse the flow of fuel (Seume *et al.*, 1997; Lee *et al.*, 2000) or a premixed secondary fuel–air stream (Billoud *et al*, 1992), oscillate the composition of a fuel–air mixture (Sivasegaram *et al.*, 1995a), and pulse the ignition of a mixture of secondary fuel and air (Wilson *et al.*, 1992). Control by oscillated removal of heat by pulsing the injection of water was found to be as effective as the pulsing of liquid fuel, with the added benefit of a reduction in NO_x emission (Sivasegaram *et al.*, 1995b).

Pulsed fuel flow is subject to fluid-dynamic damping as well as effects of delay in the evaporation and burning of the fuel so that the actual oscillated input, unlike that of acoustic drivers, could be much less than what corresponds to the rate of pulsing. Although injectors are available to pulse fuel with frequencies close to 1 kHz, damping of the oscillated input becomes increasingly severe with frequency, and ensuring a large oscillated input with frequency greater than 200 Hz would require the addition of pulsed fuel close to the location where it is intended to be consumed (Sivasegaram *et al.*, 1995a). Pulsing the flow of liquid fuel as well as that of a spray of water would require fine atomization of the liquid so that the time delay in the evaporation would not be greater than the period of oscillation. Finer atomization of injected fuel sprays is not practical inside industrial burners, and fuel injectors with fine atomization (mean droplet diameter around 10 μm) will be necessary. Pulsing the flow rate or the fuel content of a secondary air stream carrying pre-vaporized or atomized fuel could also be an option.

Fuel ignition pulsation (Wilson *et al.*, 1992) is sensitive to fuel characteristics and requires a gaseous fuel with a high flame speed such as ethylene or acetylene and, besides requiring modifications to the flow arrangement, could give rise to additional

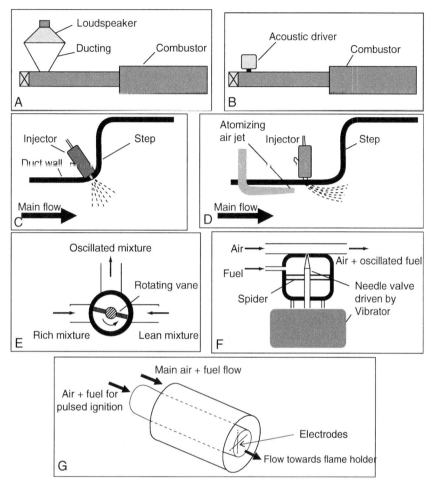

Figure 7.16 Some actuators used in active control: (A) loudspeaker, (B) acoustic driver, (C) pulsed fuel injector, (D) pulsed injector with cross jet of air to atomize fuel spray, (E) arrangement to oscillate secondary air-fuel mixture composition, (F) arrangement to pulse fuel content of secondary air fuel supply, and (G) arrangement to pulse the ignition of fuel.

frequency modes that may be associated with large amplitudes (Sivasegaram *et al.*, 1995a).

The use of an actively tuneable Helmholtz resonator attached to the combustor has been proposed by Wang and Dowling (2002) to make the resonator adapt to variations in the dominant frequency. However, the slow speed of response of the system to variations in frequency restricts the use of the resonator as an actuator for active control.

7.3.2. ACTIVE CONTROL STRATEGIES

Active control has mostly concerned the amelioration of a single acoustic frequency of the combustor using a closed-loop control circuit (see Figure 7.17). The feedback is often a fluctuating wall static pressure signal, and free-field sound signals could be

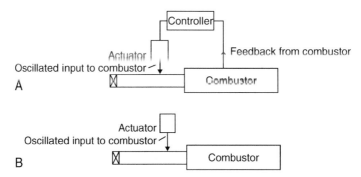

Figure 7.17 Control circuits for active and passive control: (A) a simple active (closed-loop) control circuit and (B) passive (or open-loop) control circuit for imposed oscillations.

equally effective. The use of feedback in the form of fluctuation in light emission (Sreenivasan *et al.*, 1985) or the chemi-luminescence of CH or OH radical (Higgins *et al.*, 2001a, b) has been demonstrated in the laboratory. However, optical access to the flame could be a problem in some combustors. Generally, the feedback is band-pass filtered to provide a clean signal at the dominant discrete frequency, except where rival frequencies exist, in which case filtering needs to be arranged with separate windows for the competing frequencies. The oscillated input is provided at the dominant frequency of the feedback, but with a difference in phase and amplitude based on that of the feedback.

The optimum phase difference between the feedback and the oscillated input could be predetermined when the input is a pressure oscillation. With oscillation of heat release, however, the optimum phase difference needs to be determined empirically because of the uncertainty in the time lag between the addition of the pulsed fuel and its burning.

Modulating the oscillated input to match the feedback is a complex task, and attempts to derive a universal transfer function to relate the feedback from the combustor to the oscillated input have been unsuccessful because of the non-linear nature of the problem and difficulties in extrapolating results obtained in combustors with small heat release rates to larger heat release rates (Gulati and Mani, 1990). A variety of approaches have been used to increment the oscillated input according to the amplitude of the feedback. A relatively simple method is to have the amplitude of the feedback directly proportional to that of the input (Bloxsidge *et al.*, 1987). Other approaches include incrementing the input according to the variation in amplitude of the feedback during the preceding cycle, and a pseudo-adaptive strategy where the input is varied according to an empirically predetermined modulated waveform (Sivasegaram *et al.*, 1995a). The rms pressure fluctuations of around 10 kPa with heat release on the order of 100 kW have been attenuated by between 12 and 15 dB (a factor of 4–5) using different control strategies, and the effectiveness of control was not very sensitive to the control parameters or phase within 20° (Sivasegaram *et al.*, 1995a).

Adaptive filters have been used successfully to attenuate rms pressures of around 5 kPa by between 15 and 20 dB (a factor of 5–10), with heat release rates of up to 100 kW (Hendricks *et al.*, 1992). However, for adaptive control, where the controller provides an appropriate oscillated input in anticipation of the change in feedback, to be effective, a large database would be required. Hence, practical adaptive control systems are likely to be specific to a burner design and to a range of operating conditions.

Several adaptive control strategies are being developed, as for example by Polifke and Paschereit (1998), Evesque and Dowling (2001), Park *et al.* (2003); Waschman *et al.* (2003), and Yang *et al.* (2002), to address the problems of non-linearity in the control of combustion oscillation, but so far the emphasis has been on near-stoichiometric conditions.

Although active control can successfully control oscillations dominated by a single discrete frequency, there are instances where suppression of the dominant frequency has led to the excitation of a dormant rival frequency mode, as for example in the combustors of Bhidayasiri *et al.* (2002), with complex flame-holding arrangements comprising the combinations of a bluff body and a step, and swirl and a step. Control of the dominant frequency could also lead to the excitation of its harmonic, as in the bluff-body stabilized flame of Sivasegaram and Whitelaw (1992). Thus, there is a need for control systems that may require a combination of strategically located actuators to deal individually with rival frequencies, which could either coexist or occur alternately, and to respond to each frequency as it arises.

Bulk-mode oscillations, especially when associated with poor flame stabilization, are less amenable than acoustic frequencies to active control, and levels of attenuation have been modest, typically less than 4 dB (a factor of 1.6) compared with over 12 dB (a factor of 4) for acoustic oscillations with comparable amplitudes and heat release rates. Addition of oscillations at an empirically determined alternative frequency, usually an acoustic frequency of the combustor associated with small amplitudes of oscillation or its sub-harmonic, has been found to be effective (Sivasegaram and Whitelaw, 1992; Wilson *et al.*, 1992). This control strategy is referred to here and by Emiris *et al.* (2003) and De Zilwa *et al.* (1999), among others, as passive control since the input is not linked to the feedback from the combustor and is provided in the manner that swirl or fuel may be added to improve flame stability. It is also referred to as open-loop control (Wilson *et al.,* 1992; McManus *et al.*, 1993) since there is no control circuit linking the input signal to the feedback, and has been found to be more effective than active (closed-loop) control, with attenuation around 10 dB (a factor of 3) in the suppression of bulk-mode oscillations in ducts with an exit nozzle (Wilson *et al.*, 1992; Sivasegaram *et al.*, 1995a).

Active control also performed poorly in the control of acoustic oscillations (the ¾-wave frequency of the combustor duct) in plane and round sudden expansions with area expansion ratios of around 2.5 or less (De Zilwa *et al.*, 1999), as a result of the modulation in amplitude as well as frequency caused by the movement in flame position during control, leading to a loss of lock between the feedback and the input (Sivasegaram and Whitelaw, 1992; Yu *et al.*, 2002). The problem is further aggravated when the amplitude and frequency of the naturally occurring oscillations themselves suffer modulation due to a periodic shift in the position of the flame and a variation in the distance over which heat release occurs (De Zilwa *et al.*, 1999). Also, as shown in Figure 7.18, passive control by adding oscillations at around 420 Hz (the next harmonic of the ¾-wave frequency of the entire duct length) with an acoustic driver resulted in more effective suppression of the dominant 220 Hz frequency (¾-wave in the whole duct) with attenuation of the rms pressure fluctuation between 5.5 and 12 dB (factors of 1.8 and 4) compared with 3 and 7 dB (factors of 1.4 and 2.2) by active control. Passive control by pulsing the fuel flow at values of alternate frequencies less than 100 Hz was as effective as using acoustic drivers at 420 Hz.

Oscillations close to the lean flammability limit involve a broadband low frequency accompanied by acoustic or bulk-mode oscillations, whose amplitude and frequency are

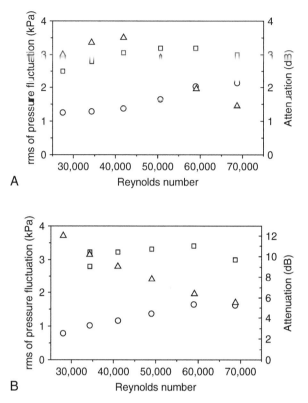

Figure 7.18 Active and passive control by imposed oscillation (De Zilwa *et al.*, 1999). Duct geometry as in Figure 7.4; upstream duct length = downstream duct length = 850 mm; control at $\phi = 1.0$, dominant frequency = 220 Hz: (A) active control and (B) passive control, oscillations imposed at 420 Hz. □ – amplitude without control, ○ – amplitude with control, △ – attenuation.

subject to modulation. Thus the task of control is more complex than that of oscillations close to stoichiometry because of the presence of two frequencies whose amplitudes and frequency values are subject to modulation, and because of the need to establish control within a few cycles of the onset of oscillation to avert the risk of global extinction.

7.3.3.   CONTROL OF OSCILLATIONS CLOSE TO THE LEAN FLAMMABILITY LIMIT

Oscillations at values of ϕ close to the lean flammability limit are, as explained earlier, driven by cycles of extinction and relight with a broadband low frequency generally of the order of 10 Hz compared to acoustic and bulk-mode frequencies of the order of 100 and 50 Hz, respectively. The low frequency oscillation is transient in nature, with the flame weakening and the amplitude of oscillation and the period of the cycles increasing with every cycle of extinction and relight. This is accompanied by a natural frequency of the duct, an acoustic or bulk-mode frequency, depending on the duct geometry.

Thus, to prevent blow-off close to the lean flammability limit, active control is confronted with the need to determine the low frequency within a few, probably less than ten, cycles from the onset of oscillations, whose period could vary by as much as

33%, generally decreasing but not always systematically. As a result, control strategies developed to deal with acoustic oscillations are, in themselves, unlikely to be effective in the control of the very low frequency associated with extinction and relight. Oscillations at a natural frequency of the combustor accompanying the low frequency are also subject to modulation in amplitude and frequency owing to the cyclical variation in flame position and heat release at the low frequency. Active control is likely to face problems of loss of lock between the input and the feedback, so passive control by the addition of oscillations at an alternative frequency is likely to be more effective. Thus it may be necessary to use a combination of strategies to deal with the two frequencies individually.

Oscillated input in the form of imposed pressure oscillations is simpler to install and to implement than the pulsing of fuel flow. Large oscillated inputs would, however, require the pulsing of fuel, and imposed pressure oscillations, except when they significantly attenuate the naturally occurring oscillations, will lead to increased strain rates (Emiris and Whitelaw, 2003) and an increase in ϕ_{lean} as well as levels of NO_x emission. Thus, the oscillated inputs for control close to the lean flammability limit need to be much smaller in magnitude than at higher equivalence ratios.

Since the low frequency is a direct result of cycles of extinction and relight, strategies for its control could include, in part or whole, the addition of swirl or fuel as a means of making the flame more robust and less prone to local extinction. Excessive swirl, however, entails the risk of flashback.

Active control of oscillations close to extinction was examined in the uniformly premixed flows of Emiris and Whitelaw (2003), and Emiris *et al.* (2003) and was found to be ineffective. Passive control by the addition of oscillations with a pair of acoustic drivers at an alternate frequency was ineffective in sudden-expansion flows with an unconstricted exit. In sudden expansions with an exit nozzle, passive control reduced the amplitudes of the bulk-mode frequency and, consequently, the broadband low frequency by a factor of three (Figure 7.19) as well as reduced the lean flammability limit (ϕ_{lean} around 0.66 compared with 0.69 without control) owing to the reduction in amplitude of oscillation leading to lower strain rates (De Zilwa *et al.*, 2001).

Injection of fuel into the shear layer helped to remove the low frequency in some cases, and it has been shown by Emiris and Whitelaw (2003) that radial injection of small quantities of fuel upstream of the step in a round sudden expansion improved flame stabilization and reduced the amplitude of extinction-and-relight oscillations in unconstricted as well as constricted ducts. No significant lowering in the lean flammability limit was demonstrated, although injection of fuel in high concentration could have adverse implications for NO_x emissions. Sensitivity to fuel injection rate near the lean flammability limit has been further examined by Emiris *et al.* (2003) to show that the benefit of fuel injection is limited to a narrow range of values of injection rate, with a 35% reduction in amplitude, from 0.08 to 0.05 kPa by the injection of 3.8% of the total fuel supply close to the step (Figure 7.20). Larger injection rates led to poorer flame stability and larger amplitudes. Additionally, the position of the injectors, the bulk equivalence ratios and flow rates are important for *a priori* decisions relating to the use of fuel injection in the control of oscillations.

Addition of swirl, as discussed in Section 7.2.1, offers the benefit of improving flame stabilization and therefore reducing the amplitude of the low frequency. The added strain rate due to the introduction of swirl could, however, offset this benefit to increase ϕ_{lean}, except over a limited range of flow conditions in ducts with exit nozzles.

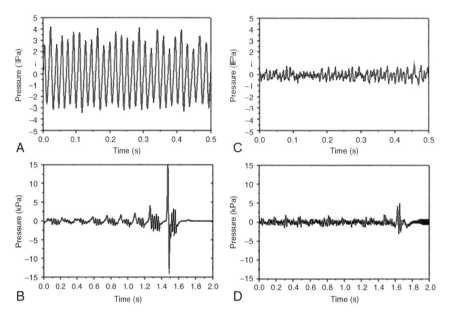

Figure 7.19 Control at the lean flammability limit (De Zilwa *et al.*, 2001). Uniformly premixed flow in round duct with 25 mm exit nozzle, other dimensions as in Figure 7.4, *Re* = 34 000: (A) $\phi = 1$, no imposed oscillations, (B) oscillations added at 220 Hz with acoustic drivers, (C) $\phi = 0.69$, no imposed oscillations, and (D) $\phi = 0.66$, oscillations added at 220 Hz with acoustic drivers.

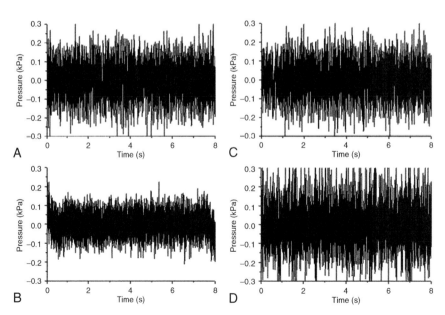

Figure 7.20 Effect of fuel injection rate on the time-resolved pressure signals (De Zilwa *et al.*, 2003). Duct diameters as in Figure 7.4, $X = L = 1030$, methane–air, *Re* = 26 500, $\phi = 0.6$. Fuel injection given as fraction of total fuel: (A) 0%, (B) 1.9%, (C) 3.8%, and (D) 5.0%.

The stratified flows of Luff *et al.* (2004), and Luff (2005), with fuel concentration increasing with proximity to the center of the duct, gave rise to cycles of extinction and relight and large amplitudes of oscillation over a fairly wide range of values of ϕ_{overall} adjoining the lean flammability limit. The broadband low frequency of the extinction and relight cycle was accompanied by the longitudinal acoustic ¾-wave frequency of the combustor duct. Control of the acoustic frequency proved ineffective because of the modulation in amplitude and frequency; and passive control by the addition of an alternative frequency with a pair of acoustic drivers was effective over a wide range of flow conditions (Luff, 2005).

Experiments with methane and ethylene as fuels identified an alternative frequency of around 70 Hz, corresponding to the ¼-wave frequency of the entire duct, and another around 420 Hz, close to the second harmonic of the ¾-wave frequency of the whole duct, to be dominant for ϕ_{overall} close to unity. For smaller values of ϕ_{overall}, the imposed frequency reduced the amplitude of the acoustic ¾-wave and thereby the instantaneous strain rates and proneness to extinction so that the low frequency oscillations were averted over a wide range of values of ϕ_{overall}, except near the lean flammability limit where the amplitude of the low frequency was large and cycles of local extinction and relight could not be prevented by suppressing the acoustic frequency.

Large amplitudes of the low frequency in the stratified flow are due to local extinction in the lean outer region of the flow being relit by the richer flow nearer the center, with the fuel concentration in the core flow influencing the location of relight and the amplitude and frequency of extinction and relight cycles. Local extinction could be avoided by enriching the outer flow, as has been demonstrated in uniformly premixed

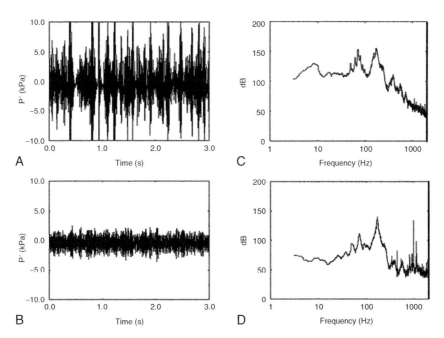

Figure 7.21 Control by imposed alternate frequency in stratified flow (Luff, 2005). Methane–air flame, duct dimensions as in Figure 7.11, $Re = 62\ 000$, $\phi_{\text{core}} = 1.0$, $\phi_{\text{annulus}} = 0.55$: (A) and (B) without control, (C) and (D) acoustic forcing at 1000 Hz; input power around 40 W.

flows (Emiris and Whitelaw, 2003). In the stratified flow of Luff (2005) this was achieved by enhancing fine-scale mixing between the core and the annular flows by adding oscillations at a frequency of the order of 1 kHz using an acoustic driver. Figure 7.21 shows that the addition of oscillations at 1 kHz suppressed the low frequency oscillations, thereby reducing the peak-to-peak amplitude of over 12 kPa by a factor of 4, the 3.5 kPa rms pressure by a factor of 3, and the amplitudes of the low as well as the acoustic frequency by around 30 dB (a factor of 30). This reduction occurred mainly because enhanced mixing between the core and annular flows enriched the latter and suppressed local extinction.

The strategy of enriching the outer flow by imposed oscillation could, sometimes, lead to large amplitudes of the acoustic 3/4-wave frequency, and control may require the combination of a high frequency to enhance mixing with a smaller alternative frequency to suppress the acoustic frequency. This strategy was not always effective, as a result of the inadequate strength of the oscillated input needed for a large difference between ϕ_{core} and $\phi_{annulus}$, for $\phi_{overall}$ close to the lean limit (Luff, 2005).

Enhancing the mixing between the core and annular flows by the addition of pressure oscillations will not be feasible in practical combustors but could be achieved by the addition of a moderate amount of swirl to the core flow. In this case, control of oscillations will require the pulsing of fuel flow, desirably at the longitudinal acoustic frequency of the entire combustor or a sub-harmonic.

7.4. CONCLUDING REMARKS

Although the causes of combustion oscillations are not always very clear, it is known that acoustic coupling between the combustion instability and a natural frequency of the combustor generally gives rise to large amplitudes of oscillation. Studies close to the flammability limit in laboratory combustors with flame stabilization behind sudden-expansions and bluff bodies have shown that cycles of local extinction and relight could cause a progressive weakening of the flame and large amplitudes of low frequency oscillation, leading to global extinction. It was also found that constriction of the combustor exit and stratification of the fuel distribution could lead to larger amplitudes of oscillation than in ducts without an exit constriction. In all cases, the dominant broadband low frequency was accompanied by a natural frequency of the combustor, with the amplitude and frequency mode dependent on the flow boundary conditions. Oscillations associated with cycles of extinction and relight have also been observed in opposed-jet flames and swirl burners and are thus likely to occur in a wider range of combustor geometries than investigated so far.

The amplitude of oscillations close to the lean flammability limit is likely to be larger in practical combustors than in the laboratory due to heat release rates being on the order of a hundred times larger, and to augmentation of the oscillations by acoustic coupling between the cycles of extinction and relight and a natural frequency of the combustor. It was estimated in De Zilwa *et al.* (2001) that the frequency of the extinction and relight cycle could be as high as 100 Hz in practical systems, compared with on the order of 10 Hz in laboratory combustors as a result of the higher temperatures and flow speeds in practical combustors. Therefore, the prospects are high for acoustic coupling with a natural frequency of the practical combustor with bulk-mode frequency around 20 Hz compared with around 50 Hz in the laboratory combustor, or a longitudinal acoustic frequency of around 100 Hz compared with around 200 Hz in laboratory combustors, leading to an increase in the lean extinction limit. Oscillations in lean premixed flames are also known to lead to increased levels of NO_x emission. Thus, besides considerations of structural integrity and stable operation of the combustor, the need for the control of oscillations is further underlined by the need to widen the flammability range and to reduce the level of NO_x emissions, in view of the adverse implications of the oscillations for the flammability limit and NO_x concentrations.

A variety of active control strategies have been developed along with a wide range of actuator designs, mainly to pulse the injection of fuel in order to control discrete frequency oscillations observed close to stoichiometric burning conditions. However, the broadband nature of the dominant low frequency close to the flammability limit, which distinguishes near-extinction oscillations from those under near-stoichiometric burning conditions, makes the low frequency less amenable to active control by imposing oscillations at the frequency of the instability but with a predetermined difference in phase. The low frequency oscillations also lead to a modulation in the amplitude as well as the frequency value of the accompanying oscillations at a natural frequency of the combustor, so that active control of the latter frequency is also made difficult. However, recent trends in control algorithms that adapt in real time to identify and match the dominant frequency, as reviewed for example by Yang *et al.* (2002), offer some promise and are the subject of continuing research.

Since the low frequency oscillations arise from cycles of local extinction and relight, control strategies need to emphasize improved flame stability in order to reduce the prospect of local extinction. The addition of a moderate amount of swirl and injection of fuel into the shear layer are known to assist flame stabilization. However, excessive swirl is accompanied by the risk of flashback and is known to increase the frequency of the extinction and relight cycles and the chances of acoustic coupling with a natural frequency of the combustor. In the case of the latter, it is necessary to ensure that the injection of fuel does not imply a significantly higher overall equivalence ratio or higher NO_x emissions. Stratification of the flow with the provision of a richer core flow, with the addition of moderate swirl if necessary, could lower the overall equivalence ratio at the lean flammability limit. However, stratification could give rise to large amplitudes of the low broadband or natural frequency of the combustor.

Control by the addition of oscillations at an alternative frequency is more effective than conventional active control in ameliorating combustion oscillations with modulated amplitude and frequency, and besides reducing the amplitude of oscillations, could be useful in extending the flammability limit. The effectiveness of this approach close to the flammability limit is yet to be determined. A combination of strategies, one to address the problem of flame stability and the extinction and relight cycles, and another to control the natural frequency has been suggested by Emiris and Whitelaw (2003) with, for example, steady fuel injection to stratify the mixture prior to injection and pulsed injection of fuel to attenuate the natural frequency of the combustor. Furthermore, the performance of any control system depends strongly on the suitability of the actuator for the task, and the strategic location of fuel injectors for pulsed as well as steady injection could be crucial to ensure optimum performance (Lee *et al.*, 2000). Attention also needs to be paid to aspects of burner design to ensure sufficiently high wall temperatures in the region of flame stabilization, since low temperatures could make the combustor more prone to local extinction and therefore larger amplitudes of oscillation (Korusoy and Whitelaw, 2004).

ACKNOWLEDGMENTS

The author acknowledges with gratitude the encouragement and support of Dr. Derek Dunn-Rankin, valuable advice and suggestions by Professor J.H. Whitelaw, and useful discussions with Professor R.P. Linsteadt and Dr. G.D. Roy. The author is also thankful to Drs. S.R.N. De Zilwa and D. Luff for helpful discussions and for softcopies of figures from their work.

REFERENCES

Ahmed, S.A. and Nejad, A.S. (1992). Swirl effects on confined flows in axisymmetric geometries. *J. Prop. Power* **8**, 339–345.

Allen, R.P. and Kovacik, J.M. (1984). Gas turbine cogeneration: principles and practice. *ASME J. Eng. Gas Turbine Power* **106**, 725–730.

Andrews, S.A. and Bradley, D. (1972). The burning velocity of methane–air mixtures. *Combust. Flame* **19**, 275–288.

Bhidayasiri, R., Sivasegaram, S., and Whitelaw, J.H. (2002). Control of oscillations in premixed gas turbine combustors. In: Roy, G.D. (Ed.). *Advances in Chemical Propulsion: Science to Technology*. CRC Press, pp. 303–322, London, UK.

Billoud, G., Galland, M.A., Huynh Huu C., and Candel, S. (1992). Adaptive active control of combustion instabilities. *Combust. Sci. Tech.* **164**, 257–283.

Bloxsidge, J.G., Dowling, A.P., Hooper, N., and Langhorne, P.J. (1987). Active Control of Reheat Buzz, AIAA Paper 87-0433.

Bowman, C.T. (1992). Control of combustion-generated nitrogen oxides emissions: technology driven by regulations. *Invited Lecture, 24th International Symposium on Combustion*, 859–876.

Bradley, D., Gaskell, P.H., Gu, X.J., Lawes, M., and Scott, M.J. (1998). Premixed turbulent flame instability and NO formation in a lean-burn swirl burner. *Combust. Flame* **115**, 515–538.

Bradley et al. (2006). Fundamentals of lean combustion. In: Dunn-Rankin, D. (Ed.). *Lean Combustion: Technology and Control*, Academic Press, Burlington, MA.

Correa, S. (1993). A review of NO$_x$ formation under gas-turbine combustion conditions. *Combust. Sci. Tech.* **87**, 329–362.

Crump, J.E., Schadow, K.C., Bloomshield, F.S., Culick, F.E.C., and Yang, V. (1985). Combustion instability in dump combustors: acoustic mode determinations. *Proceedings of the 19th JANNAF Combustion Meeting*, CPIA Publication 366.

Culick, F.E.C. (1989). Combustion Instabilities in Liquid-fuelled Propellant Systems: An Overview, AGARD-CP-450 Paper 1.

De Zilwa, S.R.N., Sivasegaram, S., and Whitelaw, J.H. (1999). Control of combustion oscillations close to stoichiometry. *Flow Turb. Combust.* **63**, 395–414.

De Zilwa, S.R.N., Uhm, J.H., and Whitelaw, J.H. (2000). Combustion oscillations close to the lean flammability limit. *Combust. Sci. Tech.* **161**, 231–258.

De Zilwa, S.R.N., Emiris, I., Uhm, J.H., and Whitelaw, J.H. (2001). Combustion of premixed air and methane in ducts. *Proc. R. Soc. Lon. A* **457**, 1915–1949.

Dowling, A.P. and Williams J.E.F. (1983). *Sound and Sources of Sound*. Ellis Horwood, London, UK.

Egolfopoulos, F.N. and Dimotakis, P.E. (2001). A comparative numerical study of premixed and non-premixed ethylene flames. *Combust. Sci. Tech.* **162**, 19–35.

Emiris, I. and Whitelaw, J.H. (2003). Control of premixed ducted flames from lean limit to stoichiometry. *Combust. Sci. Tech.* **175**, 1–28.

Emiris, I., Korusoy, E., Sivasegaram, S., and Whitelaw, J.H. (2003). Control of ducted flames by combinations of imposed oscillation and added fuel. In: Roy, G.D. (Ed.). *Combustion and Noise Control*. Cranfield University Press, pp. 3–8, Cranfield, UK.

Evesque, S. and Dowling, A.P. (2001). LMS algorithm for adaptive control of combustion oscillations. *Combust. Sci. Tech.* **164**, 65–93.

Feikema, D., Chen, R-H., and Driscoll, J.F. (1991). Blowout of non-premixed flames: maximum coaxial velocities achievable with and without swirl. *Combust. Flame* **86**, 347–358.

Gabruk, R.S. and Roe, L.A. (1994). Velocity characteristics of reacting and non-reacting flows in a dump combustor. *J. Prop. Power* **10**, 148–154.

Gore, J.P. (2002). Structure and NO_x emission properties of partially premixed flames. In: Roy, G.D. (Ed.). *Advances in Chemical Propulsion*. CRC Press, Boca Raton.

Gulati, A. and Mani, R. (1990). Active Control of Unsteady Combustion Induced Oscillations, AIAA Paper 90-2700.

Gutmark, E., Parr, T.P., Hanson-Parr, D.M., and Schadow, K.C. (1989). On the role of large and small scale structures in combustion control. *Combust. Sci. Tech.* **66**, 107–126.

Hardalupas, Y., Selbach, A., and Whitelaw, J.H. (1999). Aspects of oscillating flames. *J. Visual* **1**, 79–85.

Heitor, M.V., Taylor, A.M.K.P., and Whitelaw, J.H. (1984). Influence of confinement on combustion instabilities of premixed flames stabilised on axisymmetric baffles. *Combust. Flame* **57**, 109–121.

Hendricks, E., Sivasegaram, S., and Whitelaw, J.H. (1992). Control of oscillations in ducted premixed flames. In: Lee, R.S., Whitelaw, J.H., and Wang, T.S. (Eds.). *Aerothermodynamics in Combustors*. Springer-Verlag, pp. 215–230.

Higgins, B., McQuay, M.Q., Lacas, F., and Candel, S. (2001a). An experimental study on the effect of pressure and strain rate on CH chemiluminescence of premixed fuel-lean methane/air flames. *Fuel* **80**, 1583–1591, New York, USA.

Higgins, B., McQuay, M.Q., Lacas, F., Rolon, J.C., Darabiha, N., and Candel, S. (2001b). Systematic measurements of OH chemiluminescence for fuel-lean, high-pressure, premixed, laminar flames. *Fuel* **80**, 67–74.

Hubbard, S. and Dowling, A.P. (2003). Acoustic resonances of an industrial gas turbine combustion system. *J. Eng. Gas Turbine Power* **125**, 670–676.

Keller, J.O. and Hongo, I. (1990). Pulse combustion: the mechanisms of NO_x production. *Combust. Flame* **80**, 219–237.

Khezzar, L., De Zilwa, S.R.N., and Whitelaw, J.H. (1999). Combustion of premixed fuel and air downstream of a plane sudden-expansion. *Exp. Fluid* **27**, 296–309.

Korusoy, E. and Whitelaw, J.H. (2002). Extinction and relight in opposed flames. *Exp. Fluid* **33**, 75–89.

Korusoy, E. and Whitelaw, J.H. (2004). Effects of wall temperature and fuel on flammability, stability and control of ducted premixed flames. *Combust. Sci. Tech.* **176**, 1217–1241.

Kostuik, L.W., Shepherd, I.G., Bray, K.N.C., and Cheng, R.K. (1999). Experimental study of premixed turbulent combustion in opposed streams. Part II: reacting flow field. *Combust. Flame* **92**, 396–409.

Langhorne, P.J., Dowling, A.P., and Hooper, N. (1990). Practical active control system for combustion oscillations. *J. Prop. Power* **6**, 324–330.

Lee, J.G., Kim, K., and Santavicca, D.A. (2000). Effect of injection location on the effectiveness of an active control system using secondary fuel injection. *Proc. Combust. Inst.* **28**, 739–746.

Li, S.C., Ilincic, N., and Williams, F.A. (1997). Reduction of NO_x formation by water spray in strained two-stage flames. *ASME J. Eng. Gas Turbine Power* **119**, 836–843.

Luff, D. (2005). Experiments and calculations of opposed and ducted flames. Ph.D. thesis, University of London.

Luff, D., Korusoy, E., Lindstedt, P., and Whitelaw, J.H. (2003). Opposed flames with premixed air and methane, propane and ethylene. *Exp. Fluid* **35**, 618–626.

Luff, D.S., Sivasegaram, S., and Whitelaw, J.H. (2004). Combustion oscillations in round sudden expansions with stratified premixed upstream flows. *Proceedings of the Tenth Asian Congress of Fluid Mechanics*, Paper 98.

Mastarakos, E., Taylor, A.M.K.P., and Whitelaw, J.H. (1992). Extinction and temperature characteristics of turbulent counterflow non-premixed flames. *Combust. Flame* **91**, 55–64.

McManus, K.R., Poinsot, T., and Candel, S.M. (1993). A review of active control of combustion instabilities. *Prog. Energy Combust. Sci.* **19**, 1–29.

Milosavljevic, V.D. (2003). Perturbation in combustor near-field aerodynamics as a main source of thermoacoustic instabilities in modern industrial dry low NO_x gas turbine combustion systems. In: Roy, G.D. (Ed.). *Combustion and Noise Control*. Cranfield University Press, pp. 55–60, Cranfield, UK.

Milosavljevic, V.D., Persson, M., Lindstedt, R.P., and Vaos, E.M. (2003). Evaluation of the thermoacoustic performance of an industrial performance an industrial gas turbine combustion chamber with a second moment method. In: Roy, G.D. (Ed.). *Combustion and Noise Control*. Cranfield University Press, pp. 72–77, Cranfield, UK.

Moore, M.J. (1997). NO_x emission control in gas turbines for combined cycle gas turbine plant. *Proc IMechE* **211**(Part A), 43–52.

Moran, A.J., Steele, D., and Dowling, A.P. (2000). Active control of combustion and its applications. *Proceedings of the RTO AVP Symposium on Active Control Technology for Enhanced Performance Operation Capabilities of Military Aircraft, Land Vehicles and Sea Vehicles*, Braunschweig.

Neumeir, Y. and Zinn, B.T. (1996). Experimental demonstration of active control of combustion instabilities using real-time modes observation and secondary fuel injection. *Proc. Combust. Inst.* **26**, 2811–2818.

Norster, E.R. and De Pietro, S.M. (1996). Dry low emissions combustion system for EGT small gas turbines. *Trans. Inst. Diesels Gas Turbine Engine* **495**, 1–9.

Park, S., Pang, P., Yu, K., Annaswamy, A.M., and Ghoneim, A.F. (2003). Performance of an adaptive posicast controller in a liquid fuelled dump combustor. In: Roy, G.D. (Ed.). *Combustion and Noise Control.* Cranfield University Press, pp. 78–83, Cranfield, UK.

Pascherit, C.O., Gutmark, E., and Weisenstein, W. (1998). Structure and control of thermoacoustic instabilities in a gas turbine combustor. *Combust. Sci. Tech.* **175**, 1–28.

Poppe, C., Sivasegaram, S., and Whitelaw, J.H. (1998). Control of NO_x emissions in confined flames by oscillations. *Combust. Flame* **113**, 13–26.

Perez-Ortiz, R.M., Sivasegaram, S., and Whitelaw, J.H. (1993). Ducted Kerosene Spray Flames, AGARD-CP-536 Paper 34.

Polifke, W. and Paschereit, C.O. (1998). Determination of thermo-acoustic transfer matrices by experiment and computational fluid dynamics. *ERCOFTAC Bull.* **39**, 1–10.

Poinsot, T.J., Bourienne, F., Candel, S., Esposito, E., and Lang, W. (1989). Suppression of combustion instabilities by active control. *J. Prop. Power* **5**, 14–20.

Putnam, A.A. (1971). *Combustion Driven Oscillations in Industry.* American Elsevier Publishing Co. New York, USA.

Rayleigh, J.W. (1896). *The Theory of Sound*, Vol. 2. MacMillan & Co., London, p. 232.

Sardi, E. and Whitelaw, J.H. (1999). Extinction timescales of periodically strained lean counterflow flames. *Exp. Fluid* **27**, 199–209.

Sardi, E., Taylor, A.M.K.P., and Whitelaw, J.H. (2000). Extinction of turbulent counterflow flames under periodic strain. *Combust. Flame* **113**, 13–26.

Schadow, K.C. and Gutmark, E. (1992). Combustion instabilities related to vortex shedding in dump combustors and their passive control. *Prog. Energy Combust. Sci.* **18**, 117–132.

Seume, J.R., Vortmeyer, N., Krause, W., Hermann, J., Hantschik, C.C., Zangl, P., Gleis, S., Vortmeyer, D., and Orthmann, A. (1997). Application of active combustion control to a heavy duty gas turbine. *Trans. ASME J. Eng. Gas Turbine Power* **120**, 721–726.

Sivasegaram, S. and Whitelaw, J.H. (1987a). Oscillations in axisymmetric dump combustors. *Combust. Sci. Tech.* **52**, 423–426.

Sivasegaram, S. and Whitelaw, J.H. (1987b). Suppression of oscillations in confined disk stabilised flames. *J. Prop.* **3**, 291–295.

Sivasegaram, S. and Whitelaw, J.H. (1988). Combustion oscillations in dump combustors with a constricted exit. *Proc. IMechE C* **202**, 205–210.

Sivasegaram, S. and Whitelaw, J.H. (1991). The influence of swirl on oscillations in ducted premixed flames. *Combust. Flame* **85**, 195–205.

Sivasegaram, S. and Whitelaw, J.H. (1992). Active control of oscillations in combustors with several frequency modes. *Active Control of Noise and Vibration*. ASME, DSC-38, pp. 69–74.

Sivasegaram, S., Tsai, R.F., and Whitelaw, J.H. (1995a). Control of combustion oscillations by forced oscillation of part of the fuel supply. *Combust. Sci. Tech.* **105**, 67–83.

Sivasegaram, S., Tsai, R.F., and Whitelaw, J.H. (1995b). Control of oscillations and NO_x concentrations in ducted premixed flames by spray injection of water. *Proc. ASME, HTD* **317**(2), 169–174.

Sreenivasan, K.R., Raghu, S., and Chu, B.Y. (1985). The Control of Pressure Oscillations in Combustion and Fluid Dynamical Systems, AIAA Paper 85-0540.

Touchton, G.L. (1985). Influence of gas turbine combustion design and operating parameters on the effectiveness of NO_x suppression by injected steam or water. *ASME J. Eng. Gas Turbine Power* **107**, 706–713.

Wang, C.-H. and Dowling, A.P. (2002). Actively tuned passive control of combustion instabilities. In: Roy, G.D. (Ed.). *Advances in Chemical Propulsion: Science to Technology.* CRC Press, pp. 16–21, Boca Raton, Florida, USA.

Warnatz, J. (1984). Chemistry of high temperature combustion of alkanes up to octane. *Proc. Combust. Inst.* **20**, 845–856.

Waschman, S., Park, S., Annaswamy, A.M., and Ghoneim, A.F. (2003). Experimental verification of model-based control strategies using a backward facing step combustor. In: Roy, G.D. (Ed.). *Combustion and Noise Control.* Cranfield University Press, pp. 28–34, Cranfield, UK.

Whitelaw, J.H., Sivasegaram, S., Schadow, K.C., and Gutmark, E. (1987). Oscillations in Non-Axisymmetric Dump Combustors, AGARD CP-450 Paper 15.

Wilson, K., Gutmark, E., and Schadow, K.C. (1992). Flame kernel pulse actuator for active combustion control. *Active Control of Noise and Vibration.* ASME, DSC-38, pp. 75–81.

Yang, V. and Schadow, K.C. (1999). AGARD workshops on active control for propulsion systems. *Proceedings of the RTO AVP Symposium on Gas Turbine Engine Combustion, Emissions and Alternative Fuels,* Lisbon, RTO-MP-14, pp. 35.1–35.16.

Yang, V., Hong, B.S., and Ray, A. (2002). Robust feedback control of combustion instabilities with model uncertainty. In: Roy, G.D. (Ed.). *Advances in Chemical Propulsion; Science to Technology.* CRC Press, pp. 361–380, Boca Raton, Florida, USA.

Yu, K.H., Wilson, K.J., Parr, T.P., and Schadow, K.C. (2002). Liquid fuelled active control for ramjet combustors. In: Roy, G.D. (Ed.). *Advances in Chemical Propulsion: Science to Technology.* CRC Press, pp. 303–322.

Chapter 8

Lean Hydrogen Combustion

Robert W. Schefer, Christopher White, and Jay Keller

Nomenclature

BMEP	Brake mean effective pressure
BTE	Brake thermal efficiency
$[CH_4]$	Concentration of methane (molecules/cm^3)
CI	Compression ignition
CNG	Compressed natural gas
CR	Compression ratio
DI	Direct injection
E	Activation energy
EGR	Exhaust gas recirculation
EZEV	Equivalent zero emissions vehicle
H2ICE	Hydrogen-fueled internal combustion engine
[HC]	Total molar concentration of all non-methane hydrocarbon species
HEV	Hybrid electric vehicle
ICE	Internal combustion engine
ITE	Indicated thermal efficiency
IVC	Inlet valve closing
K_{ext}	Normalized strain rate at extinction
LFL	Lean flammability limit
L-H2ICE	Liquid hydrogen fueled internal combustion engine
mpg	Miles per gallon
n_{H2}	Volume fraction of hydrogen in the fuel mixture
NG	Natural gas
$[O_2]$	Concentration of oxygen (molecules/cm^3)
PEM	Proton exchange membrane
PEMFC	Proton exchange membrane fuel cell
PFI	Port-fuel-injection
RFL	Rich flammability limit
RON	Research octane number
S_L	Laminar flame speed
S_T	Turbulent flame speed
SI	Spark ignition

SOI	Start of injection
SULEV	Super ultra low emissions vehicles
T	Temperature (K)
TWC	Three-way catalyst
u'_{rms}	Velocity fluctuations
U_{mean}	Bulk inflow velocity
ε	Mole fraction of hydrogen in total fuel
τ_{CH4}	Autoignition delay time of pure CH_4
τ_{H2}	Autoignition delay time of pure H_2
τ_{ign}	Autoignition delay time
ϕ	Equivalence ratio
ϕ_{local}	Local equivalence ratio

8.1. INTRODUCTION

Hydrogen shows considerable promise as a primary energy carrier for the future. First, it can be produced directly from all primary energy sources, enabling energy feedstock diversity for the transportation sector. These alternative energy resources include wind, solar power, and biomass (plant material), which are all renewable fuel sources. Electricity produced from nuclear fission, or fusion, has also been mentioned with increasing frequency as a possible source of H_2 production through electrolysis of water or thermochemical cycles. A major benefit of increased H_2 usage for power generation and transportation is that all of these sources minimize our dependence on non-renewable fossil fuels and diversify our energy supply for utilization in end-use energy sectors. Alternatively, H_2 can be produced through coal gasification, or by "steam reforming" of natural gas (NG), both of which are non-renewable fossil fuels but are abundantly available in the United States and throughout the world. Combining the latter technologies with carbon capture and storage (CCS) would provide a significant increase in sources of clean burning H_2 while at the same time eliminating greenhouse gas emissions. Second, since all conventional fossil fuels contain carbon atoms in addition to hydrogen atoms, carbon dioxide (CO_2) is a major product gas formed during the conversion of the fuel to energy. The release of stored chemical energy in H_2 to useful heat produces only water as a product, thus eliminating CO_2, a significant contributor to climate change as a significant greenhouse gas.

The concept of H_2 as an energy carrier is often discussed. This concept can best be described in the context of H_2 produced directly from water using electrolysis. While this process is not economically attractive at current costs, if the electricity required to convert H_2O to H_2 is provided by wind or solar power, then the H_2 is produced without creating any CO_2. Given the intermittent nature of wind and solar power sources, surplus energy produced during very windy or bright sunny days could be used to produce H_2 that is stored for later use. Under these conditions the stored H_2 becomes an energy carrier that can be used later to produce power where it is needed, either in conjunction with a fuel cell to produce electricity or in a combustor to produce power (internal combustion engine (ICE) or turbine). It should be noted that there are many different ways to produce H_2 from a primary energy feedstock which are usually much more efficient than simple electrolysis, but this serves as a good example of how one might store energy in hydrogen as a carrier for use later.

The energy stored in hydrogen can be converted to useful energy through either fuel cells to directly produce electricity or combustion to produce power. Fuel cells are a proven technology used by NASA to produce electricity since the 1960s. However, much more work is needed to improve their cost effectiveness for everyday use in the general population. For example, fuel cells were recently estimated to be ten times more expensive to produce than a power-comparable ICE. Cost of a proton exchange membrane (PEM) fuel cell is projected to be about \$300/kW for mass production whereas the cost for an H2ICE is about \$30/kW. The lifetime and reliability of fuel cells are also issues that must be addressed for longer-term use, as are the requirements of high-purity hydrogen. Current lifetimes for PEM fuel cells are on the order of 1000 h, whereas the requirement for transportation is about 5000 h and that for stationary applications is 50 000 h.

Another energy conversion technology that is also well proven is combustion. Direct combustion of conventional fossil fuels has been used for centuries and has been refined considerably in recent years due to increasingly stringent pollutant emissions limits and higher-efficiency requirements brought on by recent fuel shortages. As will be described in greater detail in the following sections, recent research indicates that while some technical challenges exist, H_2 can be successfully burned in conventional combustion systems with minimal design changes. In particular, with regard to systems based on dilute premixed combustion technology, the unique characteristics of H_2 offer several advantages over conventional hydrocarbon fuels. For example, lean operation of hydrocarbon-based gas turbines is often limited by the onset of combustion instabilities. As will be shown, the wider flammability limits of H_2 extend the stable combustion regime to leaner mixtures thus allowing stable operation at the lower temperatures needed to reduce oxides of nitrogen (NO_x). We have already mentioned that increasing the H_2 content in a fuel replaces carbon-containing hydrocarbon fuel molecules and, at the limit of pure H_2, CO_2 emissions are eliminated. In the following section, the combustion characteristics of H_2 will be described and compared with relevant conventional hydrocarbon fuels. In the remainder of this chapter the use of hydrogen in two applications, ICEs and gas turbine combustion systems, will be described. In both of these applications lean premixed combustion technology using hydrogen-based fuels has already been successfully demonstrated.

8.2. HYDROGEN COMBUSTION FUNDAMENTALS

The use of H_2 in combustion systems is attractive because it has a very wide flammability range, it is easy to ignite, and has a large flame propagation velocity and small quenching distance. Hydrogen properties of interest are listed in Table 8.1 (Drell and Belles, 1958; Reed, 1978; Lewis and von Elbe, 1987). The same properties for typical gasoline (Reed, 1978; Heywood, 1988) and compressed natural gas (CNG) (Reed, 1978; Segeler, 1978) are listed in columns 3 and 4 to provide a basis for the discussion of hydrogen as a fuel in ICEs in Section 8.3. The corresponding properties for methane in column 5 provide a basis for the discussion of hydrogen as a gas turbine fuel.

Noteworthy in Table 8.1 are the flammability limits of hydrogen, which are much wider than those of hydrocarbons. The limits, by volume, are from 4% (lean flammability

Table 8.1

Fuel properties at 25°C and 1 atm

Property	Hydrogen	LNG	Gasoline	Methane
Density (kg/m^3)	0.0824	0.72	730[a]	0.651
Flammability limits (volume % in air)	4–75	4.3–15	1.4–7.6	5.5–15
Flammability limits (ϕ)	0.1–7.1	0.4–1.6	~0.7–4	0.4–1.6
Autoignition temperature in air (K)	858	723	550	813
Minimum ignition energy in air (mJ)[b]	0.02	0.28	0.24	0.29
Flame velocity (m/s)[b]	1.85	0.38	0.37–0.43	0.40
Adiabatic flame temperature (K)[b]	2480	2214	2580	2226
Quenching distance (mm)[b]	0.64	2.1[c]	~2	2.5
Stoichiometric fuel/air mass ratio	0.029	0.069	0.068	0.058
Stoichiometric volume fraction (%)	29.53	9.48	~2[d]	9.48
Lower heating value (MJ/m^3)	9.9	32.6	–	32.6
Lower heating value (MJ/kg)	119.7	45.8	44.79	50.0
Heat of combustion (MJ/kg air)[b]	3.37	2.9	2.83	2.9

[a]Liquid at 0°C.
[b]At stoichiometry.
[c]Methane.
[d]Vapor.

limit, LFL) to 75% (rich flammability limit, RFL) for mixtures with air. These values compare with methane flammability limits of 5.5% (LFL) to 15% (RFL). Clearly, the RFL for H_2 is significantly greater than that for conventional hydrocarbon fuels. These wide flammability limits are related to the chemical kinetics of hydrogen combustion, and the large diffusion coefficient of hydrogen (Clavin, 1985). In hydrocarbon flames such as methane, the rich limit is caused by:

$$H + CH_4 \Leftrightarrow H_2 + CH_3 \tag{8.1}$$

The highly reactive radical H atom is replaced by a weakly reactive radical CH_3. In hydrogen flames, the analogous reaction is:

$$H + H_2 \Leftrightarrow H_2 + H \tag{8.2}$$

with the consequence that the rich limit is far greater than in hydrocarbons.

The flame propagation velocity in a mixture of hydrogen and air is very large in comparison to that of hydrocarbons. This is due primarily to the faster reaction rates of the H_2/O_2 system, since the flame speed is approximately proportional to the square root of the reaction rate. The large diffusion coefficient of H_2 also plays a role in the large flame speed because of the enhanced transport of radicals and heat ahead of the flame. The flame speed in a stoichiometric mixture of hydrogen and air is 1.85 m/s while for methane it is 0.44 m/s. A further result of the faster reaction rates in hydrogen/ air flames is a much thinner flame front than for methane and other hydrocarbons.

Hydrogen also has a higher heat of combustion relative to hydrocarbon fuels on a per unit mass basis. At 298 K and 1 atm the heat of combustion (lower heating value) of H_2 is about 120 MJ/kg while that of methane is only 45 MJ/kg. It is important to note, however, that on a per unit volume basis the lower heating value for H_2 is about a factor of four less than CH_4 and CNG. Significantly larger storage capabilities are therefore required for gaseous H_2. The use of higher-pressure storage tanks (68–100 MPa) is currently under development to meet this need. Also, metal hydride storage of H_2 shows promise but still requires further improvement before being commercially viable.

The combustion characteristics in Table 8.1 also determine other aspects of hydrogen combustion that are relevant for its use in combustion systems. The quenching distance, for example, is important in safety (flashback) and micro-combustion (quenching) issues. The quenching distance is proportional to the flame thickness, and in a stoichiometric mixture of hydrogen/air at ambient pressure and temperature, the quenching distance is 0.64 mm while that for methane is approximately 2.5 mm. At higher pressure and temperatures these values are further reduced. The minimum energy for ignition (spark, flame, hot surface) is also very small in comparison with hydrocarbons.

The chemical kinetics of the H_2/air system is well known and the reader is referred to other references for a more detailed discussion of the reaction mechanism (Turns, 2000; Warnatz *et al.*, 2001). Since no carbon exists when H_2 is burned as the fuel, reactions involving hydrocarbon intermediates, CO and CO_2, are eliminated. The primary pollutant species produced during hydrogen combustion are the oxides of nitrogen (NO_x). The kinetic mechanism for the formation of NO_x has been studied extensively, is well known, and can be broken down into three mechanisms: the thermal or Zeldovich mechanism, the Fenimore or prompt mechanism, and the N_2O intermediate mechanism (Turns, 2000). The thermal mechanism consists of the two reactions:

$$O + N_2 \Leftrightarrow NO + N \tag{8.3}$$

$$N + O_2 \Leftrightarrow NO + O \tag{8.4}$$

which can be extended by adding the reaction:

$$N + OH \Leftrightarrow NO + O \tag{8.5}$$

This mechanism typically accounts for most of the NO formed at the highest temperatures and is coupled to the hydrogen combustion chemistry through the radical species O and OH. The N_2O intermediate mechanism consists of three additional reactions:

$$O + N_2 + M \Leftrightarrow N_2O + M \tag{8.6}$$

$$H + N_2O \Leftrightarrow NO + O \tag{8.7}$$

$$O + N_2O \Leftrightarrow NO + NO \tag{8.8}$$

This mechanism is important in fuel-lean combustion where the temperatures are lower. The third mechanism is the prompt NO mechanism which involves hydrocarbon intermediates:

$$CH + N_2 \Leftrightarrow HCN + N \tag{8.9}$$

As discussed later in this chapter, one strategy for introducing H_2 as a fuel into the current energy-production infrastructure involves the addition of H_2 to conventional hydrocarbon fuels. In this case, the prompt NO mechanism is relevant and is therefore presented here for completeness. Following Equation (8.9), the HCN molecule forms NO through the following sequence:

$$HCN + O \Leftrightarrow NCO + H \tag{8.10}$$

$$NCO + H \Leftrightarrow NH + CO \tag{8.11}$$

$$NH + H \Leftrightarrow N + H_2 \tag{8.12}$$

$$N + OH \Leftrightarrow NO + H \tag{8.13}$$

In the case of pure H_2 combustion, Equations (8.3) through (8.8) are adequate to account for the production of NO, while in the case of H_2 addition to a hydrocarbon fuel, the additional Equations (8.9) through (8.13) must be included.

8.2.1. RATIONALE FOR HYDROGEN IN INTERNAL COMBUSTION ENGINES

The potential for hydrogen-fueled internal combustion engines (H2ICEs) to operate as clean and efficient power plants for automobiles is now well established. In particular, H2ICEs with near-zero emissions and efficiencies in excess of conventional gasoline-fueled ICEs have been demonstrated (e.g., Swain *et al.*, 1981; Tang *et al.*, 2002; Eichlseder *et al.*, 2003). The ability for H2ICEs to operate with near-zero engine-out emissions is primarily due to the coupled effect of two characteristics unique to hydrogen: (i) in principle, nitrogen oxides (NO_x) are the only undesirable engine-out emissions[1] formed by the thermal dissociation and oxidation of N_2 in atmospheric air during combustion; and (ii) the low LFL of hydrogen allows stable combustion at highly dilute conditions. The coupled effect is that during ultra-lean operation, combustion temperatures are low enough that NO_x formation rates are too slow and engine-out emissions are near zero (for a summary, see Das, 1991). The ability to operate efficiently is in part also due to (ii) because unthrottled operation is possible at low loads.

The unique combustion characteristics of hydrogen that allow clean and efficient operation at low engine loads present difficulties at high engine loads. Here, the low ignition energies of hydrogen–air mixtures cause frequent unscheduled combustion events, and the high combustion temperatures of mixtures closer to the stoichiometric composition lead to increased NO_x production. Both effects, in practical application, limit the power densities of H2ICEs. The recent research thrust and progress on this front is the development of advanced hydrogen engines with improved power densities and reduced NO_x emissions at high engine loads.

There are several reasons for reviewing recent developments in H2ICE engine-specific research: recent progress in control strategies for NO_x reduction, development

---

[1] In practice, the burning of lubricating oil in the combustion chamber produces carbon oxides (CO_x) and hydrocarbons (HC) at near-zero levels (Das, 1991).

of advanced hydrogen engines with improved power densities, and recent progress in engine modeling and cycle analysis. Perhaps the most compelling reason is the ever-increasing likelihood that the H2ICE will serve as a transitional hydrogen power train during the initial development of a hydrogen economy. This view is based on the fact that the implementation of a production ready hydrogen proton exchange membrane fuel cell (PEMFC) is at least 10 years away (Department of Energy USA, 2004). In contrast, the H2ICE offers the potential to utilize manufacturing infrastructure already developed for petroleum-fueled ICEs, and can serve as an economical near- to mid-term option for a transportation power plant in a hydrogen economy while fuel cells undergo continued development. This view has been the accelerant behind a renewed interest and recent progress in the research and development of the H2ICE.

The literature on the H2ICE is voluminous, and dates back over 100 years (Cecil, 1822). H2ICE research prior to the early 1980s has been reviewed by Escher (Escher, 1975; Escher and Euckland, 1976; Escher, 1983), and an excellent historical perspective of the hydrogen engine and technical review of H2ICE research prior to 1990 can be found in Das (1990). For the most part, we consider here more recent H2ICE research, with an emphasis on the last 10 years. The review is intended for those who have a familiarity with ICEs but are not familiar with the hydrogen ICE. While it is not possible to be inclusive of all subject matter in the space provided here, the present chapter provides a representative overview of the "state-of-the-art" engine-specific H2ICE research. The scope of the review is as follows: H2ICE fundamentals are described in Section 8.2.2 by examination of the engine-specific properties of hydrogen. These properties will be used to show, with reference to ICE operation and control, both the advantages and the disadvantages of hydrogen compared to conventional gasoline-fueled ICEs. Past and recent studies cover some of the same ground, but the focus here is to expand understanding in the context of a survey of recent literature. Advanced hydrogen engines and their salient features are reviewed in Section 8.2.3. The intent of these engines is to overcome the limiting effects at high engine loads for naturally aspirated premixed or port-fuel-injection (PFI) H2ICEs. We conclude with some remarks regarding the overall benefit of H2ICEs and possible future work in Section 8.2.4.

8.2.2. H2ICE FUNDAMENTALS

The ability for H2ICEs to burn cleanly and operate efficiently is owed to the unique combustion characteristics of hydrogen that allow ultra-lean combustion with dramatically reduced NO_x production and efficient low engine load operation. In contrast, the same combustion characteristics impose technical challenges at high engine loads due to an increased propensity to preignite the hydrogen–air mixture and increase NO_x production. In this section, we review the benefits and technical challenges of H2ICE operation at low and high engine load through an examination of hydrogen properties relevant to engine operation and control. Hydrogen properties of interest to H2ICE are listed in Table 8.1. For direct comparison, H2ICE engine properties will be compared to those of a PFI gasoline engine.

8.2.2.1. Preignition and Knock

The high autoignition temperature of hydrogen of 858 K (see Table 8.1) means that hydrogen is most suitable as a fuel for spark ignition (SI) engines, though compression ignition (CI) has been studied in some detail (e.g., Welch and Wallace, 1990; Hedrick, 1993; Mathur *et al.*, 1993; Naber and Siebers, 1998). Despite the high autoignition temperature, the ignition energies of hydrogen–air mixtures are approximately an order of magnitude lower than those of hydrocarbon–air mixtures. This is observed in Figure 8.1 where the minimum ignition energies for hydrogen–air, propane–air, and heptane–air mixtures at atmospheric pressure are plotted as a function of equivalence ratio, ϕ, which is defined as the ratio of the actual fuel/air mass ratio to the stoichiometric fuel/air mass ratio. The low ignition energies of hydrogen–air mixtures mean that H2ICEs are predisposed toward the limiting effects of preignition. Here, preignition is defined as combustion prior to spark discharge and, in general, results from surface ignition at engine hot spots, such as spark electrodes, valves, or engine deposits. The limiting effect of preignition is that a preignition event will advance the start of combustion and produce an increased chemical heat release rate. In turn, the increased heat release rate results in a rapid pressure rise, higher peak cylinder pressure, acoustic oscillations, and higher heat rejection that leads to higher in-cylinder surface temperatures. The latter effect can advance the start of combustion further, which in turn can lead to a runaway effect, and if left unchecked will lead to engine failure (Heywood, 1988). It is therefore a necessity for practical application that preignition is avoided. The preignition-limited operating envelope defined by Tang *et al.* (2002) consists of operating conditions where 1% or fewer of the combustion cycles experience a preignition event. The preignition limit is defined as the upper bound of this envelope.

Complementary information to Figure 8.1, and relevant to the above discussion, is that studies at pressures varying from 0.2 to 1 atm and mixture temperatures varying from 273 to 373 K have found that the minimum ignition energies of hydrogen–air mixtures vary inversely with the square of the pressure and inversely with temperature (Drell and Belles, 1958). If these trends hold at high pressure and temperature, extrapolations indicate that for typical engine temperatures and pressures, the minimum ignition energies of hydrogen–air mixtures will be much lower than those shown in Figure 8.1. Similar dependencies on pressure and temperature for the minimum ignition energies of hydrocarbon–air mixtures have also been reported (Lewis and von Elbe, 1987). Knock, or spark knock (Heywood, 1988), is defined as autoignition of the

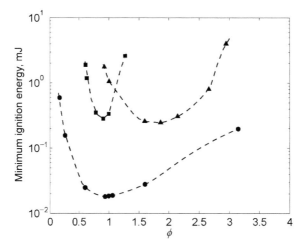

Figure 8.1 Minimum ignition energies of (●) hydrogen–air, (■) methane–air, and (▲) heptane–air mixtures in relation to ϕ at atmospheric pressure (Lewis and von Elbe, 1987).

hydrogen–air end-gas ahead of the flame front that has originated from the spark. The high autoignition temperature, finite ignition delay, and the high flame velocity of hydrogen (i.e., the latter two effects translate to less residence time for the end-gas to ignite) mean that knock, as defined, is less likely for hydrogen relative to gasoline, and hence the higher research octane number (RON) for hydrogen (RON > 120; Tang *et al.*, 2002; Topinka *et al.*, 2004) in comparison to gasoline (RON = 91–99; IIcywood, 1988). The global effect of both knock and preignition (i.e., an audible pinging or "knocking") is nearly indistinguishable, and this is the primary reason for the lack of a clear distinction between the two in the hydrogen literature (Tang *et al.*, 2002). However, a distinction is necessary since the controlling phenomena are very different: preignition can be avoided through proper engine design, but knock is an inherent limit on the maximum compression ratio (CR) that can be used with a fuel (Heywood, 1988). The differentiating factor between knock and preignition is that knock can be controlled by retarding the spark timing, while in general, preignition cannot be controlled by adjusting spark timing.

Observed in Figure 8.1 is that as the stoichiometric condition ($\phi = 1$) is approached from the lean side ($\phi < 1$), the minimum ignition energy for hydrogen is a strongly decreasing function of the equivalence ratio with the minimum at $\phi \approx 1$. This trend is responsible for the experimentally observed fact that it is extremely difficult to operate an H2ICE at or near the stoichiometric condition in the absence of frequent preignition events. Therefore, for practical application, the maximum and, consequently, peak power output can be limited by the preignition limit. Stockhausen *et al.* (2002) report a preignition limit of $\phi \approx 0.6$ for a 4-cylinder 2.0-L engine at an engine speed of 5000 rpm. Consequently, engine peak power output was reduced by 50% compared to engine operation with gasoline.

Although the preignition limit is engine specific, consistent trends with variations in engine properties and operational conditions have been found: the preignition-limited maximum ϕ decreased monotonically with increased CR (Tang *et al.*, 2002; Li and Karim, 2004; Maher and Sadiq, 2004) and with increased mixture temperature (Li and Karim, 2004). Engine speed has also been shown to have an effect (Tang *et al.*, 2002, Maher and Sadiq, 2004), but the trend is more complicated due to the coupled effect of residual mass fraction (i.e., mixture-temperature effect).

The diminished peak power output, set by the preignition limit, will decrease the performance of an H2ICE-powered vehicle in comparison to its gasoline equivalent. Therefore, determining the mechanism of preignition, practical operational limits, and control strategies has been a primary focus of many research studies (e.g., Furuhama *et al.*, 1977; Swain *et al.*, 1981; Binder and Withalm, 1982; Kondo *et al.*, 1997; Li and Karim, 2004; Maher and Sadiq, 2004). Unfortunately, despite much effort, there exist no guaranteed preventive steps. However, identification of preignition sources, such as in-cylinder hot-spots (Furuhama *et al.*, 1977; Tang *et al.*, 2002), oil contaminants (Stockhausen *et al.*, 2002), combustion in crevice volumes (Lee *et al.*, 2001), and residual energy in the ignition system (Kondo *et al.*, 1997), has provided the necessary minimizing steps. These include use of cold-rated spark plugs, low coolant temperature, and optimized fuel-injection timing.

Advanced control strategies include intake charge cooling (Fiene *et al.*, 2002), variable valve timing for effective scavenging of exhaust residuals (Berckmüller *et al.*, 2003), advanced ignition systems (Kondo *et al.*, 1997), and hydrogen direct injection (DI) (Furuhama *et al.*, 1977; Homan *et al.*, 1983). These advanced strategies can be quite effective: Berckmüller *et al.* (2003), employing a single-cylinder 0.5-L PFI-H2ICE at a CR of 12:1, used variable cam phasing and computational fluid dynamics (CFD) simulations to optimize coolant flow, injection location, and injection timing to operate at $\phi = 1$ over a speed range 2000–4000 rpm. Homan *et al.* (1983) used DI late in the compression stroke and near-simultaneous spark to eliminate preignition events. However, to prevent misfires, the hydrogen jet had to be directed toward the spark. Kondo *et al.* (1997) used an ignition system specifically designed to prevent residual energy and a water-cooled spark plug. In the absence of any advanced control, the maximum $\phi \approx 0.35$; by the elimination of residual energy in the ignition system, the maximum can be increased to $\phi \approx 0.6$; with the addition of the water-cooled spark plug, a maximum $\phi \approx 0.8$ is possible. With liquid hydrogen fueling, Furuhama *et al.* (1978) and Knorr *et al.* (1997) report that preignition can be eliminated without too much effort simply due to the cooling effect of the cold hydrogen. A further discussion of liquid-hydrogen-fueled engines follows in Section 8.2.3.2.

8.2.2.2. Flammability Range, Flame Velocity, and Adiabatic Flame Temperature

The flammability range in fuel volume fraction in air at 298 K and 1 atm for hydrogen is 4–75% and for gasoline is 1.4–7.6% (Table 8.1). For ICEs, it is more meaningful to give the flammability range in terms of equivalence ratio. Then, the flammability range of hydrogen is $0.1 < \phi < 7.1$ and gasoline is approximately $0.7 < \phi < 4$. In recasting the flammability range in terms of equivalence ratio, it is evident that the H2ICE is amenable to stable operation under highly dilute conditions, which allows more control over engine operation for both emissions reduction and fuel metering. In practical application, the diluent can either be excess air (lean operation) or recycled exhaust gas. An important distinction between the two is that the latter allows stoichiometric operation at low to medium load and the use of a three-way catalyst (TWC) for NO_x reduction. However, problems due to water condensation using recycled exhaust gas dilution in ICEs are exacerbated in H2ICEs due to the large fraction of H_2O in the exhaust stream.

Flame velocity and adiabatic flame temperature are important properties for engine operation and control, in particular thermal efficiency, combustion stability, and emissions. Laminar flame velocity and flame temperature, plotted as a function of equivalence ratio, are shown in Figures 8.2 and 8.3, respectively. These figures will be further referenced in subsections to follow.

8.2.2.3. Nitrogen Oxides

H2ICE emissions and control techniques have been thoroughly reviewed by Das (1991). Here, we provide a brief survey of more recent NO_x research.

Ultra-lean combustion (i.e., $\phi \leq 0.5$), which is adequately synonymous with low-temperature combustion, is an effective means for minimizing NO_x emissions in ICEs. As described in Section 8.2.2, H2ICEs can operate ultra-lean, and are therefore amenable to low-temperature NO_x reduction strategies. Engine-out NO_x concentration versus equivalence ratio is plotted in Figure 8.4 (closed symbols) from various studies. The claim of near-zero emissions can be made more robust by analyzing Figure 8.4 in reference to the NO_x limit of 0.07 and 0.02 g/mile to satisfy, respectively, the US Federal

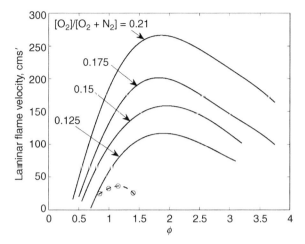

Figure 8.2 Laminar flame velocity for (—) hydrogen, oxygen, and nitrogen mixtures (Lewis and von Elbe, 1987) and (°, - -) gasoline and air mixtures (Heywood, 1988) at room temperature and atmospheric pressure. The dashed line is a least squares fit polynomial. $[O_2]/[O_2 + N_2]$ is the oxygen mole fraction in the ambient.

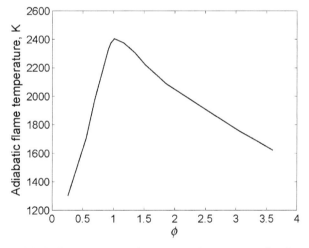

Figure 8.3 Adiabatic flame temperature for hydrogen–air mixtures (Drell and Belles, 1958).

Tier II manufacturer fleet average standard and the California Air Resources Board (CARB) Low Emissions Vehicle II (LEV II) standard for Super Ultra Low Emissions Vehicles (SULEV).[2] Conversion of NO_x concentrations in g/mile to ppm (or vice versa) depends on vehicle fuel efficiency, drive cycle, and equivalence ratio, among other factors (Van Blarigan, 1995). Here, we consider a vehicle with a fuel efficiency of 35 miles per gallon (mpg) operated at steady state. Then, as a function of equivalence ratio, the Tier II fleet average and SULEV NO_x limits are shown in Figure 8.4 as the dashed and solid lines, respectively. For different fuel efficiencies, these lines would effectively

---

[2] The SULEV NO_x standard is equal to the now-abandoned Equivalent Zero Emissions Vehicle (EZEV) standard.

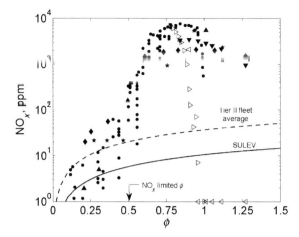

Figure 8.4 NO$_x$ emissions as a function of equivalence ratio for engine-out (closed symbols) and tailpipe with exhaust gas after-treatment (open symbols) from various studies: ● (Tang *et al.*, 2002), for CR = 12.5, 14.5, and 15.3; ▲ (Berckmüller *et al.*, 2003); ◆ (Heywood, 1988) for CR = 11:1; ▼ (Lee *et al.*, 1995); ★ (Furuhama *et al.*, 1978), for intake temperature of 298 K; ◁ (Lee *et al.*, 1995); ▷ (Heffel, 2003a), for fuel input of 1.39 kg/h. The dashed and solid lines represent the US Federal Tier II manufacturer fleet average NO$_x$ standard and CARB LEV II SULEV NO$_x$ standard, respectively, for a fuel efficiency of 35 mpg.

be translated vertically (i.e., lower for lower mpg or higher for higher mpg). However, due to the rapid increase in NO$_x$ production with increasing ϕ near $\phi \approx 0.5$, the NO$_x$-limited equivalence ratio for H2ICE operation is essentially $\phi < 0.5$.

We have also compiled in Figure 8.4 data from various sources for tailpipe emissions with exhaust after-treatment (open symbols). These data show that NO$_x$ emissions at $\phi > 0.95$ are near zero with the use of a TWC. Berckmüller *et al.* (2003) and Heffel (2003a, b) investigated charge dilution with exhaust gas recirculation (EGR) to provide a stoichiometric air/fuel ratio at low- to medium-load operation. With EGR dilution, a TWC, and $\phi = 1$, tailpipe emissions were significantly reduced for engine loads that would otherwise produce high NO$_x$ concentrations for dilution with air (i.e., $0.5 < \phi < 1.02$). For instance, with excess air at $\phi = 0.78$, Heffel (2003a) measured 7000 ppm of NO$_x$; holding the fueling constant and using EGR as the diluent, such that $\phi = 1$, NO$_x$ emissions were reduced to less than 1 ppm at the tailpipe. Lee *et al.* (1995) measured 0.04 g/mile NO$_x$ emissions for an H2ICE vehicle using a TWC on a chassis dynamometer on the US Federal Test Procedure (FTP75) drive cycle. Berckmüller *et al.* (2003) found that the maximum level of EGR while maintaining smooth engine operation was 50%. Maximum EGR levels for gasoline operation are 15–30% (Heywood, 1988). The higher EGR tolerance of the H2ICE is due to the high flame velocities of hydrogen–air mixtures as shown in the comparison of hydrogen and gasoline data in Figure 8.2. The hydrogen curve labeled $[O_2]/[O_2 + N_2] = 0.125$ is equivalent to 40% mole fraction dilution of air with nitrogen, providing an approximation of flame speeds for the case of 40% EGR dilution. This approximation ignores the effect of water contained in hydrogen-engine EGR, but at an equivalence ratio of 1, Figure 8.2 indicates a significantly higher flame speed for the diluted hydrogen mixture compared with the undiluted gasoline case.

In summary, without after-treatment, there is a tradeoff between H2ICE power output and NO$_x$ emissions. Considering that the primary benefit of an H2ICE is near-zero emissions, a practical limit of engine operation is $\phi < 0.5$ (assuming no EGR).

This restriction will translate into a large loss in the effective power density of an H2ICE. The potential to expand the power band while maintaining near-zero NO_x emissions is possible by charge dilution with EGR and use of a TWC or by improving lean power density with pressure boosting, as will be discussed in Section 8.3.1.

8.2.2.4. Power Output and Volumetric Efficiency

H2ICE peak power output is primarily determined by volumetric efficiency, fuel energy density, and preignition. For most practical applications, the latter effect has been shown to be the limiting factor that determines peak power output.

Premixed or PFI-H2ICEs inherently suffer from a loss in volumetric efficiency due to the displacement of intake air by the large volume of hydrogen in the intake mixture. For example, a stoichiometric mixture of hydrogen and air consists of approximately 30% hydrogen by volume, whereas a stoichiometric mixture of fully vaporized gasoline and air consists of approximately 2% gasoline by volume. The corresponding power density loss is partially offset by the higher energy content of hydrogen. The stoichiometric heat of combustion per standard kg of air is 3.37 and 2.83 MJ, for hydrogen and gasoline, respectively. It follows that the maximum power density of a premixed or PFI-H2ICE, relative to the power density of the identical engine operated on gasoline, is approximately 83% (Furuhama *et al.*, 1978). For applications where peak power output is limited by preignition, H2ICE power densities, relative to gasoline operation, can be significantly below 83%. Furuhama *et al.* (1978) and Tang *et al.* (2002) report preignition-limited power densities of 72% and 50%, respectively, relative to operation with gasoline.

Presently, without exception, H2ICEs are modified conventional gasoline (or diesel) engines, with varying degrees of modification. Swain *et al.* (1996) designed an intake manifold to take advantage of the characteristics of hydrogen. The important feature is the use of large passageways with low-pressure drop, which is possible with hydrogen fueling since high intake velocities required for fuel atomization at low engine speeds are not necessary (Swain *et al.*, 1996). With the use of a large diameter manifold, Swain *et al.* reported a 2.6% increase in peak power output compared to that for a small diameter manifold. However, the improvement was lower than the estimated 10% that was expected. One possible explanation for the less-than-expected performance improvement was that the intake flow dynamics with hydrogen fueling are more complex than for gasoline-fueled engines. In this context, Sierens and Verhelst (2003) found that the start and duration of injection influences volumetric efficiency due to the interaction between the injected hydrogen and the intake pressure waves

While PFI-H2ICEs suffer from low volumetric efficiency at high loads, at low- to medium-load H2ICEs offer the benefit of being able to operate unthrottled. As described in Chapter 4, the advantage is that the pumping loss due to the pressure drop across the throttle plate is eliminated and fuel efficiency is improved. The ability for the H2ICE to operate unthrottled is owed to the low LFL and high flame velocity of hydrogen. However, due to increasing amounts of unburned hydrogen at ultra-dilute conditions, some throttling is required at idle conditions (Swain *et al.*, 1981).

8.2.2.5. Thermal Efficiency

The high RON and low LFL of hydrogen provide the necessary elements to attain high thermal efficiencies in an ICE. Brake thermal efficiency (BTE) versus brake mean effective pressure (BMEP) for various sources is plotted in Figure 8.5. For direct comparison between the various studies, we have normalized BMEP by maximum BMEP for equivalent gasoline operation. Dashed lines through the data are best-fit polynomials and the solid line is the BTE of a 4-cylinder 1.6-L gasoline engine with a CR of 9.0:1 (Swain *et al.*, 1983). The data of Tang *et al.* (2002) at a CR of 14.5:1 are illustrative of the increase in BTE with higher CR that is possible with hydrogen. Tang *et al.* (2002) and Nagalingam *et al.* (1983) found a CR of approximately 14.5:1 to be optimal due to heat transfer losses at higher CRs.

Aside from the increase in BTE by increasing CR, H2ICEs have higher efficiencies than gasoline ICEs at similar CRs. This is observed by comparing the gasoline and hydrogen data sets of Swain *et al.* (1983). Compared to gasoline operation (i.e., solid curve), the BTE with hydrogen operation (triangles) is higher across the entire operating range, with the relative increase maximized at medium loads. The drop-off in the relative difference in BTE between gasoline and hydrogen at low loads is due to the need for some throttling, as discussed in Section 8.2.2.4. The drop-off at high loads is likely due to increasing heat transfer losses. Shudo *et al.* (2001) showed that for an H2ICE the relative fraction of the heat release lost by heat transfer to the cylinder walls increases monotonically with increasing equivalence ratio. The trend is explained as a consequence of increasing flame velocity, increasing flame temperature, and decreasing quenching distance with increasing equivalence ratio that leads to narrow thermal boundary layers. Shudo *et al.* (2001) reported that at $\phi = 0.4$ the energy lost by heat transfer to the wall accounted for 25% of the total heat release, and at $\phi = 1$ this percentage increased to 45%. An important conclusion is that improvements in H2ICE efficiencies will require strategies to minimize heat transfer losses to the cylinder walls (e.g., charge stratification).

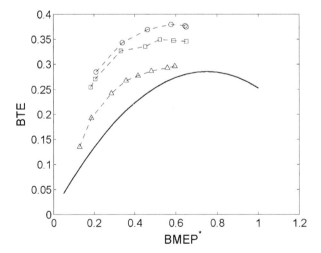

Figure 8.5 Brake thermal efficiency (BTE) as a function of normalized BMEP, where BMEP* = BMEP/ BMEP$_{max. gasoline}$. ○, □ (Tang *et al.*, 2002) for CR = 14.5 and 12.5, respectively; △, — (Swain *et al.*, 1983) for hydrogen and gasoline, respectively, at CR = 9:1. The dashed lines through the hydrogen data are least squares fit polynomials.

8.2.3. ADVANCED HYDROGEN ENGINE STRATEGIES

In the previous section, we have shown that the power density of a naturally aspirated premixed or PFI-H2ICE is inferior to the power density of the identical engine fueled with PFI gasoline. In practical application, the loss in power density can be as high as 50% (Stockhausen *et al.*, 2002). It is not surprising then that much effort has been put forth in the development of advanced hydrogen engines with improved power densities. In this section, we review several examples. Particular attention is paid to power density, NO$_x$ emissions, and thermal efficiency. For the most part, we consider only advanced reciprocating engine concepts.

8.2.3.1. Pressure-Boosted H2ICE

Intake air pressure boosting is an effective and proven strategy for increasing peak engine power in conventional petroleum-fueled ICEs. For premixed or PFI-H2ICEs, pressure boosting is likely necessary to achieve power densities comparable to gasoline engines and, fortunately, application is straightforward. Although research examining the strategy has been active for decades, optimization of boosted H2ICEs is far from complete.

Early work testing boosted H2ICEs has been carried out by Nagalingam *et al.* (1983), Furuhama and Fukuma (1986), and Lynch (1983). Nagalingam *et al.* worked with a single-cylinder research engine and simulated turbocharged operation by pressurizing inlet air to 2.6 bar and throttling the exhaust to mimic a turbine's backpressure. Researchers at the Musashi Institute of Technology turbocharged a liquid-hydrogen, two-stroke diesel engine and tested its performance on the bench and in a vehicle (Furuhama and Fukuma, 1986). In early tests of turbocharged hydrogen engines in commercial vehicles, Lynch (1983) converted gasoline and diesel engines to spark-ignited hydrogen operation at maximum inlet pressures of 1.5 bar absolute.

More recently, substantial development has been brought about by research efforts from BMW (Berckmüller *et al.*, 2003) and Ford (Natkin *et al.*, 2003; Jaura *et al.*, 2004). Berckmüller *et al.* (2003) have reported results from a single-cylinder engine supercharged to 1.8 bar that achieves a 30% increase in specific power output compared to a naturally aspirated gasoline engine. Natkin *et al.* (2003) report results for a supercharged 4-cylinder 2.0-L Ford Zetec engine and a 4-cylinder 2.3-L Ford Duratec engine that is used for conventional and hybrid vehicles (Jaura *et al.*, 2004). Two Nissan engines tested for hydrogen hybrid vehicle use at Musashi Institute of Technology showed a similar 35% increase in power due to boosting while holding NO$_x$ emissions at 10 ppm (Escher, 1975).

Because boosting pressure increases charge pressure and temperature the problems of preignition, knock, and NO$_x$ control are heightened during boosted operation. In addition, Nagalingam *et al.* (1983) reported that the preignition-limited equivalence ratio decreased from 1 down to 0.5 when they increased intake pressure from 1 to 2.6 bar. In their work, water injection was used to mitigate the effects of higher charge temperatures associated with boosted operation. Berckmüller *et al.* (2003) used other methods to mitigate increased charge temperatures including optimizing coolant flow to the exhaust valve seats and spark plugs and varying cam phasing to improve residual scavenging. They reported a decrease in the preignition-limited equivalence ratio from 1 to 0.6 when inlet pressure was increased from 1 to 1.85 bar. Intercooling is a commonly used strategy for addressing this problem as well: Ford's experimental hydrogen engines incorporate dual intercoolers to maximize cooling of the supercharged air (Natkin *et al.*, 2003). With proper ignition-timing control, they were able to operate free of preignition at equivalence ratios up to 0.8.

As with preignition, NO_x production is sensitive to charge temperature and pressure boosting aggravates the problem. Berckmüller *et al.* (2003) required equivalence ratios lower than 0.45 to operate below the knee of the NO_x curve (occurring at $NO_x \approx 100$ ppm, by their definition). Similarly, Nagalingam *et al.* (1983) published results showing NO_x levels below 100 ppm for equivalence ratios less than 0.4 when operating at supercharged intake pressures of 2.6 bar absolute. Much the same, Natkin *et al.* (2003) reported 90 ppm NO_x emissions at $\phi = 0.5$. To achieve emissions of 3–4 ppm (levels likely required to attain SULEV standards, see Figure 8.4), Ford's supercharged engine was run at a leaner ϕ of 0.23 (Natkin *et al.*, 2003). The Musashi tests mentioned above achieved NO_x levels below 10 ppm for equivalence ratios leaner than 0.4 (Furuhama and Fukuma, 1986). The scatter in the NO_x-limited equivalence ratio between the various studies is a consequence of the dependency of NO_x formation rates on CR, intercooling, and in-cylinder mixing. In general, the scatter is equivalent to that observed in Figure 8.4 for naturally aspirated H2ICEs. Multi-mode operating strategies are often adopted for boosted H2ICEs. Berckmüller *et al.* (2003) recommended the following strategy for increasing loads: (1) unthrottled lean operation at low enough loads to avoid NO_x production, then (2) unthrottled stoichiometric operation with EGR dilution to the naturally aspirated full-load condition, and finally (3) supercharged stoichiometric operation with EGR dilution. The lean operating mode is clean enough to require no after-treatment, while the $\phi = 1.0$ modes enable the use of low-cost TWC to deal with the associated higher NO_x production. Natkin *et al.* (2003) used the expedient, but less efficient approach of throttling their supercharged H2ICE at low loads while holding fixed at 0.5. For high loads, they too proposed an EGR strategy enabling $\phi = 1.0$.

8.2.3.2. Liquid-Hydrogen-Fueled Internal Combustion Engine (L-H2ICE)

The use of liquid hydrogen as an automotive fuel has been recently reviewed by Peschka (1998). The L-H2ICE label means that hydrogen is stored as a liquid, but not necessarily injected as a liquid. While the primary benefit of the L-H2ICE is the higher stored-energy density of hydrogen available with liquefaction, it is not the sole benefit. The charge-cooling effect of the cold hydrogen provides for several advantages compared to conventional gaseous PFI.

Intake charge cooling improves volumetric efficiency, minimizes preignition, and lowers NO_x emissions. The increase in volumetric efficiency and, subsequently, power density is a simple consequence of intake mixture density varying inversely with temperature. Furuhama *et al.* (1978) calculated that with a hydrogen temperature of 120 K, the peak power output of an L-H2ICE can equal that of the identical engine fueled with gasoline. Similarly, Wallner *et al.* (2004) estimated that with intake charge cooling to 210 K, the power density of an L-H2ICE will be 15% higher compared to fueling with PFI gasoline. Furthermore, the lower charge temperature will also mitigate preignition events (similar to water injection) and thereby increase the preignition-limited maximum equivalence ratio. For example, stoichiometric operation in the absence of preignition events has been reported by Furuhama *et al.* (1978) and Knorr *et al.* (1997). The net effect is that not only are high power densities possible with L-H2ICE they are also attainable in practice. An improvement in BTE and NO_x emissions are also realized with charge cooling: Furuhama *et al.* (1978) report that, relative to gaseous hydrogen fueling, BTE increases and specific NO_x emissions decrease with decreasing injection temperatures. The latter effect was described as being primarily due to the fact that with liquid hydrogen fueling a leaner mixture could be used to produce the same power that would otherwise require a richer mixture with gaseous hydrogen fueling.

For practical application an L-H2ICE fueling system typically requires a vacuum-jacketed fuel line, heat exchanger and cryogenic pumps, and injectors. A detailed description of an L-H2ICE fueling system is given by Peschka (1992). The practical difficulties of liquid storage include the energy penalty of liquefaction, evaporation during long-term storage, and the cost of onboard cryogenic dewars.

Several experimental and prototype L-H2ICE-powered vehicles have been demon-strated over approximately the last 25 years: students at Musashi University converted a gasoline-powered automobile to operate on liquid hydrogen (Furuhama *et al.*, 1978).[3] The automobile was competed in a student road rally in California totaling roughly 1800 miles. The reported fuel economy was the gasoline equivalent of 36.3 mpg. MAN, over a 2-year test period (1996–1998), operated a dual-fueled liquid hydrogen and gasoline bus in regular public transport service (Knorr *et al.*, 1997). The bus was powered by a 6-cylinder 12-L engine with a CR of 8:1. The liquid hydrogen storage capacity was 570 L, of which 6% of this total was reported as lost to evaporation per day. BMW has developed a small fleet of dual-fueled liquid hydrogen and gasoline-powered vehicles under the model name 750 hL. The engine has 12 cylinders and a displacement of 5.4 L. Liquid hydrogen storage capacity is 140 L providing a range of 400 km. Acceleration is respectable at 0–60 mph in 9.6 s.

8.2.3.3. Direct-Injection Hydrogen-Fueled Internal Combustion Engine (DI-H2ICE)

The direct-injection H2ICE has long been viewed as one of the most attractive advanced H2ICE options (for a summary, see Homan, 1978). The view is based on high volumetric efficiency (since hydrogen is injected after intake valve closing) and the potential to avoid preignition. The latter effect is controlled by timing injection to both minimize the residence time that a combustible mixture is exposed to in-cylinder hot-spots (i.e., late injection) and to allow for improved mixing of the intake air with the residual gases. The improved volumetric efficiency (equal to PFI gasoline or higher) and the higher heat of combustion of hydrogen compared to gasoline provide the potential for DI-H2ICE power density to be approximately 115% that of the identical engine operated on gasoline. This estimate is consistent with that measured by Eichlseder *et al.* (2003). In particular, they measured a 15% increase in IMEP for engine operation with DI hydrogen compared to engine operation with PFI gasoline.

The challenge with DI-H2ICE operation is that in-cylinder injection requires hydrogen–air mixing in a very short time. For early injection (i.e., coincident with IVC), maximum available mixing times range from approximately 20 to 4 ms across

---

[3] See Furuhama (1997), for a summary of Musashi University's 20 years of L-H2ICE research.

the speed range 1000–5000 rpm, respectively. In practice, to avoid preignition, start of
injection (SOI) is retarded with respect to IVC, and mixing times are further reduced.
To evaluate the plausibility of complete mixing in a DI-H2ICE, Homan (1978), using
experimental correlations for air-entrainment rates in free-turbulent jets and order-of-
magnitude expressions for turbulent mixing times, estimated that a free-hydrogen jet
with sonic velocity at the orifice issuing into air will entrain a stoichiometric amount
of air in approximately 1 ms. However, contrary to this optimistic estimate, the
overwhelming experimental evidence (Homan, 1978; Homan *et al.*, 1983; Glasson
and Green, 1994; Kim *et al.*, 1995; Jorach *et al.*, 1997) demonstrates that complete
mixing in an engine takes approximately 10 ms. Homan *et al.* (1983) conjectured that
the order-of-magnitude difference in mixing times between the estimate for a free jet
and that measured in an engine is a result of fluid flow interaction between in-cylinder
flow and the hydrogen jet. A strong interaction will cause the free-jet analysis to break
down. As evidence, Homan *et al.* (1983) demonstrated, in an engine with side
injection and side spark (separated by 180°), that the frequency of misfire for late
injection decreased significantly when the injected hydrogen was in coflow with the
swirl instead of in crossflow.

Given the high probability of incomplete mixing with late injection, much effort has
been devoted to understanding the effect of injection timing on DI-H2ICE properties.
The effect of SOI on NO_x emissions has been investigated by Homan *et al.* (1983),
Glasson and Green (1994), and Eichlseder *et al.* (2003). The data of NO_x versus SOI for
these studies are shown in Figure 8.6, where it is observed that the effect of SOI on NO_x
emissions is not simple: NO_x emissions increase with retardation of SOI in several data
sets and decrease in others. These conflicting trends can be explained by separating the
data sets of Glasson and Green (1994) and Eichlseder *et al.* (2003) into two groups:
(i) data sets with a global equivalence ratio at or below the NO_x-limited equivalence
ratio ($\phi < 0.5$, open symbols); and (ii) data sets with a global equivalence ratio above
the limit (▼, ★, +). For all data sets, retarding SOI is assumed to increase mixture

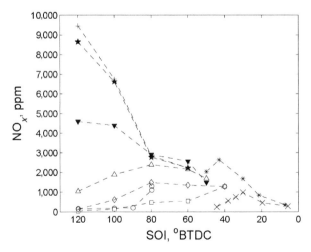

Figure 8.6 NO_x concentration versus SOI: ○ (Glasson and Green, 1994) at $\phi = 0.5$; (□, ◇, △, ▼, ★, +)
from (Eichlseder *et al.*, 2003) at $\phi = 0.35, 0.43, 0.51, 0.60, 0.69$, and 0.79, respectively; (×, □) from (Homan
et al., 1983) at $\phi = 0.35$ and 0.52, respectively. Dashed lines through the data are for visual aid.

inhomogeneity. For the data that fall below the NO_x-limited equivalence ratio, the increase in mixture inhomogeneity leads to locally rich regions with high combustion temperatures and subsequent higher NO_x production. For the data sets above the limit, with increasing mixture inhomogeneity, rich regions with $\phi_{local} > 1$ are produced with a local reduction in NO_x formation. Subsequently, the overall heat release rate is decreased (in the locally rich regions most of the excess H_2 will eventually burn, but in the slower mixing-controlled mode). Then, the overall reduction in NO_x emissions is thought to be due to the coupled effect of a slower heat release rate and in-cylinder heat losses that lead to lower peak in-cylinder temperatures and slower NO_x formation rates.

The data of Homan *et al.* (1983) are unique in that a modified diesel head was used to locate the spark plug close to the injector. This arrangement allowed DI-H2ICE operation with late injection and near-simultaneous spark. Homan *et al.* termed this operating scheme LIRIAM (late injection, rapid ignition and mixing). The observed increase, then decrease in NO_x production with SOI retard is also explained as an effect of charge stratification. However, due to the engine geometry and late-injection strategy, it is not surprising that the data follows a different trend with ϕ.

Similar to the NO_x emissions described above, the effect of SOI on thermal efficiency is not straightforward. Eichlseder *et al.* (2003) found that at low loads (or similarly, low equivalence ratio), indicated thermal efficiency (ITE) increased with retardation of SOI. The increase was shown to be due to a decrease in the compression work caused by differences in mixture gas properties and charge mass with retarded SOI. At high loads, the same authors found that ITE first increases and then decreases with retardation of SOI. The reversing trend is assumed to be a consequence of an unfavorable mixture formation. However, Lee *et al.* (2002) report results contradictory to Eichlseder *et al.* (2003), where they find that, for both low and high load, thermal efficiency decreases monotonically with retardation of SOI. These contradictory findings may be a result of differences in mixture formation. The view is based on the recent results of Shudo *et al.* (2003), in which it was shown that with charge stratification such that the local mixture fraction in the near-wall region is expected to be lean compared to the bulk mixture, cooling losses can be decreased significantly and thermal efficiency increased. In order to take full advantage of the potential of the DI-H2ICE, a high-pressure, high flow-rate hydrogen injector is required for operation at high engine speeds and to overcome the in-cylinder pressure for injection late in the compression stroke. Here, we define high pressure as greater than 80 bar to ensure sonic injection velocities and high enough mass flow rates for SOI throughout the compression stroke. The need for rapid mixing necessitates the use of critical flow injectors, and the short time duration with late injection requires high mass flow rates. The development of high-pressure injectors has been reported by Green and Glasson (1992) and Jorach *et al.* (1997). The development of hydrogen injectors for injection pressures lower than 80 bar has been reported by Homan *et al.* (1983) and Varde and Frame (1985). In addition, a DI-H2ICE injector has been developed by Furuhama *et al.* (1987). To prevent valve leakage at the valve seat a seal made of an elastomer material has been used with success (Homan *et al.*, 1983; Green and Glasson, 1992). Typical flow rates required are 1–10 g/s.

Multi-mode operating strategies have also been proposed for DI-H2ICEs (Lee *et al.*, 2002; Rottengruber *et al.*, 2004), similar to pressure-boosted H2ICEs. Here, a dual-injector strategy would be used to take advantage of the high thermal efficiencies at low to medium loads with PFI fueling and the high peak power with DI fueling. Rottengruber *et al.* (2004) propose the following strategy: (1) unthrottled lean operation with external mixture at loads below the NO_x formation limit, then (2) throttled stoichiometric operation with external mixture, and finally (3) stoichiometric internal mixture formation. Lee *et al.* (2002) report, for a dual-injector H2ICE, improvements in thermal efficiency of approximately 15–30% at low to medium loads with PFI fueling compared to DI fueling, and an increase in peak power output of approximately 60–70% with DI fueling compared to PFI fueling. The disadvantage of the dual-injector strategy is the added cost.

8.2.3.4. H2ICE Electric Hybrid

A hybrid electric version of an H2ICE offers the potential for improved efficiencies and reduced emissions without the need for after-treatment. In a hybrid electric vehicle (HEV), the ICE operates either in series or in parallel with an electric motor. There are advantages and disadvantages for both configurations. Recent hydrocarbon-fueled HEVs have incorporated a configuration that combines features of both series and parallel.[4] The series hybrid configuration for H2ICEs has been investigated by Van Blarigan and Keller (1998) and Fiene *et al.* (2002). In this configuration, the H2ICE is used to drive an alternator that generates electricity. The electricity is used either to charge the batteries or power the electric motor that powers the vehicle drivetrain. The advantage is the ability to operate and optimize the H2ICE for single-speed operation at maximum power. For this purpose, Van Blarigan and Keller (1998) have extensively studied the optimization of H2ICE thermal efficiencies and NO_x emissions for single-speed operation. Peak ITEs of 44–47% were reported (Van Blarigan and Keller, 1998). To assess the potential for H2ICE vehicles to meet EZEV (i.e., SULEV, see Footnote 2) emission standards, Aceves and Smith (1997) modeled a non-hybrid H2ICE vehicle, and parallel and series H2ICE HEVs. They reported that for the input parameters selected, all three vehicles can satisfy EZEV emissions standards. An attractive feature of the HEV is that the peak power output of the ICE can be significantly lower than that required for a non-hybrid ICE without any sacrifice in vehicle performance. Then, with respect to H2ICE HEVs, the obvious operating strategy would be to operate the H2ICE lean enough such that engine-out NO_x emissions are near zero (see Figure 8.4). It follows that since lean power density can be improved with intake pressure boosting, a boosted H2ICE HEV would be an attractive option, as demonstrated by the Ford H2RV (Jaura *et al.*, 2004). The H2RV is an HEV that uses a supercharged 4-cylinder 2.3-L H2ICE. The boosted H2ICE has a peak power of 110 HP at 4500 rpm and the electric motor provides an additional 33 HP, and is used primarily for power assist. Acceleration is 0–60 mph in 11 s. Fuel economy is 45 miles/kg of H_2 (gasoline equivalent of 45 mpg) and driving range is 125 miles.

The potential for improved efficiencies and reduced emissions with H2ICE hybridization appears quite promising. In particular, in comparing an H2ICE HEV to a PEMFC vehicle, Keller and Lutz (2001) found that the H2ICE HEV compares favorably.

---

[4] For example, see Toyota Hybrid System (THS).

8.2.4. CONCLUDING REMARKS

The last decade has produced significant advancements in the development of the H2ICE and H2ICE-powered vehicles. Undoubtedly aided by the technological advancements of the ICE, simple H2ICE options are convenient and economically viable in the near term. Consequently, there is little doubt that the H2ICE can serve as a near-term option for a transportation power plant in a hydrogen economy. This is best illustrated in the fact that at least one H2ICE-powered vehicle, the BMW 745 h, is expected to reach the market in the next few years.

However, the long-term future of the H2ICE is less certain and hard to predict, as is the future of the hydrogen economy itself. The uncertainty is in part due to the multiple H2ICE options available, as described in Section 8.2.3. There are good prospects for increased efficiencies, high power density, and reduced emissions with hybridization, multi-mode operating strategies, and advancements in ICE design and materials. The commercial viability of these advanced H2ICE options requires: continued advancements in fundamental H2ICE research, reduction of NO_x formation and emissions, research and development of advanced engine components, and highly advanced control and optimization strategies. Provided that these efforts produce an H2ICE option that is highly efficient, with near-zero emissions, and a drivability that surpasses present day gasoline-fueled ICEs, then competition will dictate the transition to the PEMFC.

8.3. HYDROGEN IN GAS TURBINE ENGINES

As described in Section 8.1, the primary uses of gas turbines are in stationary power generation and aircraft operation. A third class is aeroderivative power generating gas turbines. With regard to the use of hydrogen as a fuel, the primary differences between these applications are related to combustor operating conditions. Typical operating condition ranges for the three classes of gas turbine combustors are shown in Table 8.2.

Generally, operating pressures are lowest in land-based power generation combustors and increase by about a factor of two to three in high-performance aircraft systems. The higher pressures in aircraft systems allow reductions in combustor size and provide higher efficiencies and thrust for aircraft applications. The higher combustor inlet temperatures in aircraft engines also provide higher efficiencies. These operating conditions have the greatest impact on technical challenges such as autoignition and flashback, which depend on temperature and pressure through their effect on chemical kinetic rates. These effects will be discussed in a later section. Turbulence intensity is

Table 8.2

Operating conditions in gas turbine combustors

Application	Land-Based Power Generation	Aeroderivative Power Generation	Aircraft
Inlet temperature (K)	670–1020[a]	670–1020	670–1020
Operating pressure (atm)	9–20	20–40	20–42
Turbulence intensity (u'/U_{mean})	30–50% ($u' = 10$–$15\,\mathrm{m/s}$)[b]	30–50%	30–50%

[a]Richards *et al.* (2001).
[b]Based on bulk inflow velocity of 30 m/s.

also important in determining the combustion rate and flame speed and is typically defined as the local velocity fluctuations (rms) divided by the local mean velocity. The turbulence intensity appears in most expressions for the turbulent flame speed, which is relevant to flame location and phenomena such as flashback. However, since the existence of multiple recirculation zones in most gas turbine combustors results in regions where the mean velocity is near zero, a meaningful turbulence intensity based on the local mean velocity is often difficult to determine. Thus, the turbulence intensities given in Table 8.2 are based on the expected maximum rms velocity fluctuations normalized by the bulk inflow velocity. They will be used in the discussion that follows to estimate turbulent flame speeds in the combustor.

Other differences between power generation and aircraft applications are related to the source of hydrogen and its storage. For stationary power generation several options are feasible. Piping of hydrogen directly to the plant location as is done with conventional NG gas turbines should be straightforward once the necessary infrastructure is in place. Alternatively, an attractive option is the transport of hydrocarbon feedstock such as NG to the plant site where it is reformed into hydrogen and CO_2. This approach facilitates the subsequent sequestration of the CO_2 or its transport and use in industrial plants located elsewhere. Hydrogen for aircraft operation will likely require large-scale storage capabilities for H_2 at the airport as well as solutions for onboard H_2 storage. Somewhat surprisingly, considerable efforts were put into design and feasibility studies in the 1970s with regard to H_2 utilization for aircraft, with the general conclusion that these potential problems were not insurmountable (Brewer, 1991; Scott, 2004). This subject will be discussed in a later section.

8.3.1. Rationale for Hydrogen in Gas Turbine Engines

8.3.1.1. Emissions

Gas turbines are dominant for both power generation and aircraft propulsion technology in the United States. While gas turbines currently use a wide variety of hydrocarbon fuels, most new power generation installations will be fueled using NG. Jet fuel is typically used in aircraft gas turbines. It is currently recognized that a transition to hydrogen as a primary fuel offers numerous advantages over conventional hydrocarbon fuels. Principal among these advantages are reduced greenhouse gas (CO_2) emissions and independence from foreign oil sources since hydrogen can be produced from coal, photochemistry, and biomass or nuclear sources. The advantages of a transition to hydrogen fuel in the gas turbine sector become particularly apparent when considered in the context of projected growth figures for gas turbine utilization. Figure 8.7 shows that gas turbines currently account for 15% of the total CO_2 emissions. This number is expected to grow to nearly 28% by the year 2020. Passenger cars, by comparison, currently account for 11% of total CO_2 production and will increase to only 13% by 2020. An immediate transition to hydrogen fuel use in gas turbines would significantly impact future CO_2 emission issues.

NO_x emissions are also major concerns for air quality. Lean premixed combustion is an effective approach to minimize NO_x emissions from gas turbine engines (Lefebvre, 1983; Department of Energy, Office of Fossil Energy, 1999). Operation under premixed, fuel-lean conditions makes burner operation at the lower temperatures needed to limit

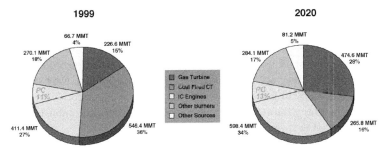

Figure 8.7 Contribution of various combustion sources to total US CO₂ emissions (Schefer, 2003). (See color insert.)

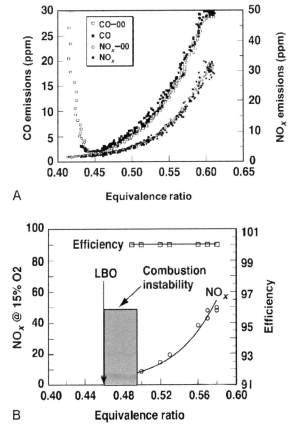

Figure 8.8 (A) Tradeoff between NO_x and CO emissions under premixed fuel-lean conditions (data provided courtesy of Dr Dan Maloney at National Energy Technology Center in Morgantown, WV) and (B) tradeoff between NO_x emissions and efficiency (data provided courtesy of Bill Proscia at United technologies Research Center in East Hartford)

NO_x emissions possible. Unfortunately, this approach is limited by both flame instability and by the increase in CO emissions that occurs under ultra-lean conditions. This tradeoff in NO_x and CO emissions is shown in Figure 8.8a where emissions from a laboratory scale, premixed gas turbine combustor burning NG are shown as a function of equivalence ratio, ϕ. It can be seen that at increasingly lean conditions the NO_x emissions monotonically decrease while CO emissions rapidly increase for $\phi < 0.44$. The region where combustion instabilities become significant in Figure 8.8b typically occurs between ϕ of 0.46 and 0.49 and continues to the lean blowout limit (where complete flame extinction occurs), which is in the range 0.40 to 0.44. As discussed below, hydrogen and hydrogen blended with traditional hydrocarbon fuels significantly improves flame stability during lean operation and allows stable combustion at the low temperatures needed to minimize NO_x production.

8.3.1.2. Alternative Fuel Utilization

The increased use of gas turbines for power generation is accompanied by the need to use more cost-effective alternate fuels with a wide range of heating values (Meier *et al.*, 1986; Richards *et al.*, 2001; Zhang *et al.*, 2005). For example, available and developing technologies for the production of syngas from coal gasification and biomass result in a fuel gas mixture typically consisting of 24–44% (by volume) of hydrogen in CO, with smaller amounts of CO_2 and nitrogen. Typical syngas compositions at a number of current installations are given in Table 8.3. Complete removal of all carbon prior to combustion would be prohibitively costly from an economic and energy loss standpoint. However, combusting the mixture of hydrogen and CO could be attractive and, with sequestration of the CO_2 formed in the product gases, would minimize greenhouse gas emissions into the atmosphere.

Other sources of fuels with high hydrogen content are the low- and medium-heating value fuels containing hydrogen, carbon monoxide, and inert gases that are produced in integrated coal-gasification combined cycle (ICGCC) installations. Refineries and chemical plants produce by-product gases, usually rich in hydrogen, that are often flared. Biomass-derived fuels and sewage/landfill off-gases represent another economical energy source. The ability to effectively burn these fuels in gas turbines will provide substantial cost advantages, while minimizing adverse effects on the environment. The economical and environmental benefits are an immediate motivator to develop fuel flexible gas turbine systems capable of using many different fuels containing various blends of hydrogen and hydrocarbons.

8.3.1.3. Transition Strategies for Future Hydrogen Utilization

Ultimately, it would be desirable to replace conventional hydrocarbon fuels with pure hydrogen, thus entirely eliminating the source of combustion generated CO_2. This

Table 8.3

Syngas compositions in operating gasification power plants (Todd, 2000)

Syngas	PSI	Tampa	El Dorado	ILVA	Schwartz
H_2	24.8	37.2	35.4	8.6	61.9
CO	39.5	46.6	45.0	26.2	26.2
CH_4	1.5	0.1	0.0	8.2	6.9
CO_2	9.3	13.3	17.1	14.0	2.8
$N_2 + Ar$	2.3	2.5	2.1	42.5	1.8
H_2O	22.7	0.3	0.4	–	–
H_2/CO	0.63	0.8	0.79	0.33	2.36

strategy assumes that any CO_2 produced as a by-product through, for example, hydrocarbon reforming to produce hydrogen is sequestered or otherwise eliminated. While this scenario may be achieved in the longer term, several considerations favor the shorter-term use of fuel blends consisting of hydrogen- and carbon-containing species. One consideration is the current lack of an infrastructure for the production, storage, and distribution of hydrogen. An infrastructure comparable to that currently in place for conventional fuels that could provide sufficient hydrogen to replace all hydrocarbon fuels is not likely for years in the future. One shorter-term approach that shows promise is to blend hydrogen with conventional fuels, thus reducing the quantity of hydrogen needed to a more sustainable level. Purely economic considerations also favor limiting the amount of hydrogen to something less than 20–40% in NG for power generation. At these addition levels, the costs are comparable to other NO_x control strategies for achieving emission levels of 3 ppm or less (TerMaath *et al.*, 2006).

The use of "designer fuels", formed by blending hydrogen with a hydrocarbon feedstock to optimize combustor operation, is well suited to power generating gas turbine systems (Boschek *et al.*, 2005; Strakey *et al.*, 2006). Blending the hydrogen with hydrocarbons provides a near-term technology, enabling stable combustion at ultra-lean premixed conditions and hence, lowering temperatures to significantly reduce NO_x emissions. In the medium to long term, HC and CO_2 emissions can be reduced by systematically increasing the hydrogen concentration in the hydrocarbon feedstock. In the limit of a pure hydrogen combustion system, HC and CO_2 emissions are eliminated entirely. Not only does this approach provide a solution to the immediate need for NO_x control, it also offers a transition strategy to a carbon-free energy system of the future. An ultra-lean combustion system for gas turbine power generation with hydrogen enrichment will provide fuel flexibility and reduced NO_x emissions in the near term, and CO_2 reduction in both current and future systems.

8.3.2. Benefits of Hydrogen in Gas Turbine Applications

In this section, we will address issues associated with the expected benefits of H_2 addition to gas turbine combustors. Recent research findings will be presented that demonstrate the extent of these benefits. While there are many expected benefits of H_2 as a gas turbine fuel, there are some challenging technical issues that must be addressed. These issues will be addressed in Section 8.3.3, and potential solutions for overcoming them will be explored. Both the use of pure hydrogen and hydrogen blended with conventional carbon-containing fuel components will be considered where appropriate.

8.3.2.1. Wider Flammability Limits for Lean Operation

As seen in Table 8.1, the flammability limits of H_2 are significantly wider than conventional hydrocarbon fuels. In particular, the lean limit equivalence ratio of H_2 occurs at 0.1, which compares with a lean limit for NG of 0.4. With reference to Figure 8.8, it is therefore expected that this lower limit will allow operation of gas turbine combustors at significantly lower fuel/air ratios and thus at the low temperatures required to produce sub-ppm range NO_x emissions. Several early studies in large-scale gas turbine combustors have shown this to be the case. For example, Clayton (1976) carried out an early study of the effect of hydrogen addition on aircraft gas turbines.

Up to 15% hydrogen addition to JP-5 or JP-6 produced leaner blowout limits and corresponding reductions in NO_x (less than 10 ppm at 15% O_2) while maintaining acceptable CO and HC emissions levels. Tests conducted in a single combustor test stand at full pressure and temperature at the GE Corporate Research and Development Center and in the field in an NG-fired cogeneration plant at Terneuzen, The Netherlands, demonstrated improved flame stability with hydrogen addition to NG fuel (Morris *et al.*, 1998). Blends of up to 10% hydrogen by volume showed reduced CO emissions under lean conditions with increased hydrogen, and lower NO_x emissions for a given CO level. Increasing the hydrogen did not affect the observed combustion dynamics. Conceptual studies on hydrogen addition in large-scale, gas turbine-based power plants have concluded that hydrogen addition can be used to improve gas turbine combustor performance (Bannister *et al.*, 1998; Phillips and Roby, 1999). The combustion of low BTU gasified fuels with up to 35% hydrogen content has also been demonstrated in an existing, low emission gas turbine with minimal modification (Ali and Parks, 1998).

Figure 8.9 shows details of the flame blowout characteristics under fuel-lean conditions in a premixed centerbody-stabilized dump combustor (Schefer *et al.*, 2002). The curve labeled $n_{H2} = 0$ corresponds to pure CH_4 mixed with air, where n_{H2} is the volume fraction of hydrogen in the fuel mixture. The velocity is based on the bulk average velocity at the combustor inlet plane. The curve indicates the equivalence ratios at which the flame extinguishes. Thus, for a fixed velocity, values of equivalence ratio to the right of the curve produce a stable flame, while leaner values of equivalence ratio to the left of the curve result in flame blowout. Generally, the flame blowout velocity decreases as the fuel/air ratio becomes leaner. To characterize the effect of hydrogen addition on lean burner stability, various amounts of hydrogen were added to the

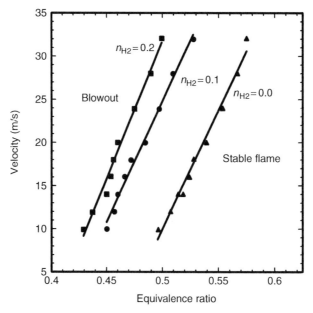

Figure 8.9   Effect of hydrogen addition on swirl burner flame blowout under fuel-lean conditions (Schefer *et al.*, 2002).

methane/air mixture and the flame stability measurements were repeated. Also shown in Figure 8.9 are the resulting lean blowout curves for 10% and 20% hydrogen dilution, labeled $n_{H2} = 0.10$ and 0.20, respectively. It can be seen that hydrogen addition significantly extends the lean stability limits. For example, at a velocity of 20 m/s, the methane flame blows out at an equivalence ratio of about 0.48. The addition of 10% and 20% of hydrogen extends the equivalence ratio at blowout to 0.43 and 0.40, respectively. These values represent a 10.4% and 16.6% reduction in the lean stability limit. Results obtained in a similar burner with multiple inlet holes carrying premixed methane and air with hydrogen addition between 70% and 100% showed stable flames were maintained at equivalence ratios as low as 0.1 (Schefer *et al.*, 2003).

The ability of H_2 addition to extend the lean flame stability limits has been attributed to improved resistance to strain that is imposed on the flame by the fluid dynamics (Gauducheau *et al.*, 1998; Vagelopoulos *et al.*, 2003). The effect of H_2 addition on lean flame blowout behavior is shown in Figure 8.10 where the normalized strain rate at extinction, K_{ext}, is shown for various amounts of H_2 addition (Vagelopoulos *et al.*, 2003). Here, K_{ext} is defined such that the extinction strain rate is normalized by the extinction strain rate for pure CH_4. These calculated results were obtained for a premixed counter flow flame configuration in which back-to-back premixed flames are established by two opposed streams of premixed fuel and air that interact to form a stagnation plane. Depending on the velocity of the opposing streams, the levels of strain imposed on the flame can be increased until the flame extinguishes. As described in Chapter 2, strain is defined as the velocity gradient normal to the flame at the location of maximum heat release. Results are shown for equivalence ratios of 0.50, 0.55, and 0.60. In general, the extinction strain rate increases with the addition of H_2 to the CH_4, indicating increased resistance to flame stretch. Comparing the results for the three

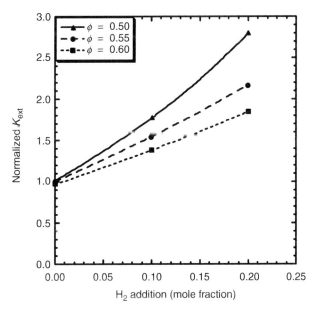

Figure 8.10 Extinction strain rate as a function of H_2 addition to a counterflow premixed CH_4/air flame (Vagelopoulos *et al.*, 2003).

equivalence ratios further shows that H_2 addition is more effective at increasing strain resistance as the equivalence ratio is reduced.

Similar conclusions were reached by Gouldsheau et al. (1998) based on the results of stretched flame calculations at elevated 30 atm pressure. The improved lean flame stability was attributed to hydrogen's higher flame speed and increased resistance to strain. It was further noted that flame properties such as temperature and radical concentrations (O, H, and OH) are little changed by hydrogen addition in unstrained flames, but radical concentrations increased significantly with hydrogen addition in strained flames. Ren *et al.* (2001) performed a sensitivity analysis that showed the chain branching reaction $H + O_2 \Leftrightarrow OH + O$ is important with respect to extinction due to the production of H radicals. More recent work by Jackson *et al.* (2003) showed that the primary benefit of H_2 addition is an improved resistance to strain and that improvements to the flame speed and flammability limits are secondary.

8.3.2.2. Reduced NO_x Emissions

As described previously, one of the primary advantages of using H_2 or H_2-blended hydrocarbon fuels is the ability to operate at leaner conditions where the lower flame temperatures limit NO_x formation. Calculations in a premixed counterflow flame configuration by Sankaran and Im (2006) explored the effects of H_2 addition on the LFL in steady and unsteady flames. The variation of maximum flame temperature with equivalence ratio is shown in Figure 8.11. For a pure CH_4 and air mixture (curve labeled $R_H = 0$) the flame temperature decreases monotonically from a maximum at near stoichiometric conditions to a value of about 1620 K at $\phi = 0.72$. This equivalence ratio is the lowest value at which a stable flame can be maintained. Further reductions in ϕ result in total flame extinction. The curves for 18% and 30% hydrogen addition (curves labeled $R_H = 0.05$ and 0.1) clearly show a decrease in the minimum equivalence ratio at which extinction occurs to 0.62 and 0.52, respectively. The corresponding

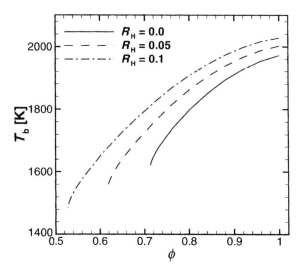

Figure 8.11 Maximum flame temperature (T_b) as a function of equivalence ratio for various amounts of H_2 addition (Sankaran and Im, 2006).

temperatures at which extinction occurs are reduced to 1560 and 1480 K, respectively. Thus, the benefits of hydrogen addition are that stable combustion can be maintained at much leaner mixtures, and that lower temperatures can be achieved while maintaining a stable flame. Since NO_x formation rates are exponentially dependent on flame temperature, reductions in temperature due to leaner operation can significantly lower NO_x emissions in gas turbine combustors.

An example of the reduced emissions achievable with hydrogen addition is shown in Figure 8.12, where the measured NO_x emissions are shown as a function of combustor equivalence ratio. These measurements were obtained in the National Energy Technology Center (NETL) SimVal laboratory-scale research combustor (Sidwell *et al.*, 2005). The combustor has a dump combustor configuration with centerbody and swirling flow. The premixed reactants enter the combustion chamber through a single annulus located in the upstream inlet plane. This facility mimics the flow field of many practical gas turbine combustors. Results are shown for pure NG, and mixtures of NG and 5–40% H_2 (by volume) at a combustor operating pressure of 6.2 atm, which is in the range of power generating gas turbine combustors. The general trend is a smooth decrease in NO_x emissions with decreasing equivalence ratio. The region labeled NG only is for pure NG and extends to an equivalence ratio of about 0.51. Flame extinction with no H_2 addition occurs slightly below this equivalence ratio, thus limiting the NO_x emissions to a minimum of between 1 and 1.5 ppm. With the addition of H_2 the trend of reduced NO_x with reduction in equivalence ratio continues, but lean extinction for 5% and 10% H_2 occurs at lower equivalence ratios of 0.45 and 0.445, respectively. The resulting lower flame temperatures allow sub-ppm levels of NO_x to be attained. Further, note that the combustor was not operated to lean extinction with the 20% and 40% H_2 content, which would allow even leaner operation.

It is also of interest to examine the effect of fuel composition (i.e., H_2 addition to methane) on NO_x formation in the limit of long combustor residence times where conditions approach thermodynamic equilibrium. Shown in Figure 8.13a is the

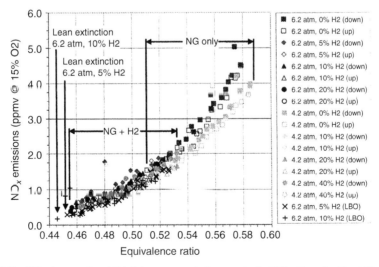

Figure 8.12 NO_x (corrected to 15% O_2) versus equivalence ratio for NG with and without H_2 addition (Sidwell *et al.*, 2005).

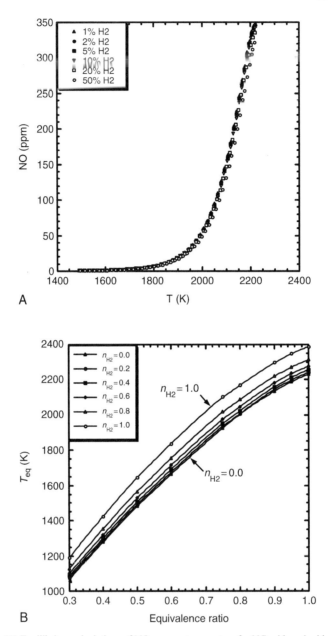

Figure 8.13   (A) Equilibrium calculations of NO_x versus temperature for NG with and without H_2 addition and (B) equilibrium calculations of NO_x versus equivalence ratio for NG with and without H_2 addition.

calculated variation of equilibrium NO_x concentration with adiabatic flame temperature. As expected, an exponential increase in NO_x with flame temperature is seen. Further, the results show that the behavior is largely independent of the amount of H_2 added to the CH_4/air mixture. The reason for this insensitivity to H_2 addition is shown in Figure 8.13b, where flame temperature is shown as a function of equivalence ratio. It can be seen that the flame temperature is minimally affected by up to 60% H_2 addition.

For example, at $\phi = 0.6$, the calculated increase in flame temperature for 60% H_2 addition is only about 50 K and for 100% H_2 the temperature increase is only about 170 K. In contrast, the flame temperature depends primarily on equivalence ratio, increasing by over a factor of two as the equivalence ratio increases from 0.3 to stoichiometric conditions. The insensitivity of temperature to H_2 addition thus accounts for the small effect of H_2 addition to NO_x emissions. NO_x emissions are thus controlled primarily by flame temperature through stoichiometry. For H_2 addition up to 60%, flame temperature is relatively insensitive to fuel composition (i.e., H_2 addition) and depends primarily on stoichiometry.

8.3.2.3. Reduced CO Emissions

The addition of H_2 to lean hydrocarbon fuels has also shown the potential to reduce CO emissions from lean gas turbine combustors. For example, Phillips and Roby (1999) found that OH and O radicals contribute to the pyrolysis of carbon-containing species. They further observed enhanced reaction rates with hydrogen that were attributed to an increase in the radical pool. The reduction in CO emissions with hydrogen addition was also attributed to the increased radical pool as higher OH concentrations promote the completion of CO oxidation to CO_2 via the OH radical.

More recently, Wicksall *et al.* (2003) and Schefer *et al.* (2002) performed experiments in a premixed swirl-stabilized burner at atmospheric pressure to quantify the effects of H_2 addition on NG and methane flames. Flame stability, blowout and emissions characteristics were investigated at fuel-lean conditions. Results demonstrated that the lean stability limit was lowered by the addition of hydrogen. A significant reduction in CO emissions and a corresponding increase in the measured OH concentration were realized as the lean stability limit was approached. For example, the addition of up to 45% H_2 to NG resulted in CO levels below 10 ppm in the central region of the combustor at downstream locations where the combustion process was nearly complete. The corresponding CO levels for pure NG ranged from 15 to 100 ppm in the central region at comparable downstream locations.

These reduced CO emission levels are attributable to two factors. First is the replacement of a carbon-containing fuel by substitution of a carbon-free fuel. This substitution effect increases as the percentage of H_2 added increases and, in the limit of pure H_2, CO emissions are eliminated. A second factor is through changes in the chemical kinetics brought on by the addition of hydrogen. Recent measurements and analysis by Juste (2006) on the use of hydrogen injection in a tubular combustion chamber, representative of gas turbines with swirling air and fueled by JP-4, showed a reduction in CO emissions of 40% with 4% H_2 addition at an equivalence ratio of 0.3. It was estimated that approximately 10% of this reduction was due to the fuel substitution effect and 30% was attributable to chemical kinetic effects. The reduction due to chemical kinetics was further attributed to increased production of radicals that promotes the oxidation of CO to CO_2. Measurements of the OH radical in a model gas turbine combustor confirm that H_2 addition in amounts up to 12% can result in a 36% increase in OH radical concentrations (Schefer *et al.*, 2002).

A reduction in CO emissions was also observed in calculations carried out in a premixed opposed jet flame (Vagelopoulos *et al.*, 2003). The calculations were carried out for premixed CH_4/air flames with varying amounts of H_2 addition. Typical results are shown in Figure 8.14 for the CO concentration (mole fraction) downstream of the

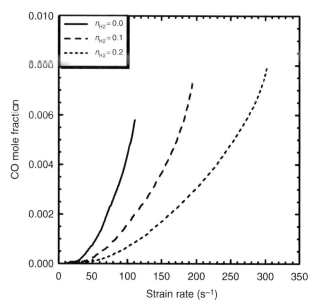

Figure 8.14 CO emission levels from premixed counterflow flame at a distance of 1.5 cm from the inlet. The fuel is CH_4 with 0%, 10%, and 20% H_2 addition. Equivalence ratio = 0.5 (Vagelopoulos *et al.*, 2003).

flame surface. Generally, the CO levels for the pure CH_4/air flame increase as the strain rate is increased. This increase is due to the lower radical concentrations that result from increased strain rate, which slows the primary reaction, accounting for the conversion of CO produced in the flame to CO_2, that is $CO + OH \Leftrightarrow CO_2 + H$. Also shown for comparison are results for 10% and 20% hydrogen added to the CH_4 fuel. Again the CO emissions increase with increased strain rate. However, for a fixed strain rate the CO emissions decrease significantly for H_2 addition. For example, at a strain rate of 100/s, the addition of 10% and 20% H_2 results in reductions of about 60% and 83%, respectively.

8.3.2.4. Energy Density

The energy density of H_2 on a per unit mass basis is about 120 MJ/kg (Table 8.1), which is nearly a factor of two higher than most conventional fuels. As such, H_2 holds considerable promise as a fuel for industrial and transportation applications. In particular, this high energy density is well suited to aircraft applications since a significant fraction of fuel costs are associated with carrying the required fuel on the plane (Scott, 2004). The higher energy density of H_2 means less fuel has to be carried, thus reducing the cost of the required fuel. However, due to the low density of H_2, its volumetric energy density is less than half that of other fuels. For example, in Table 8.1 it can be seen that the lower heating values for H_2 are 120 MJ/kg (unit mass basis) and 9.9 MJ/m^3 (unit volume basis). The corresponding values for CH_4 are 50 MJ/kg and 32.6 MJ/m^3. Thus, storing sufficient H_2 for many applications requires a large volume. Possible approaches to this storage problem are storage at very high pressures (68 MPa storage tanks are currently available and storage tanks up to 100 MPa will be available in the near future). Metal hydride storage is also promising and should be commercially feasible when storage amounts reach about DOE targets of 6% (by 2010) to 9% (by 2015) of the system weight (Department of Energy, 2005). In aircraft applications, designs developed in the 1970s proposed the use of liquid H_2 in insulated storage tanks mounted on the aircraft (Brewer, 1991; Scott, 2004). Alternatively, the use of H_2 as an additive allows one to utilize the benefits of H_2 addition to conventional fuels, while at the same time alleviating the excessive storage volume requirements.

8.3.3. TECHNICAL CHALLENGES FOR HYDROGEN IN GAS TURBINE APPLICATIONS

8.3.3.1. Flame Anchoring and Flashback

The higher flame speeds found with H_2 as a fuel could require some design modifications for optimum gas turbine combustor performance. Since flames typically stabilize, or anchor, in regions where the local flow velocity is near the local flame speed, the higher flame velocities may have to be taken into account in both combustor geometry and the manner in which the premixed reactants are introduced.

Shown in Figure 8.15 are the laminar flame speeds, S_L, for pure H_2, pure CH_4, and mixtures of H_2 and CH_4. In all cases the flame speed is highly dependent on equivalence ratio. For example, at an equivalence ratio of 0.6, S_L for pure H_2 is about 1.0 m/s, while at an equivalence ratio of 1.6 the flame speed peaks at nearly 3.0 m/s. From a gas turbine engine design standpoint, what is most relevant is that over the entire range of lean operating conditions, S_L for H_2 is significantly larger than for conventional

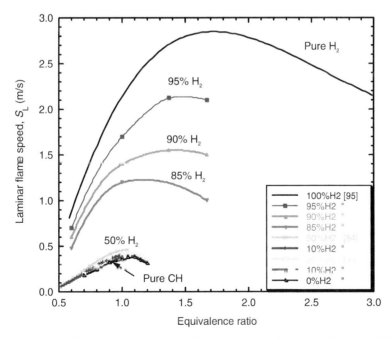

Figure 8.15 Laminar flame speeds as a function of equivalence ratio for H_2–air, CH_4–air, and blends of H_2–CH_4–air at STP. (See color insert.)

hydrocarbon fuels. In particular, at an equivalence ratio of 0.6 the laminar flame speed for methane is about 0.1 m/s, nearly an order of magnitude less than for pure H_2. As is typical for hydrocarbon fuels, the peak flame speed occurs slightly on the rich side of stoichiometric, but at a ϕ value much closer to unity than for H_2. The peak flame speed for methane is nearly a factor of seven lower than that for H_2. Also note that for mixtures of H_2 and CH_4, S_l is relatively insensitive to H_2 addition amounts up to about 50%. As the addition amount increases above 50%, S_L also increases rapidly.

As discussed in Chapter 2, the presence of turbulence significantly increases the flame speed due to increased mixing rates and, through flame surface wrinkling, the flame surface area. Figure 8.16 shows the turbulent burning velocity (normalized by S_L) for hydrogen and propane as a function of velocity fluctuations, u'_{rms}. The solid line is a proposed fit to turbulent flame speed data given by $S_T/S_L = 1 + 3.1(u'_{rms}/S_L)^{0.8}$. Turbulent flame speeds in a turbine combustor environment can be estimated from this correlation and Figure 8.16. For example, the estimated range of turbulence intensities in gas turbine combustors is 30–50%. Assuming an average bulk flow inlet velocity of 30 m/s, the resulting velocity fluctuations are $u'_{rms} = 10-20$ m/s. At the maximum laminar flame speed for H_2 of about 3 m/s, $u'_{rms}/S_L = 3-7$ and from Figure 8.16, $S_T/S_L = 8-16$. Thus, for $S_L = 3$ m/s this yields a turbulent flame speed of $S_T = 24-44$ m/s. The same calculation for CH_4 using a maximum stoichiometric $S_L = 0.45$ m/s gives $S_T/S_L = 40-60$ and $S_T = 16-25$ m/s. These velocities, of course, will decrease significantly under lean conditions where S_L falls off rapidly with ϕ. The effect of operating pressure and temperature on S_L must be taken into account. It is clear, however, that the increased turbulence intensities in gas turbine combustors result in significant increases in flame velocity that must be considered in combustor design to prevent flashback and to determine where the flame is stabilized.

It is worth noting that while S_T can be estimated using the above approach for different fuel mixtures, it is also not straightforward to establish because it depends on

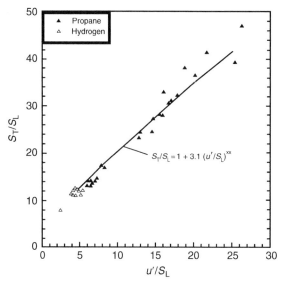

Figure 8.16   Effect of velocity fluctuations on turbulent flame speed (Abdel-Gayed and Bradley, 1981).

both the local chemical reaction rates and flow properties (turbulence intensity, pressure, and temperature) in a manner not well established over the entire range of conditions existing in gas turbine combustors. These properties vary significantly even within a single combustor geometry. Thus, the use of H_2 may require combustor modifications since the flame stabilization location is largely determined by flow mechanics (i.e., where low velocity, recirculation regions are established).

Figure 8.17 shows a comparison between calculated laminar flame speeds, S_L, and measured turbulent flame speeds, S_T, for various amounts of H_2 addition to CH_4–air mixtures (Boschek *et al.*, 2007). The calculations were carried out for freely propagating, unstretched flames. The results, for equivalence ratios of 0.50 and 0.43, show that S_L increases approximately linearly with H_2 addition over the range of conditions studied. The increase in the measured S_T with H_2 addition is comparable to the rate of increase in laminar flame speed up to about 20% H_2 addition. For H_2 addition amounts greater than 20% ($\phi = 0.50$) and greater than 30% ($\phi = 0.43$), the increase in S_T rapidly accelerates and increases non-linearly toward much higher values than S_L.

One proposed reason for the increase in flame speed is that H_2 addition increases OH, which increases the global reaction rate and flame propagation speed. This observation is entirely consistent with results in the literature described previously (Ren *et al.*, 2001; Schefer *et al.*, 2002). An additional aspect unique to hydrogen is that, unlike most hydrocarbon fuels, the Lewis number of H_2 is much less than unity. Thus, mass diffusivity is much greater than thermal diffusivity and differential diffusion increases the H_2 available in the reaction zone. As noted by Hawkes and Chen (2004), blending highly diffusive H_2 to a heavier hydrocarbon fuel such as CH_4 can significantly influence the manner in which the flame responds to stretch. The blended mixture has a higher tendency for cusp formation and flame surface area generation, which leads

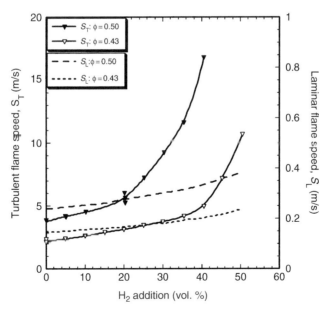

Figure 8.17 Effect of hydrogen addition on flame speed. Data points indicate experimentally measured values. Lines indicate calculated laminar flame speeds (Boschek *et al.*, 2007).

to an increased flame surface area and burning rate. The rapid acceleration of S_T with H_2 addition is likely a result of these factors.

An example of the effect of increased flame speed due to H_2 addition on flame anchoring and location is shown in Figure 8.18, where flame luminescence photographs are shown for a lean premixed swirl-stabilized flame with varying amounts of H_2 addition to the methane–air mixture. The flame with no H_2 added extends past the downstream end of the quartz combustion confinement tube and shows no visible flame in the corner recirculation zone where the quartz tube sits on the inlet faceplate. The addition of 12% H_2 results in a continuous stable flame in the corner recirculation zone and a significantly shorter flame, indicating more rapid combustion. A further increase in H_2 addition decreases the flame length, although the change is more modest. The results show that flames with H_2 addition and reaction in the corner recirculation zone are shorter and more stable. These changes in flame location are similar to changes seen with pure CH_4 fuel when the equivalence ratio is increased from very lean toward stoichiometric conditions, again reflecting the increase in reaction rates.

Flame flashback is also an issue of greater importance with premixed H_2/air mixtures, again due to the higher flame velocities and the large region over which the H_2 and air are premixed (Richards *et al.*, 2001; Kwon and Law, 2004). Flashback can occur in regions of lower velocity such as in the boundary layers near physical surfaces where

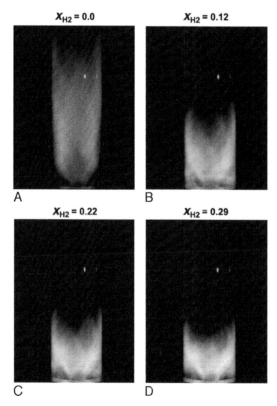

Figure 8.18   Effect of H_2 addition on global flame characteristics in lean premixed swirl burner (Schefer *et al.*, 2002). (See color insert.)

the lowest flow velocities occur. Likewise, temporary transients in the flow due to combustor oscillation may result in a sudden drop in the flow velocity (or complete flow reversal) that allows the flame to propagate upstream into the premixer where it can stabilize and result in overheating of engine components in the premixer or at the fuel injector and cause significant damage to the engine.

8.3.3.2. Autoignition/Preignition

Autoignition is also a potential problem in lean premixed combustors. Typically, the air is preheated by the compressor to temperatures in the range of 670–1020 K (Table 8.2). Thus, once the fuel is sufficiently mixed with the preheated air, the potential exists for autoignition to occur in the premixer section upstream of the combustor where the residence time is sufficiently long. As with flashback, this could lead to significant overheating and damage to the premixer and other engine hardware.

To prevent autoignition of the fuel/air mixture, premixers are designed to prevent residence times from exceeding the autoignition delay time, τ_{ign}, of the fuel mixture. This requires estimates of both the premixer residence time and τ_{ign}. Several references exist on ignition delay characteristics of the relevant fuels. For example, Spadaccini and Colkett (1994) developed an extensive database for pure methane and mixtures of methane and hydrocarbon species (ethane, propane, or butane) in air. The methane/ hydrocarbon mixtures selected were representative of variations in NG composition. They found that the data for ignition delay time were well correlated by the expression:

$$\tau_{ign} = 1.77 \times 10^{-14} \exp\left(1893/T\right)[O_2]^{-1.05}[CH_4]^{0.66}[HC]^{-0.39} \quad (8.14)$$

where T is the temperature (K), $[O_2]$ and $[CH_4]$ are the concentrations (molecules/cm^3) of oxygen and methane, respectively, and $[HC]$ is the total molar concentration of all non-methane hydrocarbon species. This expression was found to correlate the data well over pressures from 3 to 15 atm, temperatures from 1300 to 2000 K, and equivalence ratios between 0.45 and 1.25. The data show that autoignition times can vary by more than a factor of five over the expected range of NG compositions, primarily due to changes in the ethane content. The authors further note that this expression slightly underpredicts the data below 1800 K and, when compared with lower temperature flow reactor data in the 860–1000 K range, there appears to be a change in the activation energy. Thus, caution should be used in extrapolating this correlation to typical turbine inlet temperatures.

Also relevant to the present discussion are the results of Cheng and Oppenheim (1984) for the autoignition of methane/hydrogen mixtures obtained using the reflected shock technique. The experimental conditions covered a range of pressures from 1 to 3 atm, temperatures from 800 to 2400 K, and equivalence ratios from 0.5 to 1.5. The autoignition delay times for pure methane and hydrogen were found to correlate with Arrhenius-type expressions. The correlation for pure methane is given by

$$\tau_{CH4} = 1.19 \times 10^{-12} \exp\left(46.4/RT\right)[O_2]^{-1.94}[CH_4]^{0.48} \ \mu s \quad (8.15)$$

Note that the activation energy, E, in the Arrhenius term is 46.4 kcal/mole. For pure hydrogen the corresponding expression is

$$\tau_{H2} = 1.54 \times 10^{-4} \exp\left(17.2/RT\right)[O_2]^{-0.56}[H_2]^{0.14} \ \mu s \quad (8.16)$$

where $E = 17.2$ kcal/mole. Thus, the activation energy for pure H_2 is significantly lower than that of pure CH_4. They then correlated the autoignition delay time data for pure hydrogen, pure methane, and mixtures of methane and hydrogen with the formula:

$$\tau_{H2} = \tau_{CH4}{}^{(1-\varepsilon)}\tau_{H2}{}^{\varepsilon} \tag{8.17}$$

where ε is the mole fraction of hydrogen in the total fuel and τ_{CH4} and τ_{H2} are the autoignition delay times of pure CH_4 and H_2 at the same conditions. A single Arrhenius-type expression was not found to correlate the mixture data well because it was found that the activation energies for various mixtures of H_2 and CH_4 were significantly different and decreased with increasing H_2 addition.

More recently Huang *et al.* (2004) measured ignition delay times for stoichiometric hydrogen/methane/air mixtures with 15% and 35% H_2 in the fuel at pressures from 16 to 40 atm and temperatures from 1000 to 1300 K. The data were correlated with the expression:

$$\tau_{ign} = 2.1 \times 10^{-8} \exp(19.06/RT)[O_2]^{-2.83}[CH_4]^{2.31}[H_2]^{-0.0055} \text{ s} \tag{8.18}$$

The results showed that H_2 addition reduced the ignition delay time, but that the reduction was much less pronounced at lower temperatures. For example, at 1300 K and 40 atm the addition of 35% H_2 resulted in a factor of 1.5 reduction in τ_{ign}, but at temperatures below 1200 K the reduction is hardly distinguishable. At a lower pressure of 16 atm, the same trend is observed, but the measured reduction in τ_{ign} is modestly greater. Also, for less than 15% H_2 addition the reduction in τ_{ign} is almost negligible. Kinetic calculation showed that the greater effect of H_2 addition at higher temperature is due to the higher OH radical concentrations produced via the reactions:

$$OH + H_2 \Leftrightarrow H + H_2O \tag{8.19}$$

$$H + O_2 \Leftrightarrow O + OH \tag{8.20}$$

At lower temperatures, the lower branching efficiency of Equation (8.20) leads to a reduced increase in OH. The authors further speculated that the weaker effect of H_2 addition at lower temperatures could be due to the production of extra H radicals from the reaction:

$$CH_3O_2 + H_2 \Leftrightarrow CH_3O_2H + H \tag{8.21}$$

Given the observed trends, it appears that changes in ignition delay time, at least with H_2 addition up to 35%, may not be significant and thus may not require major changes in premixer design. Richards *et al.* (2001) present an excellent discussion of autoignition and notes that data on ignition delay times are generally restricted to a temperature range greater than 1200 K, which is above the inlet temperatures that exist in most gas turbine premixers. Clearly, the better establishment of autoignition delay times at conditions relevant to gas turbine inlet conditions and for a wider range of H_2/CH_4 mixture compositions is needed.

8.3.4. CONCLUDING REMARKS

Generally, the use of H_2 as a fuel in lean gas turbine combustor applications is feasible, and in many ways, is also advantageous. Because of the wider flammability limits and increased resistance to strain and extinction, the use of either pure hydrogen or hydrogen blended with conventional hydrocarbon fuel should only improve the common problems associated with unstable combustor operation under fuel-lean conditions. In addition, the wider flammability limits of H_2 will extend the lean operating limits to lower temperatures than is possible with conventional fuels, making even lower NO_x levels achievable.

The main challenges with the use of pure H_2 in gas turbines are potential flashback and autoignition due to the significantly higher flame speeds and shorter autoignition times. This will require some modifications of current design features. Also, a hydrogen infrastructure for production, storage, and delivery of large amounts of hydrogen will require development. The unique combustion characteristics of H_2 (wide flammability limits, no odor for detection, and the potential for crossover from deflagration to a more destructive detonation) make the safety considerations and the development of hydrogen safety codes and standards a priority. A strategy of blending H_2 with conventional hydrocarbon fuels offers a number of advantages. Blends of H_2 and CH_4 of up to 50% result in only moderate increases in flame speed and autoignition times while still maintaining the wider flammability limits needed to improve lean flame stability and limit NO_x production. In addition, reducing the quantities of H_2 needed reduces the demands on the needed infrastructure that would be required for complete replacement of hydrocarbon fuels. Economic analysis also shows that, at current prices for conventional fuels and emissions reduction strategies, the use of H_2 blends is economically competitive. In light of concerns over global warming and given the long-term need to minimize unburned hydrocarbon and CO_2 emissions, increased use of hydrogen in this capacity provides one possible transition strategy to a carbon-free society of the future.

REFERENCES

Abdel-Gayed, Rg. G., and Bradley, D. (1981). A two-Eddy theory of premixed turbulent flame propagation. *Phil. Trans. R. Soc. Lond.* **30**, 1–25.

Aceves, S.M. and Smith, J.R. (1997). Hybrid and Conventional Hydrogen Engine Vehicles that Meet EZEV Emissions, SAE Paper 970290.

Ali, S.A. and Parks, W.P. (1998). Renewable Fuels Turbine Project, ASME Paper 98-GT-295.

Bannister, R.L., Newby, R.A., and Yang, W. (1998). Final Report on the Development of a Hydrogen-Fueled Combustion Turbine Cycle for Power Generation, ASME Paper 98-GT-21.

Berckmüller, M., Rottengruber, H., Eder, A., Brehm, N., Elssser, G., Müller-Alander, G. *et al.* (2003). Potentials of a Charged SI-Hydrogen Engine, SAE Paper 2003–01–3210.

Binder, K. and Withalm, G. (1982). Mixture formation and combustion in a hydrogen engine using hydrogen storage technology. *Int. J. Hydrogen Energy* **7**, 651–659.

Boschek, E., Griebel, P., Erne, D., and Jansohn, P. (2005). *Eighth International Conference on Energy for a Clean Environment: Cleanair 2005*, Lisbon, Portugal.

Boschek, E., Griebel, P., and Jansohn, P. (2007). Fuel variability effects on turbulent, lean premixed flames at high pressures. *Proceedings of GT2007 ASME Turbo Expo 2007: Power for Land, Sea and Air*, Montreal, Canada.

Brewer, G.D. (1991). *Hydrogen Aircraft Technology*. CRC Press Inc., Boca Raton.

Cecil, W. (1822). On the application of hydrogen gas to produce a moving power in machinery. *Trans. Camb. Phil. Soc.* **1**, 217–240.

Cheng, R.K. and Oppenheim, A.K. (1984). Autoignition in methane–hydrogen mixtures. *Combust. Flame* **38**, 125–139.

Clavin, P. (1985). Dynamic behaviour of premixed flame fronts in laminar and turbulent flows. *Prog. Energy Combust. Sci.* **11**, 1–59.

Clayton, R.M. (1976). Reduction of Gaseous Pollutant Emissions from Gas Turbine Combustors Using Hydrogen-Enriched Jet Fuel – A Progress Report, NASA TM 33–790.

Das, L.M. (1990). Hydrogen engines: a view of the past and a look into the future. *Int. J. Hydrogen Energy* **15**, 425–443.

Das, L.M. (1991). Exhaust emission characterization of hydrogen operated engine system: nature of pollutants and their control techniques. *Int. J. Hydrogen Energy* 16, 765–775.

Department of Energy, Office of Energy Efficiency and Renewable Energy, Hydrogen, Fuel Cells and Infrastructure Technologies Development Program (2005). Multi-year RD&D Plan, http://www1.eere. energy.gov/hydrogenandfuelcells/mypp/.

Department of Energy, Office of Fossil Energy (1999). Vision 21 Program Plan, Clean Energy Plants for the 21st Century.

Department of Energy USA (2004). Hydrogen Posture Plan: An Integrated Research, Development and Demonstration Plan, www.eere.energy.gov\hydrogenandfuelcells\ posture_plan04.html.

Drell, I.L. and Belles, F.E. (1958). Survey of Hydrogen Combustion Properties, National Advisory Committee on Aeronautics, Technical Report 1383.

Eichlseder, H., Wallner, T., Freyman, R., and Ringler, J. (2003). The Potential of Hydrogen Internal Combustion Engines in a Future Mobility Scenario, SAE Paper 2003–01–2267.

Escher, W.J.D. (1975). The Hydrogen-Fueled Internal Combustion Engine: A Technical Survey of Contemporary US Projects. Escher Technology Associates, Inc., Report for the US Energy and Development Administration, Report No. TEC74/005.

Escher, W.J.D. (1983). Hydrogen as an automotive fuel: worldwide update. *Symp. Papers Nonpetr. Vehicul. Fuel.* **3**, 143–180.

Escher, W.J.D. and Euckland, E.E. (1976). Recent Progress in the Hydrogen Engine, SAE Paper 760571.

Fiene, J., Braithwaite, T., Boehm, R., and Baghzouz, Y. (2002). Development of a Hydrogen Engine for a Hybrid Electric Bus, SAE Paper 2002–01–1085.

Furuhama, S. and Fukuma, T. (1986). High output power hydrogen engine with high pressure fuel injection, hot surface ignition and turbocharging. *Int. J. Hydrogen Energy* **11**, 399–407.

Furuhama, S., Yamane, K., and Yamaguchi, I. (1977). Combustion improvement in a hydrogen fueled engine. *Int. J. Hydrogen Energy* **2**, 329–340.

Furuhama, S., Hiruma, M., and Enomoto, Y. (1978). Development of a liquid hydrogen car. *Int. J. Hydrogen Energy* **3**, 61–81.

Furuhama, S., Takiguchi, M., Suzuki, T., and Tsujita, M. (1987). Development of a Hydrogen Powered Medium Duty Truck, SAE Paper 871168.

Gauducheau, J.L., Denet, B., and Searby, G. (1998). *Combust. Sci. Tech.* **137**, 81–99.

Glasson, N.D. and Green, R.K. (1994). Performance of a spark-ignition engine fuelled with hydrogen using a high-pressure injector. *Int. J. Hydrogen Energy* **19**, 917–923.

Green, R.K. and Glasson, N.D. (1992). High-pressure hydrogen injection for internal combustion engines. *Int. J. Hydrogen Energy* **17**, 895–901.

Hawkes, E.R. and Chen, J.H. (2004). Direct numerical simulation of hydrogen-enriched lean premixed methane–air flames. *Combust. Flame* **138**, 242–258.

Hedrick, J.C. (1993). Advanced Hydrogen Utilization Technology Demonstration. Technical Report, Southwest Research Institute, Final Report for Detroit Diesel Corporation, SwRI Project No. 03–5080.

Heffel, J.W. (2003a). NO_x emission and performance data for a hydrogen fueled internal combustion engine at 1500 rpm using exhaust gas recirculation. *Int. J. Hydrogen Energy* **28**, 901–908.

Heffel, J.W. (2003b). NO_x emission reduction in a hydrogen fueled internal combustion engine at 3000 rpm using exhaust gas recirculation. *Int. J. Hydrogen Energy* **28**, 1285–1292.

Heywood, J.B. (1988). *Internal Combustion Engine Fundamentals*. McGraw-Hill, New York.

Homan, H.S. (1978). An experimental study of reciprocating internal combustion engines operated on hydrogen. Ph.D. thesis, Cornell University, Mechanical Engineering, Ithaca, NY.

Homan, H.S., de Boer, P.C.T., and McLean, W.J. (1983). The effect of fuel injection on NO_x emissions and undesirable combustion for hydrogen-fuelled piston engines. *Int. J. Hydrogen Energy* **8**, 131–146.

Huang, J., Bushe, W.K., Hill, P.G., and Munshi, S.R. (2004). Experimental and kinetic study of shock initiated ignition in homogeneous methane–hydrogen–air mixtures at engine-relevant conditions. *Combust. Flame* **136**, 25–42.

Jackson, G.S., Sai, R., Plaia, J.M., Boggs, C.M., and Kiger, K.T. (2003). *Combust. Flame* **132**, 503–511.

Jaura, A.K., Ortmann, W., Stuntz, R., Natkin, B., and Grabowski, T. (2004). Ford's H2RV: An Industry First HEV Propelled with an H_2 Fueled Engine – A Fuel Efficient and Clean Solution for Sustainable Mobility, SAE Paper 2004–01–0058.

Jorach, R., Enderle, C., and Decker, R. (1997). Development of a low-NO_x truck hydrogen engine with high specific power output. *Int. J. Hydrogen Energy* **22**, 423–427.

Juste, G.L. (2006). Hydrogen injection as an additional fuel in gas turbine combustor. Evaluation of effects. *Int. J. Hydrogen Energy* **31**, 2112–2121.

Keller, J. and Lutz, A. (2001). Hydrogen Fueled Engines in Hybrid Vehicles, SAE Paper 2001–01–0546.

Kim, J.M., Kim, Y.T., Lee, J.T., and Lee, S.Y. (1995). Performance Characteristics of Hydrogen Fueled Engine with the Direct Injection and Spark Ignition System, SAE Paper 952498.

Knorr, H., Held, W., Prumm, W., and Rudiger, H. (1997). The MAN hydrogen propulsion system for city busses. *Int. J. Hydrogen Energy* **22**, 201–208.

Kondo, T., Lio, S., and Hiruma, M. (1997). A Study on the Mechanism of Backfire in External Mixture Formation Hydrogen Engines – About Backfire Occurred by Cause of the Spark Plug, SAE Paper 971704.

Kwon, O.C. and Law, C.K. (2004). Effects of hydrocarbon substitution on atmospheric hydrogen–air flame propagation. *Int. J. Hydrogen Energy* **29**, 867–879.

Lee, J.T., Kim, Y.Y., Lee, C.W., and Caton, J.A. (2001). An investigation of a cause of backfire and its control due to crevice volumes in a hydrogen fueled engine. *Trans. ASME* **123**, 204–210.

Lee, J.T., Kim, Y.Y., and Caton, J.A. (2002). The development of a dual injection hydrogen fueled engine with high power and high efficiency. *Proceedings of ASME-ICED 2002 Fall Technical Conference* **39**, 323–324.

Lee, S.J., Yoon, K.K., Han, B.H., Lee, H.B., and Kwon, B.J. (1995). Development of a Hyundai Motor Company Hydrogen-Fueled Vehicle, SAE Paper 952764.

Lefebvre, A.H. (1983). *Gas Turbine Combustion*. Hemisphere, Washington, DC.

Lewis, B. and von Elbe, G. (1987). *Combustion, Flames, and Explosions of Gases*. Academic Press, Orlando.

Li, H. and Karim, G.A. (2004). Knock in spark ignition hydrogen engines. *Int. J. Hydrogen Energy* **29**, 859–865.

Lynch, F.E. (1983). Parallel induction: a simple fuel control method for hydrogen engines. *Int. J. Hydrogen Energy* **8**, 721–730.

Maher, A.R. and Sadiq, A.B. (2004). Effect of compression ratio, equivalence ratio and engine speed on the performance and emission characteristics of a spark ignition engine using hydrogen as a fuel. *Renewable Energy* **29**, 2245–2260.

Mathur, H.B., Das, L.M., and Petro, T.N. (1993). Hydrogen-fuelled diesel engine: a performance improvement through charge dilution techniques. *Int. J. Hydrogen Energy* **18**, 421–431.

Meier, J.G., Hung, W.S.Y., and Sood, V.M. (1986). Development and application of industrial gas turbines for medium-Btu gaseous fuels. *J. Eng. Gas Turbine. Power* **108**, 182–190.

Morris, J.D., Symonds, R.A., Ballard, F.L., and Banti, A. (1998). ASME Paper 98-GT-359.

Naber, J.D. and Siebers, D.L. (1998). Hydrogen combustion under diesel engine conditions. *Int. J. Hydrogen Energy* **23**, 363–371.

Nagalingam, B., Dübel, M., and Schmillen, K. (1983). Performance of the Supercharged Spark Ignition Hydrogen Engine, SAE Paper 831688.

Natkin, R.J., Tang, X., Boyer, B., Oltmans, B., Denlinger, A., and Heffel, J.W. (2003). Hydrogen IC Engine Boosting Performance and NO_x Study, SAE Paper 2003–01–0631.

Peschka, W. (1992). *Liquid Hydrogen. Fuel of the Future*. Springer, New York.

Peschka, W. (1998). Hydrogen: the future cryofuel in internal combustion engines. *Int. J. Hydrogen Energy* **23**, 27–43.

Phillips, J.N. and Roby, R.J. (1999). ASME Paper 99-GT-115.

Reed, R.J. (Ed.) (1978). *North American Combustion Handbook*. North American Mfg. Co., Cleveland.

Ren, J.Y., Qin, W., Egolfopoulos, F.N., and Tsotsis, T.T. (2001). *Combust. Flame* **124**, 717–720.

Richards, G.A., McMillian, M.M., Gemmen, R.S., Rogers, W.A., and Cully, S.R. (2001). Issues for low-emission, fuel-flexible power systems *Prog. Energy Combust. Sci.* **27**, 141–169.

Rottengruber, H., Berckmuller, M., Elssser, G., Brehm, N., and Schwarz, C. (2004). Direct-Injection Hydrogen SI-Engine Operation Strategy and Power Density Potentials, SAE Paper 2004–01–2927.

Sankaran, R. and Im, H.G. (2006). Effects of hydrogen addition on the markstein length and flammability limit of stretched methane/air premixed flames. *Combust. Sci. Tech.* **178**, 1585-1611.

Schefer, R.W. (2003). Hydrogen enrichment for improved lean flame stability. *Int. J. Hydrogen Energy* **28**, 1131–1141.

Schefer, R.W., Wicksall, D.M., and Agrawal, A.K. (2002). Combustion of hydrogen-enriched methane in a lean premixed swirl burner. *Proc. Combust. Inst.* **29**, 843–851.

Schefer, R.W., Smith, T.D., and Marek, C.J., Evaluation of NASA Lean Premixed Hydrogen Burner, Sandia Report SAND 2002-8609, January 2003.

Scott, D.S. (2004). Contrails against an azure sky. *Int. J. Hydrogen Energy* **29**, 1317–1325.

Segeler, C.G. (Ed.) (1978). *Gas Engineers Handbook*. Industrial Press, New York.

Shudo, T., Nabetani, S., and Nakajima, Y. (2001). Analysis of the degree of constant volume and cooling loss in a spark ignition engine fuelled with hydrogen. *Int. J. Engine Research* **2**, 81–92.

Shudo, T., Cheng, W.K., Kuninaga, T., and Hasegawa, T. (2003). Reduction of Cooling Loss in Hydrogen Combustion by Direct Injection Stratified Charge, SAE Paper 2003–01–3094.

Sidwell, T., Casleton, K., Straub, D., Maloney, D., Richards, G., Strakey, P., Ferguson, D., and Beer, S. (2005). Development and operation of a pressurized optically-accessible research combustor for simulation validation and fuel variability studies. GT2005–68752, *Proceedings of GT2005 ASME Turbo Expo 2005: Power for Land, Sea and Air*, Reno-Tahoe.

Sierens, R. and Verhelst, S. (2003). Influence of the injection parameters on the efficiency and power output of a hydrogen fueled engine. *Trans ASME* **195**, 444–449.

Spadaccini, L.J. and Colkett, M.B. (1994). Ignition delay characteristics of methane fuels. *Prog. Energy Combust. Sci.* **20**, 431–460.

Stockhausen, W.F., Natkin, R.J., Kabat, D.M., Reams, L., Tang, X., Hashemi, S. *et al.* (2002). Ford P2000 Hydrogen Engine Design and Vehicle Development Program, SAE Paper 2002–01–0240.

Strakey, P., Sidwell, T., and Ontko, J. (2006). Investigation of the effects of hydrogen addition on the lean extinction in a swirl stabilized burner. *Proceedings of Combustion Institute* **31**, 3173–3180.

Swain, M.R., Pappas, J.M., Adt Jr., R.R., and Escher W.J.D. (1981). Hydrogen-Fueled Automotive Engine Experimental Testing to Provide an Initial Design-Data Base, SAE Paper 810350.

Swain, M.R., Adt Jr., R.R., and Pappas, J.M. (1983). Experimental Hydrogen Fueled Automotive Engine Design Data Base Project. A Facsimile Report, prepared for US Department of Energy, DOE/CS/51212.

Swain, M.R., Schade, G.J., and Swain, M.N. (1996). Design and Testing of a Dedicated Hydrogen-Fueled Engine, SAE Paper 961077.

Tang, X., Kabat, D.M., Natkin, R.J., and Stockhausen, W.F. (2002). Ford P2000 Hydrogen Engine Dynamometer Development, SAE Paper 2002–01–0242.

TerMaath, C.Y., Skolnik, E.G., Schefer, R.W., and Keller, J. (2006). Emissions reduction benefits from hydrogen addition to midsize gas turbine feedstocks. *Int. J. Hydrogen Energy* **31**, 1147–1158.

Todd, D.M. (2000). Gas Turbine Improvements Enhance IGCC Viability. *2000 Gasification Technologies Conference*, October 8–11, San Francisco, CA.

Topinka, J.A., Gerty, M.D., Heywood, J.B., and Keck, J.C. (2004). Knock Behavior of a Lean-Burn, H_2 and CO Enhanced, SI Gasoline Engine Concept, SAE Paper 2004–01–0975.

Turns, S.R. (2000). An Introduction to Combustion: Concepts and Applications, Second Edition, McGraw-Hill, New York.

Vagelopoulos, C.M., Oefelein, J.C., and Schefer, R.W. (2003). Effects of Hydrogen Enrichment on Lean Premixed Methane Flames. *Fourteenth Annual US Hydrogen Conference and Hydrogen Expo USA*, Washington, DC.

Van Blarigan, P. (1995). Development of a Hydrogen Fueled Internal Combustion Engine Designed for Single Speed/Power Operation, SAE Paper 961690.

Van Blarigan, P. and Keller, J.O. (1998). A hydrogen fuelled internal combustion engine designed for single speed/power operation. *Int. J. Hydrogen Energy* **23**, 603–609.

Varde, K.S. and Frame, G.A. (1985). Development of a high-pressure hydrogen injection for SI engine and results of engine behavior. *Int. J. Hydrogen Energy* **10**, 743–748.

Wallner, T., Wimmer, A., Gerbig, F., and Fickel, H.C. (2004). The hydrogen combustion engine: a basic concept study. *Gasfahrzeuge Die passende Antwort auf die CO2-Herausforderung der Zukunft*. Expert-Verlag GmbH, Springer-Verlag, Berlin, Germany.

Warnatz, J., Maas, U., and Dibble, R.W. (2001). *Combustion*. Springer-Verlag, Berlin, Germany.

Welch, A.B. and Wallace, J.S. (1990). Performance Characteristics of a Hydrogen-Fueled Diesel Engine with Ignition Assist, SAE Paper 902070.

Wicksall, D.M., Schefer, R.W., Agrawal, A.J., and Keller, J.O. (2003). ASME Paper GT 2003–38712.

Zhang, Q., Noble, D.R., Meyers, A., Xu, K., and Lieuwen, T. (2005). GT2005–68907, *Proceedings of GT2005 ASME Turbo Expo 2005: Power for Land, Sea and Air*, Reno-Tahoe.

Index

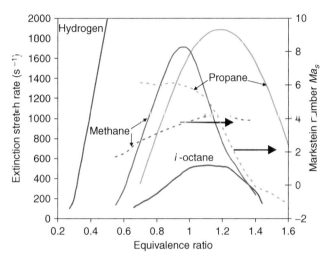

Figure 2.2 Positive stretch rates for extinction of CH$_4$-air and C$_3$H$_8$-air (Law *et al.*, 1986), hydrogen-air (Dong *et al.*, 2005), and *i*-octane-air (Holley *et al.*, 2006), under atmospheric conditions. Broken curves give Ma_s for C$_3$H$_8$-air (Bradley *et al.*, 1998a) and CH$_4$-air (Bradley *et al.*, 1996).

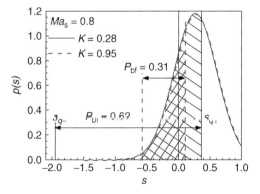

Figure 2.10 Ranges of stretch rates on $p(s)$ curve for $K = 0.28$ and 0.95. Positive and negative extinction stretch rates are s_{q+} and s_{q-}, and dotted line gives rms strain rate.

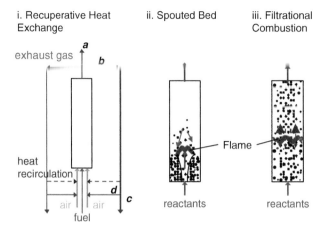

Figure 2.17 Different modes of heat recirculation. (i) For $b = c = d = 0$: conventional burner. For $b \leq a$, $d = 0$: recuperative heat recirculation, indicated by dashed line. For $b \leq a$, $d = b$, $c = 0$: both recuperator heat and burned gas recirculation; (ii) spouted bed with fountain of particles; and (iii) filtrational combustion with solid porous medium.

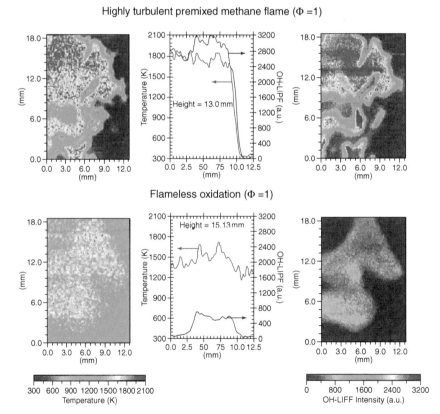

Figure 2.19 Simultaneous temperature and OH-LIPF images of premixed highly turbulent flame and flameless oxidation (Plessing *et al.*, 1998, p. 3202).

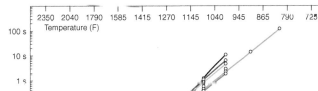

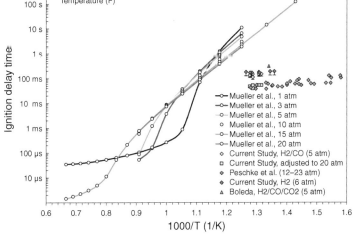

Figure 5.13 Comparison of ignition delay times for detailed mechanisms and measurements (Beerer *et al.*, 2006).

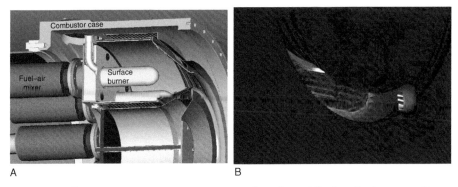

Figure 5.33 Illustration of the porous/perforated design of a surface stabilized combustion system (Images Courtesy of Alzeta Corporation): (A) schematic illustrating installation and (B) rig test.

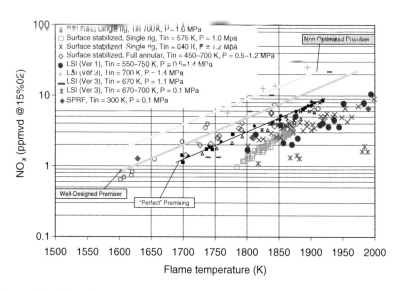

Figure 5.34 NO$_x$ Emissions for various advanced lean premixed strategies for stationary power.

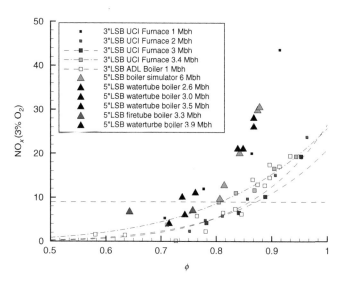

Figure 6.6 NO$_x$ emissions of LSB in furnaces and boilers of 300 kW to 1.8 MW.

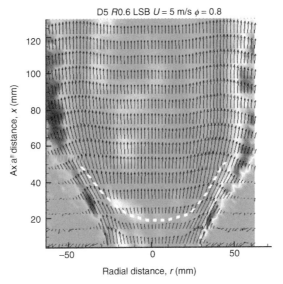

Figure 6.8 Velocity vectors and turbulence stresses for an LSB burning CH_4/air at $\phi = 0.8$ and $U_o = 5.0$ m/s.

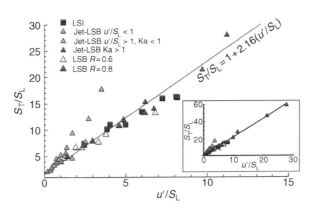

Figure 6.9 Turbulent flame speed correlation.

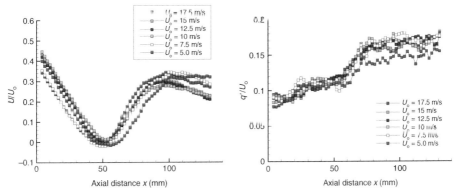

Figure 6.10 Normalized centerline profiles of non-reacting flow produced by a laboratory LSB showing self-similarity features.

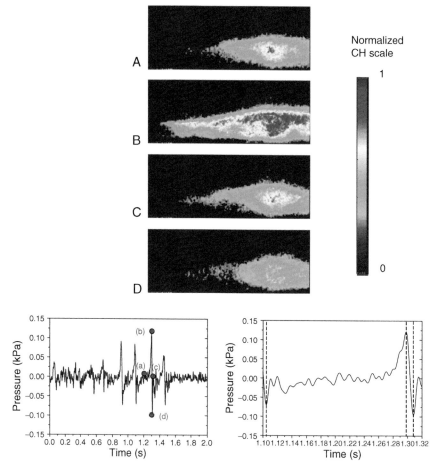

Figure 7.6 CH images of flame and pressure signal at lean flammability limit (De Zilwa *et al.*, 2001). Plane sudden expansion, duct geometry as in Figure 7.5; $Re = 57\ 000$, $\phi = 0.55$. Observation window: $40 \times 100\ \text{mm}$; exposure time $= 5\ \text{ms}$; CH values normalized by maximum; average frequency of extinction and relight cycle $= 5.6\ \text{Hz}$.

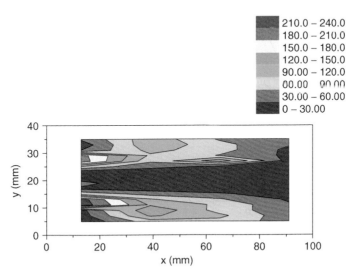

210.0 – 240.0
180.0 – 210.0
150.0 – 180.0
120.0 – 150.0
90.00 – 120.0
00.00   90.00
30.00 – 60.00
0 – 30.00

Figure 7.7 Time-averaged strain rates (De Zilwa *et al.*, 2001). Strain rate calculated as transverse gradient of axial velocity measurements of Khezzar *et al.* (1999); plane sudden expansion (duct geometry as in Figure 7.5), *Re* = 76 000, ϕ = 0.72.

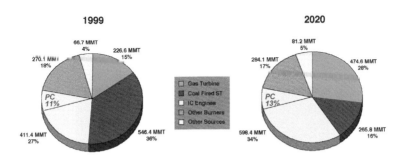

Figure 8.7 Contribution of various combustion sources to total US CO_2 emissions (Schefer, 2003).

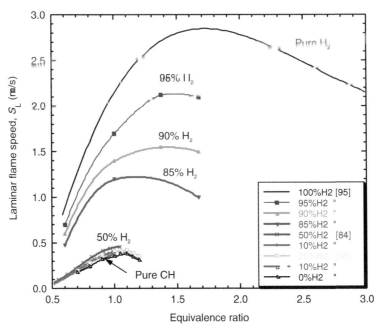

Figure 8.15 Laminar flame speeds as a function of equivalence ratio for H_2–air, CH_4–air, and blends of H_2–CH_4–air at STP.

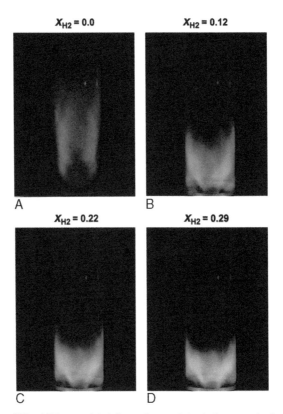

Figure 8.18 Effect of H_2 addition on global flame characteristics in lean premixed swirl burner (Schefer *et al.*, 2002).

Printed and bound by CPI Group (UK) Ltd, Croydon, CR0 4YY

03/10/2024

01040312-0012